Advances in Experimental Medicine and Biology

Volume 1442

Advances in Experimental Medicine and Biology provides a platform for scientific contributions in the main disciplines of the biomedicine and the life sciences. This series publishes thematic volumes on contemporary research in the areas of microbiology, immunology, neurosciences, biochemistry, biomedical engineering, genetics, physiology, and cancer research. Covering emerging topics and techniques in basic and clinical science, it brings together clinicians and researchers from various fields.

Advances in Experimental Medicine and Biology has been publishing exceptional works in the field for over 40 years, and is indexed in SCOPUS, Medline (PubMed), EMBASE, BIOSIS, Reaxys, EMBiology, the Chemical Abstracts Service (CAS), and Pathway Studio.

2022 CiteScore: 6.2

Meng Zhao • Pengxu Qian
Editors

Hematopoietic Stem Cells

Keystone of Tissue Development and Regenerative Medicine

Springer

Editors
Meng Zhao
Key Laboratory of Stem Cells and Tissue Engineering (Ministry of Education), Zhongshan School of Medicine
Sun Yat-sen University
Guangzhou, Guangdong, China

Pengxu Qian
Center for Stem Cell and Regenerative Medicine and Bone Marrow Transplantation Center of the First Affiliated Hospital
Zhejiang University School of Medicine
Hangzhou, Zhejiang, China

ISSN 0065-2598 ISSN 2214-8019 (electronic)
Advances in Experimental Medicine and Biology
ISBN 978-981-99-7470-2 ISBN 978-981-99-7471-9 (eBook)
https://doi.org/10.1007/978-981-99-7471-9

This Springer imprint is published by the registered company Springer Nature Singapore Pte Ltd.
The registered company address is: 152 Beach Road, #21-01/04 Gateway East, Singapore 189721, Singapore

Foreword 1

I am honored to be given the unique opportunity to write the foreword for *Hematopoietic (Blood-Forming) Stem Cells (HSCs),* by Meng Zhao and Pengxu Qian. As the two highly talented postdoctoral fellows in my laboratory, Zhao and Qian are now rising young scientists in the hematopoietic and stem cell fields. In this book, the authors put forth a brand-new view on an old subject—the hematopoietic stem cell.

In the early 1900s, Haeckel and Boveri first introduced the concept of a "Stammzelle" that has the capacity for both self-renewal and differentiation. In 1961, in their studies of bone marrow transplantation in recipient mice post-irradiation, Till and McCulloch observed the presence of spleen nodules, or the colony forming unit (S-CFU), which indicated the stem cell properties of self-renewal and the potential for multilineage differentiation. Ever since, HSCs have served as a pioneering system for novel discoveries that either lead to new concepts in the stem cell field or result in the development of new protocols for the treatment of blood disorders.

Unlike many systems, the ontogeny of the hematopoietic system is spatio-temporally dynamic, beginning in the aorta-mesonephros-gonad (AGM) region of the embryo, followed by a transition and maturation in the fetal liver, and finally moving to the bone marrow in adults. The chapter titled "HSC Development in Mammalian Embryos" is led by B. Liu and Y. Lan, two leading scientists in the study of the emergence of HSCs during embryonic development. However, HSCs are not randomly distributed in the body but instead reside in a special site in the bone marrow (termed the *niche*) that maintains HSCs in an undifferentiated state. S. Pinho and M. Zhao, each of whom has made critical contributions to the study of the niche concept, go on to summarize the chapter titled "HSCs and Their Bone Marrow Niches."

Although active-cycling HSCs are required to support daily hematopoiesis, only quiescent HSCs can support long-term hematopoiesis. Thus, early studies mainly focused on the cell cycling state of HSCs. Decades later, the field realized that a stem cell cannot simply be switched from a quiescent state to proliferative state the way that one flips a switch turn a light on and off, but rather this change in state is regulated by epigenetic changes and controlled through metabolic activities. The chapter titled "Emerging Roles of Epigenetic Regulators in Maintaining HSC Homeostasis" is led by P. Qian, who revealed that methylation at specific DNA locus (the imprinted Dlk1-Gtl2 locus) regulating the mTOR pathway plays an essential role in determining when HSCs transition from a reserve state (quiescent plus low metabolic activity) to an active state. The chapter titled "Metabolism in Hematopoiesis and Malignancy" was collectively contributed to by C. Man and X. Zeng, who discovered that regulation of the lactate/proton symporter could adjust pH levels and carbon flux and thus impact myeloid leukemia-initiating cells, as well as Y. Wang, who found that mitobiogenesis can be manipulated in normal and malignant hematopoietic cells through malic enzyme 2. In addition to epigenetic regulation at the DNA and protein levels, epigenetic regulation at the RNA level has also emerged as an active area of research. J. Chen, who revealed the critical role of N^6-methyladenosine RNA (M6A) in hematopoiesis, summarizes the chapter titled "N^6-Methyladenosine RNA Modification in Normal and Malignant Hematopoiesis."

In humans, it is estimated that there are on average 20,000–200,000 single hematopoietic stem and progenitor (HSPC)-derived hematopoietic clones after birth, and that this number declines to around 20 clones throughout the aging process, a phenomenon known as *clonal hematopoiesis*. Early onset clonal hematopoiesis due to epigenetic alterations and genetic mutations often leads to the development of myeloid dysplastic syndrome (MDS), a preleukemic condition with a high risk of leukemogenesis. J. Sun, who made the initial observation that long-lived progenitors, but not HSCs, contribute the larger portion of hematopoietic clones, leads the chapter titled "The Origin of Clonal Hematopoiesis." Age-related decline in the heterogeneity of hematopoietic clones also leads to reduced immunity, inversely increasing the risk of leukemogenesis. J. Perry, who uncovered the mechanism underlying the immune privilege of leukemia stem cells (LSCs), summarizes the chapter titled "Immune System Influence on Hematopoietic Stem Cells and Leukemia." Understanding the epigenetic basis of aging opens the opportunity to develop approaches for its therapeutic prevention. The chapter titled "Aging Causes and Rejuvenation of HSCs" was collectively contributed to by Z. Chen and Z. Ju, who revealed several mechanisms that contributed to the aging of HSCs (e.g., chronic inflammation), as well as Y. Sun, who uncovered the critical role of the intrinsic stress-response factor ATF4 in the prevention of HSC aging.

Given the lifesaving nature of HSC transplantation in the clinic, the ability to effectively expand a rather limited number of HSCs has been considered as the "holy grail" of hematopoietic research. Over the past decades, many approaches have been explored but none thus far has led to the development of a clinically successful protocol. B. Gao, Y. Chen, and X Huang, all of whom were trained by Dr. Hal Broxymeyer (the "father" of umbilical cord blood transplantation), have made contributions to the ex vivo expansion of HSCs and wrap up the chapter titled "Ex vivo Expansion of and Homing of Human UBC-HSCs."

The development of new technology always promotes scientific advancement. Y. Wu, who developed a highly efficient method for conducting genome editing in HSCs, leads off the chapter titled "Gene Editing in HSCs." Research in different model organisms can make a similar impact, often providing unexpected views that can result in important discoveries. Z. Wen leads off the chapter titled "Learning from Zebrafish

Hematopoiesis." Of note, the chapter titled "Multi-lineage Differentiation from HSCs," was collectively contributed to J. Yu, who made several important contributions involving non-coding RNAs and M6A-mediated mRNA regulation.

Finally, I hope readers are not bored by this relatively long foreword, but instead inspired by the exciting topics, intriguing discoveries, and fundamental insights that have helped to unleash the mysteries of the HSC. Undoubtedly, ongoing current and future research will continue moving us closer to the holy grail—the ability to take full advantage of HSCs, putting them to use as novel therapies capable of curing a variety of blood disorders, improving the immune system, and preventing aging. After all, rejuvenating life might not just be a dream in the future.

Stowers Institute for Medical Research, Kansas City, MO, USA
Linheng Li

Foreword 2

Since the early 1960s, hematopoietic stem cell (HSC) research has built the foundation of stem cell biology and pioneered the field of regeneration medicine. Many fundamental concepts in stem cell biology and regeneration medicine such as stem cell self-renewal, multi-potency, quiescence, homing, and mobilization are originally proposed, experimentally demonstrated, or clinically applied through HSC research.

During the last 25 years, stem cell research has been rapidly developed and expanded into a multi-disciplinary field. Many other types of stem cells have been identified and understood to a more definitive level and yet to be translated into therapeutic agents in the clinic. Especially, research advances in pluripotent stem cells offer an enormous potential in brining many types of "off-the-shelf" cell products for future regenerative medicine. Nonetheless, many principles in stem cell medicine can still be obtained from HSC research. Meanwhile, HSC research is still facing great challenges. Hurdles in HSC expansion, regenerative efficiency, clonal transformation, gene editing/therapies, etc. remain. In recent years, single cell technologies and single cell biology offer unprecedented opportunities to further study HSC ontogeny, heterogenicity, epigenetics, metabolism, microenvironment, aging, and leukemogenesis, to further enhance our abilities to therapeutically manipulate HSC, and to possibly expand the applications of HSC therapies beyond the hematopoietic diseases. Therefore, it is demanded to have an updated review book on HSC.

This book edited by Dr. Meng Zhao et al. covers a range of topics on new advances in HSC research (as partially mentioned above). Each chapter is written by expert investigators in the field. I am very pleased to see that most authors are the young investigators who have become or are becoming independent with an outstanding track record in HSC research. The content of this book is a timely collection of important advances in HSC biology and translational research. It should be quite valuable and an excellent reference for the researchers, clinicians, graduate students, postdoctoral fellows, and other relevant scholars in HSC research, general stem cell biology, and regenerative medicine. Therefore, I strongly endorse this book to our colleagues in the field.

State Key Laboratory of Experimental Hematology,
Institute of Hematology & Blood Diseases Hospital,
Chinese Academy of Medical Sciences & Peking
Union Medical College, Beijing, China

Tao Cheng

Preface

Hematopoiesis is a paradigm model for comprehending the development, maintenance, regeneration, aging, and malignant transformation of mammalian organs. Siting at the apex of the hematopoietic hierarchy, hematopoietic stem cells (HSCs) masterfully regulate their proliferation, self-renewal, and differentiation to generate all blood cell lineages throughout an organism's lifetime. This process exemplifies the most compelling example of somatic stem cell research.

The book "Hematopoietic Stem Cells: Keystone of Tissue Development and Regenerative Medicine" provides a comprehensive insight into HSCs, covering their embryonic development, adult maintenance, and aging, as demonstrated through studies conducted on both zebrafish and mammals. Additionally, this book delves into how niche components and cell-intrinsic regulatory mechanisms, such as epigenetics, metabolism, and RNA modification, govern the regulation of HSCs in both normal and diseased conditions. Furthermore, this book highlights the role of HSCs in regenerative medicine by introducing gene therapy and *ex vivo* expansion techniques for HSC transplantation.

This book offers a comprehensive history of HSC studies, detailing how research on HSCs has significant clinical implications. By exploring the insights gained from these studies, the book serves as a valuable resource for translating research findings from the laboratory to clinical applications.

Academic researchers, research scientists, and graduate students in universities, industry, and government would find this book of interest.

Hangzhou, Zhejiang, China — Pengxu Qian
Guangzhou, Guangdong, China — Meng Zhao

Contents

About the Editors

Meng Zhao a distinguished stem cell biologist and hematologist, is currently serving as a professor at the Zhongshan School of Medicine, Sun Yat-sen University. His research primarily focuses on unraveling the intricate mechanisms governing blood stem cells and developing therapeutic strategies for hematopoietic stem cell transplantation and leukemia treatment. In his laboratory, he utilizes genetic models and clinical specimens to investigate both the intrinsic cellular control and microenvironmental influence on normal and leukemic stem cells. Noteworthy contributions from him include identifying megakaryocytes as a hematopoietic stem cell niche, characterizing leukemic stem cells in T-cell acute lymphoblastic leukemia, and discovering amino acid catabolism within hematopoietic stem cells. The primary objective of his laboratory is to gain comprehensive insights into the complex metabolic intricacies that govern both normal and leukemic stem cells within the dynamic bone marrow microenvironment.

Pengxu Qian obtained his bachelor degree in 2006 and his Ph.D. degree in 2012 from University of Science and Technology of China (USTC). During his postdoctoral training in Prof. Linheng Li's lab at Stowers Institute for Medical Research from 2012 to 2017, he was funded by the American Society of Hematology (ASH) Scholar Award to study the roles of DNA methylation and imprinting genes in hematopoietic stem cells. In November 2017, he started his lab as principal investigator in Zhejiang University School of Medicine to study the roles and mechanisms of epigenetic regulators in normal hematopoiesis and leukemogenesis. The major contributions from him include illustrating the roles and mechanisms of the Dlk1-Gtl2 imprinting locus, the differentially methylated region (DMR) DERARE in the Hoxb cluster, and the m6A reader YTHDF2 in normal hematopoiesis and leukemogenesis. The primary objectives of his laboratory are to decipher the epigenetic regulatory mechanisms underlying HSC homeostasis and leukemogenesis and to develop novel therapeutic approaches against hematopoietic malignancies.

Hematopoietic Stem Cell Development in Mammalian Embryos

1

Siyuan Hou, Chen Liu, Yingpeng Yao, Zhijie Bai, Yandong Gong, Chaojie Wang, Jian He, Guoju You, Guangyu Zhang, Bing Liu, and Yu Lan

Abstract

Hematopoietic stem cells (HSCs) are situated at the top of the adult hematopoietic hierarchy in mammals and give rise to the majority of blood cells throughout life. Recently, with the advance of multiple single-cell technologies, researchers have unprecedentedly deciphered the cellular and molecular evolution, the lineage relationships, and the regulatory mechanisms underlying HSC emergence in mammals. In this review, we describe the precise vascular origin of HSCs in mouse and human embryos, emphasizing the conservation in the unambiguous arterial characteristics of the HSC-primed hemogenic endothelial cells (HECs). Serving as the immediate progeny of some HECs, functional pre-HSCs of mouse embryos can now be isolated at single-cell level using defined surface marker combinations. Heterogeneity regrading cell cycle status or lineage differentiation bias within HECs, pre-HSCs, or emerging HSCs in mouse embryos has been figured out. Several epigenetic regulatory mechanisms of HSC generation, including long noncoding RNA, DNA methylation modification, RNA splicing, and layered epigenetic modifications, have also been recently uncovered. In addition to that of HSCs, the cellular and molecular events underlying the development of multiple hematopoietic progenitors in human embryos/fetus have been unraveled with the use of series of single-cell technologies. Specifically, yolk sac-derived myeloid-biased progenitors have been identified as the earliest multipotent hematopoietic progenitors in human embryo, serving as an important origin of fetal liver monocyte-derived macrophages. Moreover, the development of multiple hematopoietic lineages in human embryos such as T and B lymphocytes, innate lymphoid cells, as well as myeloid cells like monocytes, macrophages, erythrocytes, and megakaryocytes has also been depicted and reviewed here.

Siyuan Hou, Chen Liu and Yingpeng Yao contributed equally with all other contributors.

S. Hou · B. Liu
Key Laboratory for Regenerative Medicine of Ministry of Education, Institute of Hematology, School of Medicine, Jinan University, Guangzhou, Guangdong, China

State Key Laboratory of Experimental Hematology, Haihe Laboratory of Cell Ecosystem, Institute of Hematology, Fifth Medical Center of Chinese PLA General Hospital, Beijing, China

C. Liu · Z. Bai · Y. Gong · J. He · G. Zhang
State Key Laboratory of Experimental Hematology, Haihe Laboratory of Cell Ecosystem, Institute of Hematology, Fifth Medical Center of Chinese PLA General Hospital, Beijing, China

Y. Yao · C. Wang · Y. Lan (✉)
Key Laboratory for Regenerative Medicine of Ministry of Education, Institute of Hematology, School of Medicine, Jinan University, Guangzhou, Guangdong, China

G. You
State Key Laboratory of Primate Biomedical Research, State Key Laboratory of Experimental Hematology, School of Medicine, Tsinghua University, Beijing, China

M. Zhao, P. Qian (eds.), *Hematopoietic Stem Cells*, Advances in Experimental Medicine and Biology 1442, https://doi.org/10.1007/978-981-99-7471-9_1

Keywords

Hematopoietic stem cells · Hemogenic endothelial cells · Development · Single-cell RNA sequencing · Heterogeneity · Epigenetic regulation

The formation of adult-like hematopoietic stem cells (HSCs) during embryonic development has been intensively studied for nearly 30 years. In this chapter, we will focus mainly on recent novel conceptual advances of the emerging HSCs in mouse embryos, including their precise vascular origin, functional heterogeneity, and specific regulatory mechanisms, as uncovered by a series of new technologies such as single-cell transplantation, single-cell RNA sequencing, and genetic lineage tracing. Particularly, progresses regarding the development of HSCs in human embryos will also be described in detail.

1.1 Endothelial Origin and Functional Heterogeneity of Emerging HSCs in Mouse Embryos

The first adult-repopulating HSCs are detected at embryonic day (E) 10.5 in the ventral part of the dorsal aorta in aorta-gonad-mesonephros (AGM) region of mouse embryos (Medvinsky and Dzierzak 1996; Dzierzak and Speck 2008; Yokomizo and Dzierzak 2010; de Bruijn et al. 2000). It is generally accepted that HSCs are derived from a specialized subset of endothelial cells, known as hemogenic endothelial cells (HECs), and that HECs undergo a process known as endothelial-to-hematopoietic transition (EHT) to form HSCs (Zovein et al. 2008; Boisset et al. 2010). In recent years, with the development of single-cell technologies, the cellar evolution and molecular events throughout the EHT and even in the early stages of vascular development have been deeply elucidated. HSCs arise from arterial endothelial cells and undergo successive developmental intermediate precursors through pre-HEC, HEC, and pre-HSC, accompanying with a gradual decrease in endothelial and arterial features and sustained increase in hematopoietic features (Zhu et al. 2020; Zhou et al. 2016; Baron et al. 2018; Fadlullah et al. 2021; Hou et al. 2020, 2022; Lan 2022).

In embryonic proper, HECs are located in the endothelial layer presenting endothelial features, begin to express key hemogenic transcription factor Runx1, and have hemogenic potential. The number of HECs is scarce, reaching the peak at embryonic day (E) 10.0 and almost becoming undetectable at E11.0, and the transformation process is transient. Compared to arterial endothelial cells, HECs are more active in biological processes such as cell cycle and ribosome biogenesis (Hou et al. 2020). Researchers have discovered several markers and established corresponding reporter mouse lines to enrich HECs, represented by Runx1, Gfi1, and Neurl3 (Hou et al. 2020; Swiers et al. 2013; Thambyrajah et al. 2016). Using a combination of surface markers ($CD31^{+}CD41^{-}CD43^{-}CD45^{-}Procr^{+}Kit^{+}CD44^{+}$, PK44, approximately 100 per embryo), the HSC-competent HECs are efficiently enriched. Of note, PK44 population cannot produce hematopoietic progenies under colony-forming culture conditions, validating its identity as endothelial cells rather than committed hematopoietic progenitors. Having initiated their intrinsic hemogenic program featured by Runx1 expression, HECs represented by PK44 or $Neurl3^{EGFP}$ endothelial cells have been transcriptomically and functionally proven to be a continuum of cellular states from endothelial-biased characteristics to hematopoietic-biased characteristics prior to acquiring hematopoietic function (Hou et al. 2020; Li et al. 2021a). Specifically, individual cells of a small proportion of HECs display dual endothelial-hematopoietic potential, suggesting they are undergoing hemogenic specialization, and have not been committed to either endothelial or hematopoietic cell fate (Hou et al. 2020; Li et al. 2021a; Howell and Speck 2020). Interestingly, the immunophenotypic PK44 population is also detected in the yolk sac, showing short-term multi-lineage reconstitution capacity after in vitro hematopoietic induction. PK44 cells in the yolk sac are transcriptomically different from those in the AGM region, including expression of yolk sac but not dorsa aorta endothelial cell markers Lyve1 and Stab2 (Li et al. 2021a). HECs in the AGM region undergo EHT to give rise to the intra-aortic hematopoietic clusters (IAHCs), a heterogeneous population containing pre-HSCs, HSCs, and HSC-independent progenitors such as lympho-myeloid-biased progenitors (LMPs) (Zhou et al. 2016; Rybtsov et al. 2011; Vink et al. 2020). It has been reported that Cxcr4 expression can distinguish between HSC-competent and multipotent progenitor-competent HECs, suggesting that the fate of HSCs or non-HSC progenitors is committed at the HEC stage (Dignum et al. 2021).

Pre-HSCs cannot directly repopulate irradiated recipients, but can mature into adult-repopulating HSCs in vivo or in vitro (Zhou et al. 2016; Rybtsov et al. 2011; Boisset et al. 2015). Pre-HSCs still co-express endothelial cell markers and manifest certain arterial features. According to the expression of CD45, pre-HSCs could be separated into two consecutive subsets, type I ($CD45^-$) and type II ($CD45^+$). The surface marker combinations $CD31^+CD45^-CD41^{low}Kit^+CD201^{hi}$ (approximately 11 per embryo) and $CD31^+CD45^+Kit^+CD201^{hi}$ (approximately 18 per embryo) functionally enrich type I and type II pre-HSCs, respectively, at single-cell level (Zhou et al. 2016). Type I pre-HSCs in both G0 and S/G2/M phase exhibit reconstitution potential, differing from HSCs in adult bone marrow and E14 fetal liver, which are mostly in G0 and G1 phase, respectively (Zhou et al. 2016; Bowie et al. 2006). Recently, $CD45^+$ pre-HSCs and LMPs in IAHCs have been transcriptomically distinguished, and the corresponding feature genes *Eya2*, *Procr*, *Cd27*, and *Mecom* for pre-HSCs and *Myc*, *Il7r*, *Fcer1g*, and *Gata1* for HSC-independent progenitors are identified, respectively (Zhu et al. 2020). As the signature genes for pre-HSCs (Zhou et al. 2016), Procr and Hlf expression-marked critical populations in mid-gestational embryos not only have HSC potential but also physiologically contribute to and sustain the size of the HSC pool in adult bone marrow evidenced by genetic lineage tracing studies (Tang et al. 2021; Yokomizo et al. 2019; Zheng et al. 2019). Through iterative single-cell analyses, the immunophenotype of $CD31^{hi}SSC\text{-}A^{lo}Kit^{hi}Gata2^{med}CD27^{med}$ has been recently identified to enrich all functional HSCs in IAHCs, which are localized to aortic clusters containing one to two cells (Vink et al. 2020).

From the view of lineage differentiation bias and reconstitution capacity, HSCs in adult bone marrow and fetal liver have functional heterogeneity (Dykstra et al. 2007; Benz et al. 2012; Crisan and Dzierzak 2016). During development, HSC heterogeneity exists from pre-HSC stage, showing predominantly the myeloid-deficient (γ) type and the lymphomyeloid-balanced (β) type; the latter has a stronger capacity for reconstruction and self-renewal (Ye et al. 2017). Whether the pre-HSCs are fated to become a specific heterogeneous subtype intrinsically at the HEC stage or they are in response to extrinsic cues from the microenvironment remains to be further investigated.

1.2 HSC-Dependent and HSC-Independent Origin of Lymphopoiesis and Myelopoiesis

Generation of T and B lymphocyte is generally believed to represent a hallmark of HSC-dependent definitive hematopoiesis. HSCs generated in the AGM region migrate to fetal liver and finally reside in bone marrow, where they continually generate lymphoid lineages. Nevertheless, emerging evidences have elucidated the existence of lymphoid precursors before embryonic HSC emergence. B lymphocyte and T lymphocyte potential can be detected both in the extraembryonic yolk sac and within the embryo at E8.5 and E9.5 (Liu and Auerbach 1991; Cumano et al. 1993; Huang et al. 1994). In vivo and in vitro study suggested that lymphoid potential is predominantly detected in the arterial vessel (Boisset et al. 2010; Uenishi et al. 2018). A further report demonstrated that both the nonarterial and arterial yolk sac endothelial cells have adult-repopulating lymphoid potential (Wang et al. 2021a). Progenitors isolated from yolk sac and para-aortic splanchnopleura (P-Sp) region before the onset of HSCs or in the embryos lacking HSCs could generate innate-like B-1a cells, an IgM-expressing subset eliciting T cell-independent antibody response, but not B-2 cells (Yoshimoto et al. 2011; Kobayashi et al. 2014; Ghosn et al. 2016; Hadland et al. 2017). Although HSCs could differentiate into B-1a cells (Kristiansen et al. 2016; Beaudin et al. 2016; Elsaid et al. 2021), most B-1a cells persisting into adult life are believed to arise from HSC-independent progenitors with an endothelial origin (Sawai et al. 2016; Sawen et al. 2018; Kobayashi et al. 2021). By using $Ncx1^{-/-}$ mouse model devoid of blood circulation, researchers found yolk sac and P-Sp cells at E9.5 have the potential to differentiate into both αβ and γδ T cells (Yoshimoto et al. 2012). Consistent with these findings, fate mapping studies confirmed that γδ dendritic epidermal T lymphocytes (DETCs) are generated independently of HSCs and instead originate from yolk sac hematopoiesis, which can self-renew clonally in the adult mice (Gentek et al. 2018a). It is now acknowledged that except IAHCs, yolk sac could also produce LMPs, a subset of HSC-independent progenitor downstream hemogenic endothelial cells (Ghosn et al. 2019). LMPs have been first identified as $Lin^{-}Kit^{+}Rag\text{-}1^{+}IL\text{-}7R\alpha^{+}$ in the E9.5 yolk sac, and these cells could contribute to fetal lymphopoiesis and myelopoiesis in vivo (Boiers et al. 2013). Besides, these LMPs were found to colonize the thymus rudiment at E11.25 independent of Notch signaling (Luis et al. 2016). However, a later report found E9.5 yolk sac-derived progenitors showing a transient expression of lymphoid transcripts fails to generate lymphoid progeny (Elsaid et al. 2021). Another study showed Flt3-Cre fate mapping embryonic multipotent progenitors (eMPPs) independent of HSCs are the predominant source of lymphoid lineage in adult mice (Patel et al. 2022). Although the physiological role of HSC-independent multipotent progenitors has been elucidated gradually, the relative contributions of these cells to lymphopoiesis, as well as the lineage relationship of them with downstream HSC-independent T- and B-committed progenitors, need to be further explored.

Myeloid cells, which include granulocytes, monocytes, macrophages, mast cells, and dendritic cells, are involved in important and diverse biological processes, such as organogenesis, tissue homeostasis, and innate immunity during development and after birth (Rae et al. 2007). For the past decades, researchers have been exploring the origin of myeloid cells through a combination of different approaches such as bone marrow transplantation, parabiosis, and genetic lineage tracing, revealing that both HSC-dependent and HSC-independent hematopoiesis play a role in the myeloid system in both embryonic and adult tissues (Dzierzak and Speck 2008; Hashimoto et al. 2013; Hoeffel and Ginhoux 2018; Liu et al. 2019; Neo et al. 2021). In mice, HSC-independent embryonic origins of some of these myeloid lineages have been well studied, such as in tissue-resident macrophages. All tissues at birth are populated with fetal macrophages derived from HSC-independent hematopoietic progenitors. Of note, only some tissue macrophages are replaced with time via the recruitment of HSC-derived circulating $Ly6C^{hi}$ monocytes and their subsequent differentiation into tissue macrophages, including the intestine (with the exception in certain anatomical parts of the intestine), dermis, heart, and pancreas macrophages (Scott et al. 2014; Viola and Boeckxstaens 2021). Some tissue-resident macrophage populations, such as microglia, epidermal Langerhans cells, liver Kupffer cells, and alveolar macrophages, exhibit negligible need for replacement in adulthood (Neo et al. 2021; Ginhoux and Guilliams 2016). The origin of macrophages in some adult tissues still remains elusive, such as in the peritoneum, fat, and arteries, as different genetic lineage tracing mouse models might achieve inconsistent conclusions (Weisberg et al. 2006; Yona et al. 2013; Amano et al. 2014; Sheng et al. 2015; Ensan et al. 2016; Weinberger et al. 2020). Similarly, tissue-resident mast cells in adults also have multiple sources of origin, and the mast cells in connective tissues (except adipose tissue and chest cavity) are mainly derived from erythroid-myeloid progenitors produced by yolk sac (Gentek et al. 2018b; Li et al. 2018). Up to now, granulocytes, such as basophils, neutrophils, and eosinophils, and dendritic cells are considered to be the progenies of adult bone marrow HSCs. Although erythroid-myeloid progenitors may be the precursors of the first granulocytes in embryonic stage (McGrath et al. 2015), it has not been reported how long they will last and whether they will be replaced or not.

In the past 10 years, it has become clear that HSC-independent hematopoietic cells have previously unanticipated roles in both embryos and adults. However, there are still many open questions about their exact role, origins, and contributions. The overlapping and transient nature of HSC-dependent and HSC-independent hematopoietic waves during development and the lack of specific fate tracing models to distinguish them make it challenging to determine their physiological contributions to embryonic organogenesis and adult hematopoietic system. Notably, fetal HSCs minimally contribute to the generation of progenitors and functional blood cells before birth (Yokomizo et al. 2022). In conclusion, it is now well established that HSC-independent hematopoiesis is essential for embryonic organogenesis and its progeny can, and does, persist after birth. This has opened up a new and fascinating field of hematopoiesis.

1.3 Epigenetic Regulation of HSC Formation

The formation of HSC from endothelial cells is accompanied by transcriptional and epigenetic regulation including chromatin architecture, chromatin accessibility, histone modification and methylation, etc. Recently, accumulating single-cell transcriptomic datasets serve as important sources for the identification of new regulators, and the advancement of epigenetic technologies with low input allows us to decipherer the EHT process with unprecedented precision.

As one of the most pivotal transcription factors for embryonic hematopoiesis, Runx1 is required for the EHT and consequent HSC formation but not thereafter (Chen et al. 2009; Gao et al. 2018; Yzaguirre et al. 2018; de Bruijn and Dzierzak 2017). Based on a recent single-cell transcriptome dataset involving the whole EHT process in AGM region, Runx1 dosage is proven to regulate the efficiency of the pre-HEC to HEC transition (Zhu et al. 2020). Serving as the direct downstream targets of Runx1, transcriptional factors Gfi1 and Gfi1B can silence the expression of endothelial characteristic-related genes by recruiting histone demethylase LSD1 to target genes, thus promoting HSC generation (Thambyrajah et al. 2016). Even in the case of Runx1 deficiency, forced expression of GFI1/GFI1B still leads to the downregulation of endothelial genes and the production of hematopoietic cells. Therefore, GFI1/GFI1B are considered to be key molecules in the first step of the EHT (Lancrin et al. 2012).

Taking advantage of the single-cell transcriptome dataset of HSC ontogeny, a dynamic long noncoding RNA (lncRNA) landscape of HSC development is constructed (Zhou et al. 2016, 2019). Computational and functional screening identifies lncRNA-H19 being pivotal for the endothelial-to-HSC transition in the AGM region. Mechanically, lncRNA-H19 in trans regulates the promoter demethylation and consequent activation of several critical hematopoietic transcription factors, including Runx1 and Spi1 (Zhou et al. 2019). In contrast, H19 maintains HSC quiescence in the adult bone marrow by serving as a source of miR-675 to restrict IGF2-IGF1R pathway activation, rather than functioning in the form of lncRNA (Venkatraman et al. 2013). Considering the prominent role of DNA methylation modification in regulating gene expression, the dynamic DNA methylation landscape of HSC development is further constructed. Interestingly, during HEC specification from upstream arterial endothelial cells, the differentially methylated region-associated genes are involved in hematopoiesis—but not endothelial cell-related biological processes, suggesting that DNA methylation modification participates in the hematopoietic fate priming but not endothelial feature extinguishing during this process (Li et al. 2021b). On the other hand, a dynamic RNA splicing landscape during entire HSC ontogeny is also constructed with the use of single-cell full-length transcriptome data (Zhou et al. 2016; Wang et al. 2022). EHT process is accompanied by a significant alternative splicing modality switch, which is mainly orchestrated by the splicing regulator Srsf2. Loss of Srsf2 from the endothelial stage leads to impaired HSC generation along with a profound disruption to the gene expression programs required for EHT (Wang et al. 2022).

Joint single-cell RNA sequencing and single-cell assay for transposase-accessible chromatin (ATAC) sequencing analysis reveals that pre-HEC stage is characterized by increased accessibility of chromatin enriched for SOX, FOX, GATA, and SMAD motifs. Moreover, a distal putative Runx1 enhancer displays high chromatin accessibility specifically in pre-HECs but loses accessibility thereafter (Zhu et al. 2020). With optimized low-input itChIP-seq combined with Hi-C assays, HSC regulatory regions are found to be already pre-configurated with active histone modifications as early as in the arterial endothelial stage upstream of HECs, presumably corresponding to pre-HEC stage, preceding chromatin looping dynamics within topologically associating domains (Zhu et al. 2020; Hou et al. 2020; Li et al. 2022). Specifically, half of Runx1 binding-mediated chromatin looping structures between enhancers and promoters are observed prior to HEC stage, and Runx1 and co-transcription factors together constitute a central, progressively intensified enhancer-promoter interactions along HSC ontogeny (Li et al. 2022). Moreover, with the use of a low cell number ChIP-seq protocol to profile the genome-wide histone modifications over the course of HSC ontogeny, the developmental stage-specific transcriptional regulatory networks are constructed, and two broadly expressed transcription factors, SP3 and MAZ, are predicted and validated to play a role in the formation of HECs (Gao et al. 2020).

In conclusion, with the advancing of single-cell epigenomics and spatial omics technology, researchers can accurately decipher the hematopoietic microenvironment and mechanism of HSC formation, which will provide important prospects for the regeneration of HSC in vitro.

1.4 EHT and the Generation of Human HSCs

The first HSCs in human embryos are detected in the AGM region from mid 5 weeks postconception (PCW)/Carnegie stage 14 (CS14; around embryonic day 32) and express surface proteins including CD34, CD45, VE-cadherin, Kit, THY1, and endoglin whose homologues are expressed by mouse HSCs (Hou et al. 2020; Ivanovs et al. 2011, 2014). Extreme dilution transplantation analysis showed that human embryos at 5–6 PCW (CS14–17, embryonic day 32–42) contain a mean number of less than 1 functional HSCs that can generate more than 300 daughter HSCs and reconstruct all blood lineages in the primary recipients (Ivanovs et al. 2011).

Precise dissection of dorsal aorta combined with functional assay revealed that the early hematopoietic capacity is principally distributed to the ventral part of the middle aorta (Ivanovs et al. 2014; Tavian et al. 1999). This distribution corresponds to that of IAHCs which start to appear at 4 PCW (CS12, around embryonic day 26) on the ventral floor of human AGM region (Tavian et al. 1996, 1999). IAHCs share endothelial feature markers and present a gradually increased expression of hematopoietic markers toward the top of the clusters (Tavian et al. 1999). The first HSCs in human embryos are therefore believed to locate within these IAHCs.

Resembling the conclusion drawn in mouse embryos, the IAHCs and the HSCs themselves are believed to be generated from transient hemogenic endothelial cells featured by the upregulation of key hemogenic transcription factor RUNX1 (Zovein et al. 2008; Chen et al. 2009; Oberlin et al. 2002; Zeng et al. 2019a). This kind of specialized endothelial cells goes through a series of changes of key gene expression and a morphological transition from flattened to round, which is so called EHT, to finally generate IAHCs which contain HSCs (Zeng et al. 2019a; Ivanovs et al. 2017). Plenty of molecules from SCF, BMP, Notch, WNT, and TGF-β signals are supposed to be involved in this process, both in vivo and in vitro (Zeng et al. 2019a; Marshall et al. 2000).

Given the poor investigating methods that can be used in human embryos, the lately developed single-cell molecular sequencing technologies offer an especially comprehensive view for our understanding of human EHT and HSCs. Based on single-cell RNA sequencing technology, a developmental pathway from primitive endothelial cells to hemogenic endothelial cells and finally to HSCs in human embryos is constructed, along which whole-genome molecular events are depicted by stages (Zeng et al. 2019a; Calvanese et al. 2022). The arterial feature of HSC-primed hemogenic endothelial cells is re-emphasized, especially when compared with an earlier hemogenic endothelial population irrelevant to the generation of HSCs (Zeng et al. 2019a; Calvanese et al. 2022). $WT1^+$, $DLK1^+$, and $FBLN5^+$ mesenchymal cells in the AGM region are predicted to contribute the signals required in the development of hemogenic endothelial cells, including Notch, BMP, and TGF-β signals (Zeng et al. 2019a). CD44 showing a gradual upregulation along the EHT course is screened out as a surface marker to potentially enrich HSC-primed hemogenic endothelial cells. Downstream of the $CD44^+$ hemogenic endothelial cells exist heterogeneous HSPCs featured by relatively high expression of arterial genes, cell cycle genes, and *GFI1B*, respectively, corresponding to the IAHCs that proceed to functional HSCs (Zeng et al. 2019a). The expression profile of HSCs in human early embryos is defined as $HOXA9^+MLLT3^+MECOM^+$ $HLF^+SPINK2^+$ and can be found in AGM region, placenta, and yolk sac but not in other tissues including the head and heart at CS14, implying the extraembryonic distribution of HSCs before their colonization into the liver (Calvanese et al. 2022).

About 6 days after the appearance of the first HSCs in the AGM region, HSC can be detected in the yolk sac (Ivanovs et al. 2011). Though definitive hematopoietic progenitors like yolk sac-derived myeloid-primed progenitors are derived through EHT in this site, there is no valid evidence of the independent generation of HSCs in the yolk sac (Bian et al. 2020). No HSCs with multi-lineage potential are detected in the embryonic liver till 8 PCW (CS17, around embryonic day 42), indicating that the seeding of HSCs into liver happens at later stages (Ivanovs et al. 2011). HSCs in human embryonic liver are enriched in a $CD34^+CD144^{low}CD45^{low}$ population, where angiotensin-converting enzyme (ACE) expression identifies the proliferating ones (Zhang et al. 2019). PROCR, whose homologue is expressed by mouse pre-HSCs and HSCs, is as well reported as a surface marker for enriching functional HSCs in human fetal liver (Zhou et al. 2016; Subramaniam et al. 2019; Fares et al. 2017). MLLT3, a component of super elongation complex, binds to transcription start sites of active genes in human fetal liver HSPCs and maintains stemness of HSCs (Calvanese et al. 2019, 2022). Glycophosphatidylinositol-anchored surface protein GPI-80 was also reported to define hematopoietic stem/progenitor cells with self-renewal ability in human fetal liver (Prashad et al. 2015). Beginning at 12 weeks postconception (PCW), functional HSCs arrive at fetal bone marrow and become functionally active at later stages (Zheng et al. 2022). In fetal bone marrow, pleiotrophin (PTN) derived from *CXCL12*-abundant reticular cells is required for the expansion of human HSCs. Other signals from reticular and arteriolar endothelial cells via receptors including *CD44*, *NOTHC1/2*, and *CD74* are also involved in the regulation of migration and retention of HSCs in human fetal bone marrow (Zheng et al. 2022).

There remains more to investigate about the generation of human embryonic HSCs. Though HSCs are identified in transcriptomic data, their number vastly exceeds that of functional HSCs in human AGM region suggested by transplantation assay (Ivanovs et al. 2011), indicating that knowledge obtained from transcriptomic data about human embryonic HSCs may be contaminated by hematopoietic progenitors that molecularly resemble HSCs. The temporally unrecognized counterparts of murine pre-HSCs would probably be hidden in these progenitors. Also, the lack of efficient culture system able to support the transition from endothelial or hemogenic endothelial cells to functional HSCs hampers the validation of the mediate populations as well as the confirmation of inter- and intracellular regulatory mechanisms related to EHT implied in the transcriptomic data. The continuous development of new technologies and optimization of ESC/iPSC induction system would hopefully help to complement our understanding of human embryonic HSC and its generation and to establish sufficient instructions for the generation of functional HSCs in vitro.

1.5 Development of Multiple Hematopoietic Lineages in Human Embryos

T and B lymphocytes are pivotal lymphoid outputs of hematopoiesis and represent the major arms of adaptive immunity. T lymphocyte's development initiates from the emigration of progenitors with lymphoid differentiation tendency, termed thymus seeding progenitors (TSPs), from hematopoietic sites to the thymus in 8 PCW (Farley et al. 2013; Zlotoff et al. 2008, 2010). The TSPs develop into early thymic progenitors (ETPs) and commit to T cell fate under the drive of Notch signal from the environment, followed by T cell receptor (TCR) recombination and selection (Rothenberg 2019). By 15 PCW, the first mature T cells occurred in the thymus. The precise regulatory mechanisms of T cell development stages in the thymus have been studied in depth (Hosokawa and Rothenberg 2021; Yui and Rothenberg 2014). However, less is known about the identity and characterization of T progenitors due to the rare number of TSPs and ETPs as well as lack of the appropriate technique (Rothenberg 2021; Krueger et al. 2017). By performing single-cell transcriptomic analysis with cells from human embryonic and fetal thymus and upstream hematopoietic organs, the identity and heterogeneity of ETPs were both clarified (Zeng et al. 2019b; Zhou and Rothenberg 2019; Liu et al. 2021a). Furthermore, two branches, one of which was directed to ETPs, were observed in lymphoid progenitors of human fetal liver, suggesting their TSP properties (Zeng et al. 2019b). TSPs highly expressed *TYROBP*, *IL3RA*, and *IRF8*, and the other branch was enriched by expression of B lymphocyte lineage genes like *CD79A*, *CD79B*, and *VPREB3* (Zeng et al. 2019b). This work, together with another work focused on the later stages in the thymus using TCR repertoire analysis, described a continuous development path from ETPs to multiple mature T cell types (Park et al. 2020a).

The maturation of B lymphocytes needs support from liver microenvironment in early human fetus. Their progenitors could be first observed in 7 PCW, while mature B cells only appear in 9 PCW (Popescu et al. 2019; Park et al. 2020b). During the second trimester of pregnancy, bone marrow replaces fetal liver as the main source of B cells, and a large number of B cells begin to appear in the spleen (Nuñez et al. 1996).

Innate lymphoid cells (ILCs) are known as innate counterparts of T lymphocytes considering the corresponding transcription factors driving their development and the cytokines they release (Artis and Spits 2015; Colonna 2018; Vivier et al. 2018). The hierarchical development of HSC-derived ILCs in early human fetus has also been revealed recently (Liu et al. 2021b). By computational analysis of cells from fetal hematopoietic, lymphoid and nonlymphoid tissues, from 8 to 12 PCW, together with functional validation at bulk and single-cell levels, interleukin-3 receptor (IL-3RA) was shown to be a surface marker for the lymphoid progenitors in fetal liver with T, B, ILC, and residual myeloid potentials. The lymphoid progenitors without expression of IL-3RA were mostly committed to B lymphocytes, which is consistent with previous study (Zeng et al. 2019b).

While HSC-independent lymphopoiesis has been studied thoroughly in mouse (Tavian et al. 2001; Yokota et al. 2006; Yoshimoto et al. 2012), it is unclear whether this situation exists in human embryos and, if so, to which kind of lymphocytes it contributes. Transfer experiments showed that hematopoietic progenitors with long-term uni-lineage T cell engraftment capacity were detected in AGM from mid 5 PCW (CS14), the time when the first HSCs emerge, and early 6 PCW (CS16) and then in liver at late 6 PCW

(CS17) (Ivanovs et al. 2011). Most recently, lymphoid progenitors with expression of *IL7R* were also captured in cells from AGM as early as CS14 using single-cell RNA sequencing. Notably, these IL7R-expressing progenitors showed little correlation with T progenitors from the thymus. Both evidences suggest the existence of HSC-independent lymphopoiesis in early human, consistent with that reported in mice. However, the identity of lymphoid outputs, their roles in early human, and their duration are worthy of further investigation.

Although the characteristics and development of myeloid cells in mice have been characterized, they are not well understood in humans. Advances in single-cell multi-omics approaches have brought new opportunities for the systematic understanding of myeloid development.

Macrophages are considered the earliest tissue-resident myeloid cells (TRMs); the origin and specialization of TRMs are characterized in detail in mice with two yolk sac-derived waves: a monocyte-independent primitive wave in early yolk sac and a later fetal liver monocyte-derived wave. However, early macrophage development during human embryogenesis is not well understood due to the extremely scarce cell numbers and limited access to embryonic tissues. The first hematopoietic landscape in early human embryos was provided by carrying out single-cell RNA sequencing on the $CD45^+$ hematopoietic cells of multiple tissues spanning from 3 to 8 PCW (Bian et al. 2020). Notably, a population of $CD45^+CD34^+CD44^+$ yolk sac-derived myeloid-biased progenitors (YSMPs) was identified, which appears from 3 PCW (CS11). Compared to erythromyeloid progenitors (EMPs) in mice, these progenitors exhibit much weaker erythroid but stronger myeloid signature and differentiation potential. Furthermore, prior to the appearance of the first HSCs in the human fetal liver, both two yolk sac-derived hematopoietic waves contribute to the TRM pools, which mirrors the process seen in mice. One is from $HBE1^+$ yolk sac-derived primitive macrophages, mainly represented by microglia. The other one is from YSMPs, which migrate into the fetal liver to differentiate into monocytes and then specialize into macrophages in diverse developing tissues. However, due to the lack of available lineage tracing techniques in humans, the amount of contribution of two-wave yolk sac hematopoiesis to diverse TRMs remains unclear and requires further study utilizing new technologies.

After the colonization of tissues and organs, macrophages will undergo functional specialization with the support of the local microenvironment (Varol et al. 2015). Different TRMs do not specialize at the same rate, with the specialization toward microglia in the brain and Kupffer cells in the liver occurring earlier (Bian et al. 2020). Besides, their molecular characteristics also change over time; early TRMs mainly show stronger expression related to proliferation, chemotaxis, and pro-inflammatory signals, such as TNF and NF-KB, which have been implicated in organogenesis and angiogenesis (Jeucken et al. 2019; Yang et al. 2006; Heidemann et al. 2003), while macrophages in later stages of gestation highly express characteristics related to antigen presentation and immune response (Suo et al. 2022), suggesting that macrophages play different biological functions along the different stages of human embryonic development (Suo et al. 2022; Wood and Martin 2017).

Human monocytes are first seen in the fetal liver from as early as about 4 PCW (CS12), featured by the expression of *CCR2* and *CSF1R*, and then could be detected in the fetal bone marrow and peripheral tissues (Bian et al. 2020). Notably, the migration and colonization of monocytes in different sites are accompanied by the phenotypic transition. The $CXCR4^{hi}$ monocytes are enriched in the fetal bone marrow, and the monocytes in peripheral organs highly express *CCR2* and then *IL1B* (Suo et al. 2022). Unlike monocytes, neutrophils remain absent in the fetal liver, until the appearance of mature neutrophils at around 12 PCW in bone marrow (Popescu et al. 2019; Slayton et al. 1998; Ohls et al. 1995; Jardine et al. 2021), which is in contrast to the earliest appearance of neutrophils in the fetal liver in mice. Monocytes and neutrophils are considered to originate from the granulocyte-monocyte progenitors (GMPs), which are first detected in the fetal liver and subsequently in the fetal bone marrow. As the embryo develops, GMPs, as a heterogeneous population with different transcriptome profiles, express higher levels of characteristics of monocytes in the fetal liver but stronger granulocyte characteristics in fetal bone marrow (Jardine et al. 2021), suggesting their different differentiation potential at various hematopoietic sites and explaining the absence of mature granulocytes in the fetal liver.

Erythrocytes and megakaryocytes are thought to stem from megakaryocyte-erythroid progenitors, a bipotential precursor population during hematopoiesis (Debili et al. 1996; Tober et al. 2007). In human, nucleated erythrocytes derived from primitive hematopoiesis are the most abundant hematopoietic cells in yolk sac as early as 3 PCW, while definitive erythroid cells in fetal liver and bone marrow undergo enucleation and give rise to mature erythrocytes by proliferating and differentiation (Migliaccio et al. 1986). The important changes accompanying embryonic erythropoiesis are hemoglobin switch, from embryonic $\zeta 2\varepsilon 2$ to fetal $\alpha 2\gamma 2$ and finally to adult $\alpha 2\beta 2$ (Peschle et al. 1985). A single-cell RNA-seq study revealed the expression of embryonic globin genes of primitive erythroblasts in a gastrulating human embryo at 3 PCW, consistent with that in mouse counterparts (Tyser et al. 2021). It is still uncertain that there are multiple waves of erythropoiesis in human embryonic development like that in mouse embryos, although most of erythrocytes are thought to be HSC-independent until fetal hematopoiesis after 7 PCW (Popescu et al. 2019). Unlike the classical stages of proerythroblast (ProE), basophilic erythroblast (BasoE), polychromatophilic erythroblast (PolyE), and orthochromatic erythroblast (OrthoE) identified in adult terminal erythroid differentiation (An et al. 2014; Huang et al. 2020), erythrocytes in yolk sac, fetal liver, and umbilical cord blood exhibit prominent distinctions in cell proliferation, metabolic features in oxygen, and metal homeostasis (Xu et al. 2022). In addition to the main role of erythrocytes in oxygen transport, an immune-prone erythroid subpopulation across fetal to adult hematopoiesis with an immunophenotype of $CD71^+GYPA^+CD63^+$ exerts the immunomodulatory role on other blood cells (Xu et al. 2022).

Whereas adult megakaryocyte-poiesis takes on platelet production inside the bone marrow and blood, the many faces of megakaryocytes have been revealed during human hematopoietic development (Notta et al. 2016; Izumi 1987; Ma et al. 1996). The presence of megakaryocytes in human embryo can be traced back to the 4 PCW, the timing before which first wave or primitive hematopoiesis occurs in yolk sac (Wang et al. 2021b). The single-cell RNA-seq analyses revealed that megakaryocytes in yolk sac exhibit glycolysis-biased metabolic fingerprints, while megakaryocytes in fetal liver prefer to proliferate and form polyploids (Wang et al. 2021b). Nevertheless, it is hard to distinguish megakaryocytes that originated from primitive or definitive hematopoiesis. The expression of THBS1 marks the megakaryocytic fate of progenitors in hemogenic endothelium stage (Wang et al. 2021b). Fetal megakaryocytes display two differentiating routes and functions of proplatelet generation, niche-supporting and immune response, similar to those in adult megakaryocytes (Wang et al. 2021b; Liu et al. 2021c; Castro-Malaspina et al. 1981). After 5 WPC, the fetal liver functions as the main hematopoietic organ (Migliaccio et al. 1986; Tavian and Peault 2005), producing a large number of mature megakaryocytes (Popescu et al. 2019; Timens and Kamps 1997). Until neonatal period, megakaryocytes are derived from megakaryocytic-erythroid progenitors (MEPs) (Psaila et al. 2016) or megakaryocyte-erythroid-mast cell progenitor (MEMP) (Popescu et al. 2019), intermediate progenitors in HSC hierarchy.

As two of the earliest hematopoietic cell types in yolk sac, erythrocytes, MEPs and erythro-myeloid progenitors were identified in gastrulating human embryo at 3 PCW (Tyser et al. 2021). Functions of these cells have been less studied except for single-cell sequencing analyses in recent years. The characteristics of megakaryocytic-erythroid lineage peculiar to embryonic hematopoietic development need to be elucidated in future.

With the advance of single-cell technologies, researchers have begun to unravel the cellular and molecular events underlying hematopoietic development in human embryos/fetus (Bian et al. 2020; Popescu et al. 2019; Suo et al. 2022; Jardine et al. 2021). However, due to the ethical limitations of reverse genetics research, mechanisms regulating human hematopoietic development are mainly shaped by studies either in non-primate models other than primates or from in vitro differentiation systems using pluripotent stem cells. Given the higher similarity between humans and nonhuman primates, and the development in gene editing technology, the nonhuman primate (NHP) undoubtedly serves as a more suitable model to study molecular regulation of blood/immune cell development in vivo in primates. Additionally, the ability to cell fate trace in humans will be necessary to gain comprehensive insights into human hematopoietic development. Unfortunately, genetic barcoding or inducible labeling systems, although powerful, cannot be used to study human hematopoiesis in vivo. It is worth mentioning that lineage tracing technology, represented by endogenous genome mutations, is expected to provide important technical tools for solving these problems in the future (Ludwig et al. 2019).

References

Amano SU et al (2014) Local proliferation of macrophages contributes to obesity-associated adipose tissue inflammation. Cell Metab 19:162–171. https://doi.org/10.1016/j.cmet.2013.11.017

An X et al (2014) Global transcriptome analyses of human and murine terminal erythroid differentiation. Blood 123:3466–3477. https://doi.org/10.1182/blood-2014-01-548305

Artis D, Spits H (2015) The biology of innate lymphoid cells. Nature 517:293–301. https://doi.org/10.1038/nature14189

Baron CS et al (2018) Single-cell transcriptomics reveal the dynamic of haematopoietic stem cell production in the aorta. Nat Commun 9:2517. https://doi.org/10.1038/s41467-018-04893-3

Beaudin AE et al (2016) A transient developmental hematopoietic stem cell gives rise to innate-like B and T cells. Cell Stem Cell 19:768–783. https://doi.org/10.1016/j.stem.2016.08.013

Benz C et al (2012) Hematopoietic stem cell subtypes expand differentially during development and display distinct lymphopoietic programs. Cell Stem Cell 10:273–283. https://doi.org/10.1016/j.stem.2012.02.007

Bian Z et al (2020) Deciphering human macrophage development at single-cell resolution. Nature 582:571–576. https://doi.org/10.1038/s41586-020-2316-7

Boiers C et al (2013) Lymphomyeloid contribution of an immune-restricted progenitor emerging prior to definitive hematopoietic stem cells. Cell Stem Cell 13:535–548. https://doi.org/10.1016/j.stem.2013.08.012

Boisset JC et al (2010) In vivo imaging of haematopoietic cells emerging from the mouse aortic endothelium. Nature 464:116–120. https://doi.org/10.1038/nature08764

Boisset JC et al (2015) Progressive maturation toward hematopoietic stem cells in the mouse embryo aorta. Blood 125:465–469. https://doi.org/10.1182/blood-2014-07-588954

Bowie MB et al (2006) Hematopoietic stem cells proliferate until after birth and show a reversible phase-specific engraftment defect. J Clin Invest 116:2808–2816. https://doi.org/10.1172/JCI28310

de Bruijn M, Dzierzak E (2017) Runx transcription factors in the development and function of the definitive hematopoietic system. Blood 129:2061–2069. https://doi.org/10.1182/blood-2016-12-689109

de Bruijn MF, Speck NA, Peeters MC, Dzierzak E (2000) Definitive hematopoietic stem cells first develop within the major arterial regions of the mouse embryo. EMBO J 19:2465–2474. https://doi.org/10.1093/emboj/19.11.2465

Calvanese V et al (2019) MLLT3 governs human haematopoietic stem-cell self-renewal and engraftment. Nature 576:281–286. https://doi.org/10.1038/s41586-019-1790-2

Calvanese V et al (2022) Mapping human haematopoietic stem cells from haemogenic endothelium to birth. Nature 604:534–540. https://doi.org/10.1038/s41586-022-04571-x

Castro-Malaspina H, Rabellino EM, Yen A, Nachman RL, Moore MA (1981) Human megakaryocyte stimulation of proliferation of bone marrow fibroblasts. Blood 57:781–787

Chen MJ, Yokomizo T, Zeigler BM, Dzierzak E, Speck NA (2009) Runx1 is required for the endothelial to haematopoietic cell transition but not thereafter. Nature 457:887–891. https://doi.org/10.1038/nature07619

Colonna M (2018) Innate lymphoid cells: diversity, plasticity, and unique functions in immunity. Immunity 48:1104–1117. https://doi.org/10.1016/j.immuni.2018.05.013

Crisan M, Dzierzak E (2016) The many faces of hematopoietic stem cell heterogeneity. Development 143:4571–4581. https://doi.org/10.1242/dev.114231

Cumano A, Furlonger C, Paige CJ (1993) Differentiation and characterization of B-cell precursors detected in the yolk sac and embryo body of embryos beginning at the 10- to 12-somite stage. Proc Natl Acad Sci U S A 90:6429–6433. https://doi.org/10.1073/pnas.90.14.6429

Debili N et al (1996) Characterization of a bipotent erythro-megakaryocytic progenitor in human bone marrow. Blood 88:1284–1296

Dignum T et al (2021) Multipotent progenitors and hematopoietic stem cells arise independently from hemogenic endothelium in the mouse embryo. Cell Rep 36:109675. https://doi.org/10.1016/j.celrep.2021.109675

Dykstra B et al (2007) Long-term propagation of distinct hematopoietic differentiation programs in vivo. Cell Stem Cell 1:218–229. https://doi.org/10.1016/j.stem.2007.05.015

Dzierzak E, Speck NA (2008) Of lineage and legacy: the development of mammalian hematopoietic stem cells. Nat Immunol 9:129–136. https://doi.org/10.1038/ni1560

Elsaid R et al (2021) A wave of bipotent T/ILC-restricted progenitors shapes the embryonic thymus microenvironment in a time-dependent manner. Blood 137:1024–1036. https://doi.org/10.1182/blood.2020006779

Ensan S et al (2016) Self-renewing resident arterial macrophages arise from embryonic CX3CR1(+) precursors and circulating monocytes immediately after birth. Nat Immunol 17:159–168. https://doi.org/10.1038/ni.3343

Fadlullah MZ, Neo WH, Lie-A-ling M et al (2021) Murine AGM single-cell profiling identifies a continuum of hemogenic endothelium differentiation marked by ACE. Blood 139:343

Fares I et al (2017) EPCR expression marks UM171-expanded CD34(+) cord blood stem cells. Blood 129:3344–3351. https://doi.org/10.1182/blood-2016-11-750729

Farley AM et al (2013) Dynamics of thymus organogenesis and colonization in early human development. Development 140:2015–2026. https://doi.org/10.1242/dev.087320

Gao L et al (2018) RUNX1 and the endothelial origin of blood. Exp Hematol 68:2–9. https://doi.org/10.1016/j.exphem.2018.10.009

Gao P et al (2020) Transcriptional regulatory network controlling the ontogeny of hematopoietic stem cells. Genes Dev 34:950–964. https://doi.org/10.1101/gad.338202.120

Gentek R et al (2018a) Epidermal gammadelta T cells originate from yolk sac hematopoiesis and clonally self-renew in the adult. J Exp Med 215:2994–3005. https://doi.org/10.1084/jem.20181206

Gentek R et al (2018b) Hemogenic endothelial fate mapping reveals dual developmental origin of mast cells. Immunity 48:1160-1171 e1165. https://doi.org/10.1016/j.immuni.2018.04.025

Ghosn EE et al (2016) Fetal hematopoietic stem cell transplantation fails to fully regenerate the B-lymphocyte compartment. Stem Cell Rep 6:137–149. https://doi.org/10.1016/j.stemcr.2015.11.011

Ghosn E, Yoshimoto M, Nakauchi H, Weissman IL, Herzenberg LA (2019) Hematopoietic stem cell-independent hematopoiesis and the origins of innate-like B lymphocytes. Development 146:dev170571. https://doi.org/10.1242/dev.170571

Ginhoux F, Guilliams M (2016) Tissue-resident macrophage ontogeny and homeostasis. Immunity 44:439–449. https://doi.org/10.1016/j.immuni.2016.02.024

Hadland BK et al (2017) A common origin for B-1a and B-2 lymphocytes in clonal pre- hematopoietic stem cells. Stem Cell Rep 8:1563–1572. https://doi.org/10.1016/j.stemcr.2017.04.007

Hashimoto D et al (2013) Tissue-resident macrophages self-maintain locally throughout adult life with minimal contribution from circulating monocytes. Immunity 38:792–804. https://doi.org/10.1016/j.immuni.2013.04.004

Heidemann J et al (2003) Angiogenic effects of interleukin 8 (CXCL8) in human intestinal microvascular endothelial cells are mediated by CXCR2. J Biol Chem 278:8508–8515. https://doi.org/10.1074/jbc.M208231200

Hoeffel G, Ginhoux F (2018) Fetal monocytes and the origins of tissue-resident macrophages. Cell Immunol 330:5–15. https://doi.org/10.1016/j.cellimm.2018.01.001

Hosokawa H, Rothenberg EV (2021) How transcription factors drive choice of the T cell fate. Nat Rev Immunol 21:162–176. https://doi.org/10.1038/s41577-020-00426-6

Hou S et al (2020) Embryonic endothelial evolution towards first hematopoietic stem cells revealed by single-cell transcriptomic and functional analyses. Cell Res 30:376–392. https://doi.org/10.1038/s41422-020-0300-2

Hou S et al (2022) Heterogeneity in endothelial cells and widespread venous arterialization during early vascular development in mammals. Cell Res 32:333–348. https://doi.org/10.1038/s41422-022-00615-z

Howell ED, Speck NA (2020) Forks in the road to the first hematopoietic stem cells. Cell Res 30:457. https://doi.org/10.1038/s41422-020-0331-8

Huang H, Zettergren LD, Auerbach R (1994) In vitro differentiation of B cells and myeloid cells from the early mouse embryo and its extraembryonic yolk sac. Exp Hematol 22:19–25

Huang P et al (2020) Putative regulators for the continuum of erythroid differentiation revealed by single-cell transcriptome of human BM and UCB cells. Proc Natl Acad Sci U S A 117:12868–12876. https://doi.org/10.1073/pnas.1915085117

Ivanovs A et al (2011) Highly potent human hematopoietic stem cells first emerge in the intraembryonic aorta-gonad-mesonephros region. J Exp Med 208:2417–2427. https://doi.org/10.1084/jem.20111688

Ivanovs A, Rybtsov S, Anderson RA, Turner ML, Medvinsky A (2014) Identification of the niche and phenotype of the first human hematopoietic stem cells.

Stem Cell Rep 2:449–456. https://doi.org/10.1016/j.stemcr.2014.02.004
Ivanovs A et al (2017) Human haematopoietic stem cell development: from the embryo to the dish. Development 144:2323–2337. https://doi.org/10.1242/dev.134866
Izumi T (1987) Morphometric studies of megakaryocytes in human and rat fetal, infantile and adult hematopoiesis. I. Observations on human fetuses and blood dyscrasias. Hiroshima J Med Sci 36:25–30
Jardine L et al (2021) Blood and immune development in human fetal bone marrow and Down syndrome. Nature 598:327–331. https://doi.org/10.1038/s41586-021-03929-x
Jeucken KCM, Koning JJ, Mebius RE, Tas SW (2019) The role of endothelial cells and TNF-receptor superfamily members in lymphoid organogenesis and function during health and inflammation. Front Immunol 10:2700. https://doi.org/10.3389/fimmu.2019.02700
Kobayashi M et al (2014) Functional B-1 progenitor cells are present in the hematopoietic stem cell-deficient embryo and depend on Cbfbeta for their development. Proc Natl Acad Sci U S A 111:12151–12156. https://doi.org/10.1073/pnas.1407370111
Kobayashi M et al (2021) HSC-independent definitive hematopoietic cells persist into adult life. bioRxiv 2021.2012.2002.468909. https://doi.org/10.1101/2021.12.02.468909
Kristiansen TA et al (2016) Cellular barcoding links B-1a B cell potential to a fetal hematopoietic stem cell state at the single-cell level. Immunity 45:346–357. https://doi.org/10.1016/j.immuni.2016.07.014
Krueger A, Ziętara N, Łyszkiewicz M (2017) T cell development by the numbers. Trends Immunol 38:128–139. https://doi.org/10.1016/j.it.2016.10.007
Lan Y (2022) Deciphering the continuum of hemogenic endothelium differentiation. Blood 139:308–310. https://doi.org/10.1182/blood.2021013853
Lancrin C et al (2012) GFI1 and GFI1B control the loss of endothelial identity of hemogenic endothelium during hematopoietic commitment. Blood 120:314–322. https://doi.org/10.1182/blood-2011-10-386094
Li Z et al (2018) Adult connective tissue-resident mast cells originate from late erythro-myeloid progenitors. Immunity 49:640–653.e645. https://doi.org/10.1016/j.immuni.2018.09.023
Li YQ et al (2021a) Spatiotemporal and functional heterogeneity of hematopoietic stem cell-competent Hemogenic endothelial cells in mouse embryos. Front Cell Dev Biol 9:699263. https://doi.org/10.3389/fcell.2021.699263
Li X et al (2021b) The comprehensive DNA methylation landscape of hematopoietic stem cell development. Cell Discov 7:86. https://doi.org/10.1038/s41421-021-00298-7
Li CC et al (2022) Pre-configuring chromatin architecture with histone modifications guides hematopoietic stem cell formation in mouse embryos. Nat Commun 13:346. https://doi.org/10.1038/s41467-022-28018-z
Liu CP, Auerbach R (1991) In vitro development of murine T cells from prethymic and preliver embryonic yolk sac hematopoietic stem cells. Development 113:1315–1323. https://doi.org/10.1242/dev.113.4.1315
Liu Z et al (2019) Fate mapping via Ms4a3-expression history traces monocyte-derived cells. Cell 178:1509–1525.e1519. https://doi.org/10.1016/j.cell.2019.08.009
Liu C, Lan Y, Liu B, Zhang H, Hu H (2021a) T cell development: old tales retold by single-cell RNA sequencing. Trends Immunol 42:165–175. https://doi.org/10.1016/j.it.2020.12.004
Liu C et al (2021b) Delineating spatiotemporal and hierarchical development of human fetal innate lymphoid cells. Cell Res 31:1106–1122. https://doi.org/10.1038/s41422-021-00529-2
Liu C et al (2021c) Characterization of cellular heterogeneity and an immune subpopulation of human megakaryocytes. Adv Sci (Weinh) 8:e2100921. https://doi.org/10.1002/advs.202100921
Ludwig LS et al (2019) Lineage tracing in humans enabled by mitochondrial mutations and single-cell genomics. Cell 176:1325–1339.e1322. https://doi.org/10.1016/j.cell.2019.01.022
Luis TC et al (2016) Initial seeding of the embryonic thymus by immune-restricted lympho-myeloid progenitors. Nat Immunol 17:1424–1435. https://doi.org/10.1038/ni.3576
Ma DC, Sun YH, Chang KZ, Zuo W (1996) Developmental change of megakaryocyte maturation and DNA ploidy in human fetus. Eur J Haematol 57:121–127. https://doi.org/10.1111/j.1600-0609.1996.tb01349.x
Marshall CJ, Kinnon C, Thrasher AJ (2000) Polarized expression of bone morphogenetic protein-4 in the human aorta-gonad-mesonephros region. Blood 96:1591–1593
McGrath KE et al (2015) Distinct sources of hematopoietic progenitors emerge before HSCs and provide functional blood cells in the mammalian embryo. Cell Rep 11:1892–1904. https://doi.org/10.1016/j.celrep.2015.05.036
Medvinsky A, Dzierzak E (1996) Definitive hematopoiesis is autonomously initiated by the AGM region. Cell 86:897–906. https://doi.org/10.1016/s0092-8674(00)80165-8
Migliaccio G et al (1986) Human embryonic hemopoiesis. Kinetics of progenitors and precursors underlying the yolk sac----liver transition. J Clin Invest 78:51–60. https://doi.org/10.1172/JCI112572
Neo WH, Lie ALM, Fadlullah MZH, Lacaud G (2021) Contributions of embryonic HSC-independent hematopoiesis to organogenesis and the adult hematopoietic system. Front Cell Dev Biol 9:631699. https://doi.org/10.3389/fcell.2021.631699
Notta F et al (2016) Distinct routes of lineage development reshape the human blood hierarchy across ontogeny. Science 351:aab2116. https://doi.org/10.1126/science.aab2116

Nuñez C et al (1996) B cells are generated throughout life in humans. J Immunol 156:866–872

Oberlin E, Tavian M, Blazsek I, Peault B (2002) Blood-forming potential of vascular endothelium in the human embryo. Development 129:4147–4157

Ohls RK et al (1995) Neutrophil pool sizes and granulocyte colony-stimulating factor production in human mid-trimester fetuses. Pediatr Res 37:806–811. https://doi.org/10.1203/00006450-199506000-00022

Park JE et al (2020a) A cell atlas of human thymic development defines T cell repertoire formation. Science 367:eaay3224. https://doi.org/10.1126/science.aay3224

Park JE, Jardine L, Gottgens B, Teichmann SA, Haniffa M (2020b) Prenatal development of human immunity. Science 368:600–603. https://doi.org/10.1126/science.aaz9330

Patel SH et al (2022) Lifelong multilineage contribution by embryonic-born blood progenitors. Nature 606:747–753. https://doi.org/10.1038/s41586-022-04804-z

Peschle C et al (1985) Haemoglobin switching in human embryos: asynchrony of zeta----alpha and epsilon----gamma-globin switches in primitive and definite erythropoietic lineage. Nature 313:235–238. https://doi.org/10.1038/313235a0

Popescu DM et al (2019) Decoding human fetal liver haematopoiesis. Nature 574:365–371. https://doi.org/10.1038/s41586-019-1652-y

Prashad SL et al (2015) GPI-80 defines self-renewal ability in hematopoietic stem cells during human development. Cell Stem Cell 16:80–87. https://doi.org/10.1016/j.stem.2014.10.020

Psaila B et al (2016) Single-cell profiling of human megakaryocyte-erythroid progenitors identifies distinct megakaryocyte and erythroid differentiation pathways. Genome Biol 17:83. https://doi.org/10.1186/s13059-016-0939-7

Rae F et al (2007) Characterisation and trophic functions of murine embryonic macrophages based upon the use of a Csf1r-EGFP transgene reporter. Dev Biol 308:232–246. https://doi.org/10.1016/j.ydbio.2007.05.027

Rothenberg EV (2019) Programming for T-lymphocyte fates: modularity and mechanisms. Genes Dev 33:1117–1135. https://doi.org/10.1101/gad.327163.119

Rothenberg EV (2021) Single-cell insights into the hematopoietic generation of T-lymphocyte precursors in mouse and human. Exp Hematol 95:1–12. https://doi.org/10.1016/j.exphem.2020.12.005

Rybtsov S et al (2011) Hierarchical organization and early hematopoietic specification of the developing HSC lineage in the AGM region. J Exp Med 208:1305–1315. https://doi.org/10.1084/jem.20102419

Sawai CM et al (2016) Hematopoietic stem cells are the major source of multilineage hematopoiesis in adult animals. Immunity 45:597–609. https://doi.org/10.1016/j.immuni.2016.08.007

Sawen P et al (2018) Murine HSCs contribute actively to native hematopoiesis but with reduced differentiation capacity upon aging. eLife 7:e41258. https://doi.org/10.7554/eLife.41258

Scott CL, Henri S, Guilliams M (2014) Mononuclear phagocytes of the intestine, the skin, and the lung. Immunol Rev 262:9–24. https://doi.org/10.1111/imr.12220

Sheng J, Ruedl C, Karjalainen K (2015) Most tissue-resident macrophages except microglia are derived from fetal hematopoietic stem cells. Immunity 43:382–393. https://doi.org/10.1016/j.immuni.2015.07.016

Slayton WB et al (1998) The first-appearance of neutrophils in the human fetal bone marrow cavity. Early Hum Dev 53:129–144. https://doi.org/10.1016/s0378-3782(98)00049-8

Subramaniam A, Talkhoncheh MS, Magnusson M, Larsson J (2019) Endothelial protein C receptor (EPCR) expression marks human fetal liver hematopoietic stem cells. Haematologica 104:e47–e50. https://doi.org/10.3324/haematol.2018.198515

Suo C et al (2022) Mapping the developing human immune system across organs. Science 376:eabo0510. https://doi.org/10.1126/science.abo0510

Swiers G et al (2013) Early dynamic fate changes in haemogenic endothelium characterized at the single-cell level. Nat Commun 4:2924. https://doi.org/10.1038/ncomms3924

Tang W et al (2021) Hlf expression marks early emergence of hematopoietic stem cell precursors with adult repopulating potential and fate. Front Cell Dev Biol 9:728057. https://doi.org/10.3389/fcell.2021.728057

Tavian M, Peault B (2005) Embryonic development of the human hematopoietic system. Int J Dev Biol 49:243–250. https://doi.org/10.1387/ijdb.041957mt

Tavian M et al (1996) Aorta-associated CD34+ hematopoietic cells in the early human embryo. Blood 87:67–72

Tavian M, Hallais MF, Peault B (1999) Emergence of intraembryonic hematopoietic precursors in the pre-liver human embryo. Development 126:793–803

Tavian M, Robin C, Coulombel L, Péault B (2001) The human embryo, but not its yolk sac, generates lympho-myeloid stem cells: mapping multipotent hematopoietic cell fate in intraembryonic mesoderm. Immunity 15:487–495. https://doi.org/10.1016/s1074-7613(01)00193-5

Thambyrajah R et al (2016) GFI1 proteins orchestrate the emergence of haematopoietic stem cells through recruitment of LSD1. Nat Cell Biol 18:21–32. https://doi.org/10.1038/ncb3276

Timens W, Kamps WA (1997) Hemopoiesis in human fetal and embryonic liver. Microsc Res Tech 39:387–397. https://doi.org/10.1002/(SICI)1097-0029(19971201)39:5<387::AID-JEMT1>3.0.CO;2-E

Tober J et al (2007) The megakaryocyte lineage originates from hemangioblast precursors and is an integral component both of primitive and of definitive hematopoiesis. Blood 109:1433–1441. https://doi.org/10.1182/blood-2006-06-031898

Tyser RCV et al (2021) Single-cell transcriptomic characterization of a gastrulating human embryo. Nature 600: 285–289. https://doi.org/10.1038/s41586-021-04158-y

Uenishi GI et al (2018) NOTCH signaling specifies arterial-type definitive hemogenic endothelium from human pluripotent stem cells. Nat Commun 9:1828. https://doi.org/10.1038/s41467-018-04134-7

Varol C, Mildner A, Jung S (2015) Macrophages: development and tissue specialization. Annu Rev Immunol 33:643–675. https://doi.org/10.1146/annurev-immunol-032414-112220

Venkatraman A et al (2013) Maternal imprinting at the H19-Igf2 locus maintains adult haematopoietic stem cell quiescence. Nature 500:345–349. https://doi.org/10.1038/nature12303

Vink CS et al (2020) Iterative single-cell analyses define the transcriptome of the first functional hematopoietic stem cells. Cell Rep 31:107627. https://doi.org/10.1016/j.celrep.2020.107627

Viola MF, Boeckxstaens G (2021) Niche-specific functional heterogeneity of intestinal resident macrophages. Gut 70:1383–1395. https://doi.org/10.1136/gutjnl-2020-323121

Vivier E et al (2018) Innate lymphoid cells: 10 years on. Cell 174:1054–1066. https://doi.org/10.1016/j.cell.2018.07.017

Wang C et al (2021a) Adult-repopulating lymphoid potential of yolk sac blood vessels is not confined to arterial endothelial cells. Sci China Life Sci 64:2073–2087. https://doi.org/10.1007/s11427-021-1935-2

Wang H et al (2021b) Decoding human megakaryocyte development. Cell Stem Cell 28:535–549.e538. https://doi.org/10.1016/j.stem.2020.11.006

Wang F et al (2022) Single-cell architecture and functional requirement of alternative splicing during hematopoietic stem cell formation. Sci Adv 8: eabg5369. https://doi.org/10.1126/sciadv.abg5369

Weinberger T et al (2020) Ontogeny of arterial macrophages defines their functions in homeostasis and inflammation. Nat Commun 11:4549. https://doi.org/10.1038/s41467-020-18287-x

Weisberg SP et al (2006) CCR2 modulates inflammatory and metabolic effects of high-fat feeding. J Clin Invest 116:115–124. https://doi.org/10.1172/JCI24335

Wood W, Martin P (2017) Macrophage functions in tissue patterning and disease: new insights from the fly. Dev Cell 40:221–233. https://doi.org/10.1016/j.devcel.2017.01.001

Xu C et al (2022) Single-cell transcriptomic analysis identifies an immune-prone population in erythroid precursors during human ontogenesis. Nat Immunol 23:1109–1120. https://doi.org/10.1038/s41590-022-01245-8

Yang X et al (2006) Essential contribution of a chemokine, CCL3, and its receptor, CCR1, to hepatocellular carcinoma progression. Int J Cancer 118:1869–1876. https://doi.org/10.1002/ijc.21596

Ye H et al (2017) Clonal analysis reveals remarkable functional heterogeneity during hematopoietic stem cell emergence. Cell Res 27:1065–1068. https://doi.org/10.1038/cr.2017.64

Yokomizo T, Dzierzak E (2010) Three-dimensional cartography of hematopoietic clusters in the vasculature of whole mouse embryos. Development 137:3651–3661. https://doi.org/10.1242/dev.051094

Yokomizo T et al (2019) Hlf marks the developmental pathway for hematopoietic stem cells but not for erythro-myeloid progenitors. J Exp Med 216:1599–1614. https://doi.org/10.1084/jem.20181399

Yokomizo T et al (2022) Independent origins of fetal liver haematopoietic stem and progenitor cells. Nature 609: 779–784. https://doi.org/10.1038/s41586-022-05203-0

Yokota T et al (2006) Tracing the first waves of lymphopoiesis in mice. Development 133:2041–2051. https://doi.org/10.1242/dev.02349

Yona S et al (2013) Fate mapping reveals origins and dynamics of monocytes and tissue macrophages under homeostasis. Immunity 38:79–91. https://doi.org/10.1016/j.immuni.2012.12.001

Yoshimoto M et al (2011) Embryonic day 9 yolk sac and intra-embryonic hemogenic endothelium independently generate a B-1 and marginal zone progenitor lacking B-2 potential. Proc Natl Acad Sci U S A 108: 1468–1473. https://doi.org/10.1073/pnas.1015841108

Yoshimoto M et al (2012) Autonomous murine T-cell progenitor production in the extra-embryonic yolk sac before HSC emergence. Blood 119:5706–5714. https://doi.org/10.1182/blood-2011-12-397489

Yui MA, Rothenberg EV (2014) Developmental gene networks: a triathlon on the course to T cell identity. Nat Rev Immunol 14:529–545. https://doi.org/10.1038/nri3702

Yzaguirre AD, Howell ED, Li Y, Liu Z, Speck NA (2018) Runx1 is sufficient for blood cell formation from non-hemogenic endothelial cells in vivo only during early embryogenesis. Development 145:dev158162. https://doi.org/10.1242/dev.158162

Zeng Y et al (2019a) Tracing the first hematopoietic stem cell generation in human embryo by single-cell RNA sequencing. Cell Res 29:881–894. https://doi.org/10.1038/s41422-019-0228-6

Zeng Y et al (2019b) Single-cell RNA sequencing resolves spatiotemporal development of pre-thymic lymphoid progenitors and thymus organogenesis in human embryos. Immunity 51:930–948.e936. https://doi.org/10.1016/j.immuni.2019.09.008

Zhang Y et al (2019) VE-cadherin and ACE co-expression marks highly proliferative hematopoietic stem cells in human embryonic liver. Stem Cells Dev 28:165–185. https://doi.org/10.1089/scd.2018.0154

Zheng X et al (2019) Embryonic lineage tracing with Procr-CreER marks balanced hematopoietic stem cell fate during entire mouse lifespan. J Genet Genomics 46:489–498. https://doi.org/10.1016/j.jgg.2019.10.005

Zheng Z et al (2022) Uncovering the emergence of HSCs in the human fetal bone marrow by single-cell RNA-seq analysis. Cell Stem Cell 29:1562–1579.e1567. https://doi.org/10.1016/j.stem.2022.10.005

Zhou W, Rothenberg EV (2019) Building a human thymus: a pointillist view. Immunity 51:788–790. https://doi.org/10.1016/j.immuni.2019.10.003
Zhou F et al (2016) Tracing haematopoietic stem cell formation at single-cell resolution. Nature 533:487–492. https://doi.org/10.1038/nature17997
Zhou J et al (2019) Combined single-cell profiling of lncRNAs and functional screening reveals that H19 is pivotal for embryonic hematopoietic stem cell development. Cell Stem Cell 24:285–298.e285. https://doi.org/10.1016/j.stem.2018.11.023
Zhu Q et al (2020) Developmental trajectory of prehematopoietic stem cell formation from endothelium. Blood 136:845–856. https://doi.org/10.1182/blood.2020004801
Zlotoff DA, Schwarz BA, Bhandoola A (2008) The long road to the thymus: the generation, mobilization, and circulation of T-cell progenitors in mouse and man. Semin Immunopathol 30:371–382. https://doi.org/10.1007/s00281-008-0133-4
Zlotoff DA et al (2010) CCR7 and CCR9 together recruit hematopoietic progenitors to the adult thymus. Blood 115:1897–1905. https://doi.org/10.1182/blood-2009-08-237784
Zovein AC et al (2008) Fate tracing reveals the endothelial origin of hematopoietic stem cells. Cell Stem Cell 3: 625–636. https://doi.org/10.1016/j.stem.2008.09.018

Hematopoietic Stem Cells and Their Bone Marrow Niches

2

Sandra Pinho and Meng Zhao

Abstract

Hematopoietic stem cells (HSCs) are maintained in the bone marrow microenvironment, also known as the niche, that regulates their proliferation, self-renewal, and differentiation. In this chapter, we will introduce the history of HSC niche research and review the interdependencies between HSCs and their niches. We will further highlight recent advances in our understanding of HSC heterogeneity with regard to HSC subpopulations and their interacting cellular and molecular bone marrow niche constituents.

Keywords

Hematopoietic stem cells · Niche · Leukemia · Aging

S. Pinho (✉)
Department of Pharmacology & Regenerative Medicine, College of Medicine, University of Illinois at Chicago, Chicago, IL, USA
e-mail: spinho@uic.edu

M. Zhao (✉)
Key Laboratory of Stem Cells and Tissue Engineering (Ministry of Education), Zhongshan School of Medicine, Sun Yat-sen University, Guangzhou, Guangdong, China
e-mail: zhaom38@mail.sysu.edu.cn

2.1 The HSC Niche Concept

Hematopoietic stem cells (HSCs) can give rise to multipotent and more committed hematopoietic progenitors that produce all blood and immune cell lineages, such as red blood cells, white blood cells, and platelets, throughout life. HSCs engage in hematopoiesis within the bone marrow, where they are preserved in a microenvironment termed niche, which regulates HSC proliferation, self-renewal, localization, and differentiation. Raymond Schofield first proposed the HSC niche concept in 1978, when he discovered that HSCs that migrated to the spleen following transplantation were less efficient in reconstituting irradiated mice than HSCs harvested from the bone marrow (Schofield 1978). Therefore, he postulated the concept that stem cells rely on interactions with specific bone marrow anatomical sites or niche cells to maintain their function. Building on Schofield's hypothetical view of the stem cell niche, studies in the *Drosophila* gonads provided the first compelling experimental evidence that a niche supports the maintenance of germline stem cells (Xie and Spradling 2000). Studies in mammals, by two independent groups, reported the existence of bone lining *osteoblastic* cells and spindle-shaped N-cadherin^{+}CD45^{-} osteoblastic (SNO) cells, later identified as multipotent mesenchymal stem cells (MSCs), that contribute to the maintenance of the HSC pool size (Calvi et al. 2003; Zhang et al. 2003; Zhao et al. 2019). Subsequent studies

M. Zhao, P. Qian (eds.), *Hematopoietic Stem Cells*, Advances in Experimental Medicine and Biology 1442, https://doi.org/10.1007/978-981-99-7471-9_2

further revealed the critical role of vascular and perivascular niches in the regulation of HSC function in the bone marrow (Kiel et al. 2005; Mendez-Ferrer et al. 2010; Pinho and Frenette 2019).

2.2 The Heterogeneity of HSCs and Their Niches

HSCs are a rare and unique population but functionally and molecularly heterogeneous (Eaves 2015; Yamamoto et al. 2013). HSCs switch between quiescent and active cell cycle stages (Wilson et al. 2008) and have distinct self-renewal and reconstitution potentials upon transplantation, which further separates them as long-term HSCs (LT-HSCs), intermediate-term HSCs, and short-term HSCs (Benveniste et al. 2010; Yamamoto et al. 2013). Furthermore, HSC subsets were also identified based on distinct differentiation abilities toward the lymphoid or myeloid lineages (Dykstra et al. 2007; Muller-Sieburg et al. 2004). Platelet-/megakaryocyte-biased HSCs were also identified within quiescent LT-HSCs based on the expression of high c-Kit levels, CD41, and von Willebrand factor (vWF) (Gekas and Graf 2013; Sanjuan-Pla et al. 2013; Shin et al. 2014).

Recent studies suggest that the microanatomical complexity of the bone marrow microenvironment contributes to the regulation of HSC heterogeneity. The endosteal region and arteriolar niches maintain quiescent HSCs, with low reactive oxygen species (ROS) levels (Kunisaki et al. 2013; Nombela-Arrieta et al. 2013). Conversely, the sinusoid niche holds active HSCs for proliferation with high ROS levels (Itkin et al. 2016). A fraction of quiescent HSCs with low ROS levels is also localized in the sinusoidal niche, adjacent to megakaryocytes, a specialized type of mature blood cells that produce platelets and multiple factors that regulate HSC quiescence (Bruns et al. 2014; Zhao et al. 2014). Furthermore, vWF^+ platelet-biased HSCs are associated with and regulated by megakaryocytes, indicating that HSCs are regulated in a feedback loop by their offspring (Pinho et al. 2018). Accordingly, recent state-of-the-art imaging studies further revealed that myelopoiesis is spatially organized alongside the perisinusoidal niche (Zhang et al. 2021). On the other end, vWF^- HSCs and lymphopoiesis are associated with periarteriolar niches (Pinho et al. 2018; Shen et al. 2021). Recent advances in single-cell studies both at the transcription and protein levels further extended our understanding of the degree of heterogeneity within the bone marrow stromal compartment, in particular within MSCs and endothelial cells (Baccin et al. 2020; Baryawno et al. 2019; Severe et al. 2019; Tikhonova et al. 2019). Altogether, these studies revealed an unappreciated degree of heterogeneity and complexity within the bone marrow microenvironment through which HSC and progenitor cells navigate, suggesting that specialized niches might regulate phenotypically and functionally distinct HSC subpopulations.

2.3 The Regulation of HSCs by Bone Marrow Niche Cells

Niche cells support HSC maintenance and regeneration through paracrine signals in the bone marrow microenvironment. MSCs are known to express a broad repertoire of secreted trophic cytokines essential for tissue maintenance and regeneration (Ranganath et al. 2012). MSCs, which can be identified in the mouse bone marrow by the expression of different cell surface markers, including PDGFRα, CD51 (Pinho et al. 2013), and leptin receptor (Ding et al. 2012), are able to self-renew and differentiate into osteogenic, chondrogenic, and adipogenic progenies that support skeleton formation and maintenance (Bianco et al. 2008; Frenette et al. 2013). Importantly, MSCs also produce numerous factors that support HSC maintenance in the bone marrow niche, including stem cell factor (SCF) (Ding et al. 2012), chemokine (C-X-C motif) ligand 12 (CXCL12; also known as stromal-derived factor 1) (Ding and Morrison 2013; Greenbaum et al. 2013; Sugiyama et al. 2006), FGF2 (Itkin et al. 2012), pleiotrophin (Himburg et al. 2018), and Wnt ligands (Sugimura et al. 2012). Osteoblasts also produce factors such as CXCL12, BMPs, and osteopontin that regulate HSCs and lymphoid lineage determination (Ding and Morrison 2013; Shen et al. 2021; Stier et al. 2005). Bone marrow vascular endothelial cells, including sinusoids and arterioles, also express SCF and CXCL12, although at lower levels compared to MSCs (Asada et al. 2017). Nevertheless, endothelial cells express high levels of other niche factors, such as angiopoietin-like protein 2 to support HSC maintenance, localization, and proliferation with important roles during homeostasis and regeneration and in aging (Hooper et al. 2009; Kusumbe et al. 2016; Xu et al. 2018; Yu et al. 2022). In the perivascular niche, megakaryocytes produce transforming growth factor β (TGF-β), platelet factor 4 (PF4; also known as CXCL4), and thrombopoietin (TPO) to regulate HSC quiescence and fibroblast growth factor (FGF1) to promote HSC regeneration (Bruns et al. 2014; Nakamura-Ishizu et al. 2014; Zhao et al. 2012, 2019). Nerves are also an important component of the HSC perivascular niche. Non-myelinating Schwann cells activate TGF-β signaling to maintain HSC quiescence (Yamazaki et al. 2011), and the sympathetic nervous system regulates the aging and regeneration of HSCs through adrenergic signaling (Lucas et al. 2013; Maryanovich et al. 2018). Furthermore, cholinergic signals from the sympathetic nervous system and nociceptive neurons regulate the egress and granulocyte colony-stimulating factor (G-CSF)-mediated HSC mobilization from the bone marrow (Gao et al. 2021; Katayama et al. 2006).

Niche cells also regulate HSCs through other mechanisms independent of the activity of growth factors and cytokines. For example, niche cells can secret angiogenin, an RNase that regulates the synthesis of tRNA-derived stress-induced small RNA (tiRNA), which controls protein synthesis and overall proteostasis in HSCs (Goncalves et al. 2016). Alternatively, osteoblastic cells, mesenchymal stromal cells, and endothelial cells can also produce extracellular vesicles containing tiRNAs, microRNAs, and niche factors that contribute to the maintenance of HSCs (Gu et al. 2016; Huang et al. 2021; Kfoury et al. 2021; Stik et al. 2017). Microenvironment-secreted metabolites also regulate HSCs in the bone marrow, although the cellular sources for some of these metabolites remain undetermined. For example, vitamin A-retinoic acid, bile acid, and ascorbate signaling regulate the development of embryonic HSCs and the quiescence of adult HSCs (Agathocleous et al. 2017; Cabezas-Wallscheid et al. 2017; Chanda et al. 2013; Sigurdsson et al. 2016). Recent studies also revealed that the amino acid levels in the bone marrow microenvironment and their catabolism regulate HSC proteostasis and overall HSC maintenance and proliferation (Li et al. 2022). Valine and aspartate availabilities are essential for HSC maintenance and regeneration; interestingly, a valine-restricted diet permits nonmyeloablative HSC transplantation in mice (Qi et al. 2021; Taya et al. 2016). Finally, glucose and glutamine metabolism regulate human HSC lineage specification (Oburoglu et al. 2014). Therefore, the amino acid levels in the bone marrow microenvironment are important regulators of HSC self-renewal, regeneration, and differentiation. However, the cellular source(s) of these small metabolites in the bone marrow niche remain(s) unknown.

Recent studies also indicate that cross-organ communication contributes to HSC regulation through endocrine signals. Circulating TPO, generated by liver hepatocytes, is required for bone marrow HSC maintenance (Decker et al. 2018). Muscarinic receptor type-1 signaling in the hypothalamus of the central nervous system promotes G-CSF-induced HSC mobilization and embryonic HSC development through the hypothalamic-pituitary-adrenal axis (Kwan et al. 2016; Pierce et al. 2017). Hormones, such as luteinizing hormone, prostaglandin E2, and estrogen signaling, further contribute to HSC regulation in regeneration, expansion, and maintenance (Hoggatt et al. 2013; Nakada et al. 2014; Sánchez-Aguilera et al. 2014; Velardi et al. 2018). Recent studies further revealed that the intestinal microbiota also regulates HSC fate decisions by modulating local iron availability in the bone marrow (Zhang et al. 2022). Altogether, these studies suggest the existence of long-distance regulatory mechanisms that extend past the local bone marrow niche regulation to control HSC function and hematopoietic homeostasis.

2.4 The Aging Bone Marrow Niche

Mammalian aging is associated with reduced tissue regeneration due to the declining function of tissue-specific stem cells. In the hematopoietic system, aging is accompanied by an expansion of HSCs but with impaired engraftment, self-renewal, and lymphoid differentiation capacity (Liang et al. 2005; Pang et al. 2011; Rossi et al. 2005). Increased myeloid-lineage output is a hallmark of aged HSCs (Gekas and Graf 2013). HSC aging results from cell-intrinsic alterations, such as epigenetic changes (Chambers et al. 2007), replication stress, genomic instability and DNA damage (Flach et al. 2014; Rossi et al. 2007), increased unfolded protein responses (Mohrin et al. 2015), autophagy, and HSC-intrinsic cell signaling changes (Florian et al. 2013; Ho et al. 2017). Importantly, microenvironmental alterations also contribute to HSC aging, as seen by the lower engraftment efficiency and differentiation skewing toward the myeloid lineage of young HSCs when transplanted into old recipients compared to young ones (Ergen et al. 2012; Rossi et al. 2005). For example, telomere dysfunction in aged bone marrow stromal cells impairs HSC maintenance and function, suppresses lymphopoiesis, and increases myelopoiesis (Ju et al. 2007). Aging also leads to altered bone marrow MSC differentiation, where osteogenesis is suppressed in favor of increased adipogenesis, which impairs hematopoiesis and HSC regeneration (Ambrosi et al. 2017; Nishikawa et al. 2010). Although bone marrow MSCs expand with age, they have reduced clonogenic capacity and reduced expression of HSC niche factors such as CXCL12 and SCF (Maryanovich et al. 2018). A decline in the production of the longevity-associated molecule insulin-like growth factor 1 (IGF1) in older mesenchymal stromal cells was also recently shown to compromise HSC function in aged mice due to impaired mitochondrial activity (Young et al. 2021).

The bone marrow vascular niche is also severely altered upon aging (Kusumbe et al. 2016; Maryanovich et al. 2018; Poulos et al. 2017). Endothelial Notch signaling, which is important for the expansion of niche-forming vessels and perivascular cells, is downregulated in aged mice, which contributes to HSC aging (Kusumbe et al. 2016). Arteriolar vessels (Maryanovich et al. 2018) and type H capillaries (Kusumbe et al. 2016) are the most disturbed vascular beds, while sinusoidal niches are uniquely preserved upon bone marrow aging (Sacma et al. 2019). Nevertheless, aged sinusoidal niches are selectively disrupted upon chemotherapy due to the lack of endothelial jagged canonical Notch ligand 2 (Jag2), which subsequentially impairs HSC regeneration (Sacma et al. 2019). Interestingly, transplantation of young endothelial cells improves old HSC function in a murine model (Poulos et al. 2017).

Bone marrow aging is also associated with significant alterations in adrenergic signals, relayed to the bone marrow niche by the sympathetic nervous system, which contributes to the aging-associated features of HSCs (Ho et al. 2019; Maryanovich et al. 2018). Although there's still a lack of consensus regarding the nature of the alterations to the noradrenergic nerve fibers in the aged bone marrow, administration of a β3 adrenergic receptor (ADRβ3) agonist in physiological old mice or progeroid mice increased the in vivo function of old HSCs; on the contrary, the deletion of ADRβ3 signal led to premature HSC aging (Ho et al. 2019; Maryanovich et al. 2018). Mammalian aging is frequently associated with chronic low-grade inflammation, which likely contributes to changes in HSC function. Inflammatory factors such as interferon-α and interferon-γ are known to activate HSC proliferation but impair their self-renewal potential (Baldridge et al. 2010; Essers et al. 2009; Yamashita and Passegué 2019). Increasing evidence demonstrates that multiple inflammatory cytokines, such as interferon-γ, tumor necrosis factor alpha (TNFα), and interleukin-1 (IL-1β), are present at increased levels in the aged bone marrow microenvironment leading to increased myeloid differentiation and reduced self-renewal of HSCs (Ergen et al. 2012; Kovtonyuk et al. 2022; Meacham et al. 2022). The plasma cells in aged bone marrow express high levels of pathogen sensors and inflammatory cytokines, which also regulate the production of inflammatory factors from stromal cells to promote HSC aging (Pioli et al. 2019). Accordingly, recent studies demonstrated that repeated inflammatory exposure in early life leads to irreversible long-lived impairment of HSC self-renewal and accelerates HSC aging (Bogeska et al. 2022). Interestingly, the inhibition of IL-1β and TNFα in old mice decreased myelopoiesis (Pioli et al. 2019), suggesting that blocking inflammatory pathways may help to rejuvenate aged HSCs.

2.5 Bone Marrow Niche for Leukemia

Leukemia is a group of blood cancers characterized by the expansion of abnormal hematopoietic cells at the expense of healthy hematopoiesis. Genetic mutations in the hematopoietic system are thought to be the leading cause of leukemia; however, the bone marrow niche can also play a role in the initiation and progression of leukemia (Marchand and Pinho 2021). Mutations in niche cells occasionally initiate hematopoietic malignancies. For example, the gain-of-function mutation in the protein tyrosine phosphatase non-receptor type 11 (PTPN11) in MSCs and osteoprogenitors promotes the development and progression of myeloproliferative neoplasm (MPN) (Dong et al. 2016). Furthermore, deletion of the miRNA processing endonuclease *Dicer1* (Raaijmakers et al. 2010) or introduction of an activated β-catenin mutation in osteolineage cells (Kode et al. 2014) leads to myelodysplastic syndrome (MDS) that can evolve to acute myeloid leukemia (AML). Conversely, inhibition of Wnt signaling in the bone marrow niche prevents the development of MDS in a mouse model (Stoddart et al. 2017). These observations demonstrate that an abnormal bone marrow niche can initiate MDS/MPN disease or favor the growth of abnormal blood cells with genetic mutations and eventually lead to the development of leukemia. However, how abnormal bone marrow niche cells can introduce genetic mutations into the hematopoietic system remains unclear. Leukemic cells can also remodel the bone marrow niche to promote their growth while suppressing the function of normal HSCs and hematopoiesis (Colmone et al. 2008). Early studies in human AML revealed that leukemic cells secrete vascular endothelial growth factor (Fiedler et al. 1997) that promotes angiogenesis, leading to increased microvasculature density in the bone marrow of AML patients (Aguayo et al. 2000; Hussong et al. 2000; Padró et al. 2000). However, antiangiogenic therapy did not show antileukemic activity in clinical trials (Ossenkoppele et al. 2012; Zahiragic et al. 2007). Intravital microscopy studies in the mouse bone marrow later revealed that AML cells cause morphological and functional changes in vessels, which are more complex than the simple induction of angiogenesis. AML expands the overall vascular niche but induces specific loss of endosteal vessels, which reduces healthy HSC numbers and promotes disease progression (Duarte et al. 2018). On the contrary, the rescue of endosteal vessel defects (Duarte et al. 2018) or inhibition of vascular-derived nitric oxide (Passaro et al. 2017) improves healthy hematopoiesis and the response to chemotherapy.

Leukemia also impairs the normal HSC niche function of MSCs and their multilineage differentiation capacity. In the MLL-AF9-driven model of AML, the disease is accompanied by the proliferation of MSC and progenitors primed for osteoblastic differentiation and a reduction in the number of HSC-supporting $NG2^+$ periarteriolar MSCs (Hanoun et al. 2014). AML samples from patients also showed that leukemia induces MSC differentiation toward osteogenic lineages at the expense of adipocytic differentiation, which promotes leukemogenesis (Boyd et al. 2017). In T-cell acute lymphoblastic leukemia (T-ALL), leukemic cells selectively remodel the endosteal space, leading to the complete loss of mature osteoblastic cells but the preservation of perivascular cells (Hawkins et al. 2016).

Consistent with this, osteoblast ablation reduces healthy HSC function but promotes leukemic stem cells (LSCs) to accelerate leukemia development in chronic myelogenous leukemia (CML) (Bowers et al. 2015). Other mouse studies in AML and in a model of MPN revealed that disease progression is accompanied by sympathetic neuropathy in the bone marrow niche that promotes disease progression (Arranz et al. 2014; Hanoun et al. 2014).

The interactions between cancer cells and their niches contribute to chemoresistance and disease relapse (Straussman et al. 2012; Wilson et al. 2012). Upon chemotherapy, resistant ALL cells secrete factors to recruit and remodel Nestin^{+} MSCs to build a leukemia protective reservoir that confers chemoresistance on LSCs via growth differentiation factor 15 (GDF15) signaling (Duan et al. 2014). In T-ALL, leukemic cells migrate across the bone marrow, interact dynamically with different niche cells, and become more migratory under chemotherapy in the bone marrow (Hawkins et al. 2016). Although leukemic cells did not show any preferential association with bone marrow niche sub-compartments (Hawkins et al. 2016), whether LSCs are preserved in a specific niche in the bone marrow remains to be determined. Niche cells also contribute to the drug resistance of LSCs through cytokine-independent mechanisms. For example, endothelial cells in the bone marrow supply miR-126 to CML LSCs to support their quiescence and leukemia growth. Deletion of miR-126 in endothelial cells enhances the in vivo antileukemic effects of tyrosine kinase inhibitor treatment and strongly diminishes LSC cancer-initiating capacity (Zhang et al. 2018).

Healthy HSCs rely on glycolysis, but LSCs have high mitochondrial oxidative phosphorylation (OXPHOS) metabolism for their energy production. High OXPHOS confers survival to LSCs in response to chemotherapy in both mouse and human models (Lagadinou et al. 2013; Saito et al. 2015). Unlike glycolysis, which mainly consumes glucose, oxidative phosphorylation generates energy through catabolizing amino acids, fatty acids, and glucose which underlies the progress and chemoresistance of leukemic cells (Jones et al. 2021). Therefore, chemoresistant leukemic cells have high amino acid and fatty acid metabolism to increase their oxidative phosphorylation levels (Jones et al. 2018; Stevens et al. 2020). These observations indicate that the metabolic microenvironment determines the maintenance of healthy HSCs for hematopoiesis as well as abnormal LSCs for leukemogenesis. In the leukemia bone marrow, niche discrete anatomical locations have distinct metabolic microenvironments and ATP levels, which determine the self-renewal and homing capacity of LSCs (Hao et al. 2019; He et al. 2021). In addition, MSCs also enhance the antioxidant defense of AML cells through metabolic support or mitochondria transfer which protects leukemia cells from chemotherapy (Burt et al. 2019; Forte et al. 2020). Thus, targeting the leukemic niche in addition to chemotherapy may be an effective treatment approach, especially for hematopoietic malignancies for which standard therapies have not been effective.

2.6 Future Perspectives

Expanding functional human HSCs in vitro is still an unmet task. Lack of niche support is believed to be one of the main reasons. Niche cells lose their ability to support HSCs due to stresses, such as inflammation, myeloablation, and reactive oxygen species (Lou et al. 2022; Nakahara et al. 2019). Engineering niche cells that retain their HSC niche support ability and designing improved mouse and human bone marrow vascularized organoids that mimic the native bone marrow microenvironment are two priorities in the field.

The overall number of niche cells, such as endothelial cells and MSCs, is significantly higher than the number of HSCs. The development of single-cell technologies allowed the definition of the cellular taxonomy of the mouse bone marrow stroma (Baryawno et al. 2019). These studies revealed the high heterogeneity of niche cells in the bone marrow stroma compartment, which suggests the existence of novel uncharacterized rare subpopulations of niche cells that directly contribute to the regulation of HSC function and hematopoiesis. Another possibility is that HSCs dynamically shift their localization in the bone marrow microenvironment. To this end, further development of live tracing internal imaging systems for HSCs in the bone marrow is required to better understand their regulation in the niche during maintenance, regeneration, and aging and in disease.

Acknowledgments We thank the National Key Research and Development Program of China (2022YFA1103300, 2022YFA1104100), National Natural Science Foundation of China (82325002, 82170112), and the Sanming Project of Medicine in Shenzhen (SZSM201911004) for their generous support.

References

Agathocleous M, Meacham CE, Burgess RJ, Piskounova E, Zhao Z, Crane GM et al (2017) Ascorbate regulates haematopoietic stem cell function and leukaemogenesis. Nature 549(7673):476–481. https://doi.org/10.1038/nature23876

Aguayo A, Kantarjian H, Manshouri T, Gidel C, Estey E, Thomas D et al (2000) Angiogenesis in acute and chronic leukemias and myelodysplastic syndromes. Blood 96(6):2240–2245

Ambrosi TH, Scialdone A, Graja A, Gohlke S, Jank AM, Bocian C et al (2017) Adipocyte accumulation in the bone marrow during obesity and aging impairs stem cell-based hematopoietic and bone regeneration. Cell Stem Cell 20(6):771–784.e776. https://doi.org/10.1016/j.stem.2017.02.009

Arranz L, Sanchez-Aguilera A, Martin-Perez D, Isern J, Langa X, Tzankov A et al (2014) Neuropathy of haematopoietic stem cell niche is essential for myeloproliferative neoplasms. Nature 512(7512):78–81. https://doi.org/10.1038/nature13383

Asada N, Kunisaki Y, Pierce H, Wang Z, Fernandez NF, Birbrair A et al (2017) Differential cytokine contributions of perivascular haematopoietic stem cell niches. Nat Cell Biol 19(3):214–223. https://doi.org/10.1038/ncb3475

Baccin C, Al-Sabah J, Velten L, Helbling PM, Grünschläger F, Hernández Malmierca P et al (2020) Combined single-cell and spatial transcriptomics reveal the molecular, cellular and spatial bone marrow niche organization. Nat Cell Biol 22(1):38–48. https://doi.org/10.1038/s41556-019-0439-6

Baldridge MT, King KY, Boles NC, Weksberg DC, Goodell MA (2010) Quiescent haematopoietic stem cells are activated by IFN-gamma in response to chronic infection. Nature 465(7299):793–797. https://doi.org/10.1038/nature09135

Baryawno N, Przybylski D, Kowalczyk MS, Kfoury Y, Severe N, Gustafsson K et al (2019) A cellular taxonomy of the bone marrow stroma in homeostasis and leukemia. Cell 177(7):1915–1932.e1916. https://doi.org/10.1016/j.cell.2019.04.040

Benveniste P, Frelin C, Janmohamed S, Barbara M, Herrington R, Hyam D, Iscove NN (2010) Intermediate-term hematopoietic stem cells with extended but time-limited reconstitution potential. Cell Stem Cell 6(1):48–58. https://doi.org/10.1016/j.stem.2009.11.014

Bianco P, Robey PG, Simmons PJ (2008) Mesenchymal stem cells: revisiting history, concepts, and assays. Cell Stem Cell 2(4):313–319. https://doi.org/10.1016/j.stem.2008.03.002

Bogeska R, Mikecin AM, Kaschutnig P, Fawaz M, Buchler-Schaff M, Le D et al (2022) Inflammatory exposure drives long-lived impairment of hematopoietic stem cell self-renewal activity and accelerated aging. Cell Stem Cell 29(8):1273–1284.e1278. https://doi.org/10.1016/j.stem.2022.06.012

Bowers M, Zhang B, Ho Y, Agarwal P, Chen CC, Bhatia R (2015) Osteoblast ablation reduces normal long-term hematopoietic stem cell self-renewal but accelerates leukemia development. Blood 125(17):2678–2688. https://doi.org/10.1182/blood-2014-06-582924

Boyd AL, Reid JC, Salci KR, Aslostovar L, Benoit YD, Shapovalova Z et al (2017) Acute myeloid leukaemia disrupts endogenous myelo-erythropoiesis by compromising the adipocyte bone marrow niche. Nat Cell Biol 19(11):1336–1347. https://doi.org/10.1038/ncb3625

Bruns I, Lucas D, Pinho S, Ahmed J, Lambert MP, Kunisaki Y et al (2014) Megakaryocytes regulate hematopoietic stem cell quiescence through CXCL4 secretion. Nat Med 20(11):1315–1320. https://doi.org/10.1038/nm.3707

Burt R, Dey A, Aref S, Aguiar M, Akarca A, Bailey K et al (2019) Activated stromal cells transfer mitochondria to rescue acute lymphoblastic leukemia cells from oxidative stress. Blood 134(17):1415–1429. https://doi.org/10.1182/blood.2019001398

Cabezas-Wallscheid N, Buettner F, Sommerkamp P, Klimmeck D, Ladel L, Thalheimer FB et al (2017) Vitamin A-retinoic acid signaling regulates hematopoietic stem cell dormancy. Cell 169(5):807–823.e819. https://doi.org/10.1016/j.cell.2017.04.018

Calvi LM, Adams GB, Weibrecht KW, Weber JM, Olson DP, Knight MC et al (2003) Osteoblastic cells regulate the haematopoietic stem cell niche. Nature 425(6960):841–846. https://doi.org/10.1038/nature02040

Chambers SM, Shaw CA, Gatza C, Fisk CJ, Donehower LA, Goodell MA (2007) Aging hematopoietic stem cells decline in function and exhibit epigenetic dysregulation. PLoS Biol 5(8):e201. https://doi.org/10.1371/journal.pbio.0050201

Chanda B, Ditadi A, Iscove NN, Keller G (2013) Retinoic acid signaling is essential for embryonic hematopoietic

stem cell development. Cell 155(1):215–227. https://doi.org/10.1016/j.cell.2013.08.055
Colmone A, Amorim M, Pontier AL, Wang S, Jablonski E, Sipkins DA (2008) Leukemic cells create bone marrow niches that disrupt the behavior of normal hematopoietic progenitor cells. Science 322(5909): 1861–1865. https://doi.org/10.1126/science.1164390
Decker M, Leslie J, Liu Q, Ding L (2018) Hepatic thrombopoietin is required for bone marrow hematopoietic stem cell maintenance. Science 360(6384):106–110. https://doi.org/10.1126/science.aap8861
Ding L, Morrison SJ (2013) Haematopoietic stem cells and early lymphoid progenitors occupy distinct bone marrow niches. Nature 495(7440):231–235. https://doi.org/10.1038/nature11885
Ding L, Saunders TL, Enikolopov G, Morrison SJ (2012) Endothelial and perivascular cells maintain haematopoietic stem cells. Nature 481(7382): 457–462. https://doi.org/10.1038/nature10783
Dong L, Yu WM, Zheng H, Loh ML, Bunting ST, Pauly M et al (2016) Leukaemogenic effects of Ptpn11 activating mutations in the stem cell microenvironment. Nature 539(7628):304–308. https://doi.org/10.1038/nature20131
Duan CW, Shi J, Chen J, Wang B, Yu YH, Qin X et al (2014) Leukemia propagating cells rebuild an evolving niche in response to therapy. Cancer Cell 25(6): 778–793. https://doi.org/10.1016/j.ccr.2014.04.015
Duarte D, Hawkins ED, Akinduro O, Ang H, De Filippo K, Kong IY et al (2018) Inhibition of endosteal vascular niche remodeling rescues hematopoietic stem cell loss in AML. Cell Stem Cell 22(1):64–77.e66. https://doi.org/10.1016/j.stem.2017.11.006
Dykstra B, Kent D, Bowie M, McCaffrey L, Hamilton M, Lyons K et al (2007) Long-term propagation of distinct hematopoietic differentiation programs in vivo. Cell Stem Cell 1(2):218–229. https://doi.org/10.1016/j.stem.2007.05.015
Eaves CJ (2015) Hematopoietic stem cells: concepts, definitions, and the new reality. Blood 125(17): 2605–2613. https://doi.org/10.1182/blood-2014-12-570200
Ergen AV, Boles NC, Goodell MA (2012) Rantes/Ccl5 influences hematopoietic stem cell subtypes and causes myeloid skewing. Blood 119(11):2500–2509. https://doi.org/10.1182/blood-2011-11-391730
Essers MA, Offner S, Blanco-Bose WE, Waibler Z, Kalinke U, Duchosal MA, Trumpp A (2009) IFNalpha activates dormant haematopoietic stem cells in vivo. Nature 458(7240):904–908. https://doi.org/10.1038/nature07815
Fiedler W, Graeven U, Ergün S, Verago S, Kilic N, Stockschläder M, Hossfeld DK (1997) Vascular endothelial growth factor, a possible paracrine growth factor in human acute myeloid leukemia. Blood 89(6): 1870–1875
Flach J, Bakker ST, Mohrin M, Conroy PC, Pietras EM, Reynaud D et al (2014) Replication stress is a potent driver of functional decline in ageing haematopoietic stem cells. Nature 512(7513):198–202. https://doi.org/10.1038/nature13619
Florian MC, Nattamai KJ, Dorr K, Marka G, Uberle B, Vas V et al (2013) A canonical to non-canonical Wnt signalling switch in haematopoietic stem-cell ageing. Nature 503(7476):392–396. https://doi.org/10.1038/nature12631
Forte D, García-Fernández M, Sánchez-Aguilera A, Stavropoulou V, Fielding C, Martín-Pérez D et al (2020) Bone marrow mesenchymal stem cells support acute myeloid leukemia bioenergetics and enhance antioxidant defense and escape from chemotherapy. Cell Metab 32(5):829–843.e829. https://doi.org/10.1016/j.cmet.2020.09.001
Frenette PS, Pinho S, Lucas D, Scheiermann C (2013) Mesenchymal stem cell: keystone of the hematopoietic stem cell niche and a stepping-stone for regenerative medicine. Annu Rev Immunol 31:285–316. https://doi.org/10.1146/annurev-immunol-032712-095919
Gao X, Zhang D, Xu C, Li H, Caron KM, Frenette PS (2021) Nociceptive nerves regulate haematopoietic stem cell mobilization. Nature 589(7843):591–596. https://doi.org/10.1038/s41586-020-03057-y
Gekas C, Graf T (2013) CD41 expression marks myeloid-biased adult hematopoietic stem cells and increases with age. Blood 121(22):4463–4472. https://doi.org/10.1182/blood-2012-09-457929
Goncalves KA, Silberstein L, Li S, Severe N, Hu MG, Yang H et al (2016) Angiogenin promotes hematopoietic regeneration by dichotomously regulating quiescence of stem and progenitor cells. Cell 166(4):894–906. https://doi.org/10.1016/j.cell.2016.06.042
Greenbaum A, Hsu YM, Day RB, Schuettpelz LG, Christopher MJ, Borgerding JN et al (2013) CXCL12 in early mesenchymal progenitors is required for haematopoietic stem-cell maintenance. Nature 495(7440):227–230. https://doi.org/10.1038/nature11926
Gu H, Chen C, Hao X, Wang C, Zhang X, Li Z et al (2016) Sorting protein VPS33B regulates exosomal autocrine signaling to mediate hematopoiesis and leukemogenesis. J Clin Invest 126(12):4537–4553. https://doi.org/10.1172/jci87105
Hanoun M, Zhang D, Mizoguchi T, Pinho S, Pierce H, Kunisaki Y et al (2014) Acute myelogenous leukemia-induced sympathetic neuropathy promotes malignancy in an altered hematopoietic stem cell niche. Cell Stem Cell 15(3):365–375. https://doi.org/10.1016/j.stem.2014.06.020
Hao X, Gu H, Chen C, Huang D, Zhao Y, Xie L et al (2019) Metabolic imaging reveals a unique preference of symmetric cell division and homing of leukemia-initiating cells in an endosteal niche. Cell Metab 29(4): 950–965.e956. https://doi.org/10.1016/j.cmet.2018.11.013
Hawkins ED, Duarte D, Akinduro O, Khorshed RA, Passaro D, Nowicka M et al (2016) T-cell acute

leukaemia exhibits dynamic interactions with bone marrow microenvironments. Nature 538(7626): 518–522. https://doi.org/10.1038/nature19801
He X, Wan J, Yang X, Zhang X, Huang D, Li X et al (2021) Bone marrow niche ATP levels determine leukemia-initiating cell activity via P2X7 in leukemic models. J Clin Invest 131(4):e140242. https://doi.org/10.1172/jci140242
Himburg HA, Termini CM, Schlussel L, Kan J, Li M, Zhao L et al (2018) Distinct bone marrow sources of pleiotrophin control hematopoietic stem cell maintenance and regeneration. Cell Stem Cell 23(3):370–381.e375. https://doi.org/10.1016/j.stem.2018.07.003
Ho TT, Warr MR, Adelman ER, Lansinger OM, Flach J, Verovskaya EV et al (2017) Autophagy maintains the metabolism and function of young and old stem cells. Nature 543(7644):205–210. https://doi.org/10.1038/nature21388
Ho YH, Del Toro R, Rivera-Torres J, Rak J, Korn C, García-García A et al (2019) Remodeling of bone marrow hematopoietic stem cell niches promotes myeloid cell expansion during premature or physiological aging. Cell Stem Cell 25(3):407–418.e406. https://doi.org/10.1016/j.stem.2019.06.007
Hoggatt J, Mohammad KS, Singh P, Hoggatt AF, Chitteti BR, Speth JM et al (2013) Differential stem- and progenitor-cell trafficking by prostaglandin E2. Nature 495(7441):365–369. https://doi.org/10.1038/nature11929
Hooper AT, Butler JM, Nolan DJ, Kranz A, Iida K, Kobayashi M et al (2009) Engraftment and reconstitution of hematopoiesis is dependent on VEGFR2-mediated regeneration of sinusoidal endothelial cells. Cell Stem Cell 4(3):263–274. https://doi.org/10.1016/j.stem.2009.01.006
Huang D, Sun G, Hao X, He X, Zheng Z, Chen C et al (2021) ANGPTL2-containing small extracellular vesicles from vascular endothelial cells accelerate leukemia progression. J Clin Invest 131(1):e138986. https://doi.org/10.1172/jci138986
Hussong JW, Rodgers GM, Shami PJ (2000) Evidence of increased angiogenesis in patients with acute myeloid leukemia. Blood 95(1):309–313
Itkin T, Ludin A, Gradus B, Gur-Cohen S, Kalinkovich A, Schajnovitz A et al (2012) FGF-2 expands murine hematopoietic stem and progenitor cells via proliferation of stromal cells, c-kit activation, and CXCL12 down-regulation. Blood 120(9):1843–1855. https://doi.org/10.1182/blood-2011-11-394692
Itkin T, Gur-Cohen S, Spencer JA, Schajnovitz A, Ramasamy SK, Kusumbe AP et al (2016) Distinct bone marrow blood vessels differentially regulate haematopoiesis. Nature 532(7599):323–328. https://doi.org/10.1038/nature17624
Jones CL, Stevens BM, D'Alessandro A, Reisz JA, Culp-Hill R, Nemkov T et al (2018) Inhibition of amino acid metabolism selectively targets human leukemia stem cells. Cancer Cell 34(5):724–740.e724. https://doi.org/10.1016/j.ccell.2018.10.005
Jones CL, Inguva A, Jordan CT (2021) Targeting energy metabolism in cancer stem cells: progress and challenges in leukemia and solid tumors. Cell Stem Cell 28(3):378–393. https://doi.org/10.1016/j.stem.2021.02.013
Ju Z, Jiang H, Jaworski M, Rathinam C, Gompf A, Klein C et al (2007) Telomere dysfunction induces environmental alterations limiting hematopoietic stem cell function and engraftment. Nat Med 13(6):742–747. https://doi.org/10.1038/nm1578
Katayama Y, Battista M, Kao WM, Hidalgo A, Peired AJ, Thomas SA, Frenette PS (2006) Signals from the sympathetic nervous system regulate hematopoietic stem cell egress from bone marrow. Cell 124(2):407–421. https://doi.org/10.1016/j.cell.2005.10.041
Kfoury YS, Ji F, Mazzola M, Sykes DB, Scherer AK, Anselmo A et al (2021) tiRNA signaling via stress-regulated vesicle transfer in the hematopoietic niche. Cell Stem Cell 28(12):2090–2103.e2099. https://doi.org/10.1016/j.stem.2021.08.014
Kiel MJ, Yilmaz OH, Iwashita T, Terhorst C, Morrison SJ (2005) SLAM family receptors distinguish hematopoietic stem and progenitor cells and reveal endothelial niches for stem cells. Cell 121(7): 1109–1121. https://doi.org/10.1016/j.cell.2005.05.026
Kode A, Manavalan JS, Mosialou I, Bhagat G, Rathinam CV, Luo N et al (2014) Leukaemogenesis induced by an activating β-catenin mutation in osteoblasts. Nature 506(7487):240–244. https://doi.org/10.1038/nature12883
Kovtonyuk LV, Caiado F, Garcia-Martin S, Manz EM, Helbling P, Takizawa H et al (2022) IL-1 mediates microbiome-induced inflammaging of hematopoietic stem cells in mice. Blood 139(1):44–58. https://doi.org/10.1182/blood.2021011570
Kunisaki Y, Bruns I, Scheiermann C, Ahmed J, Pinho S, Zhang D et al (2013) Arteriolar niches maintain haematopoietic stem cell quiescence. Nature 502(7473):637–643. https://doi.org/10.1038/nature12612
Kusumbe AP, Ramasamy SK, Itkin T, Mäe MA, Langen UH, Betsholtz C et al (2016) Age-dependent modulation of vascular niches for haematopoietic stem cells. Nature 532(7599):380–384. https://doi.org/10.1038/nature17638
Kwan W, Cortes M, Frost I, Esain V, Theodore LN, Liu SY et al (2016) The central nervous system regulates embryonic HSPC production via stress-responsive glucocorticoid receptor signaling. Cell Stem Cell 19(3): 370–382. https://doi.org/10.1016/j.stem.2016.06.004
Lagadinou ED, Sach A, Callahan K, Rossi RM, Neering SJ, Minhajuddin M et al (2013) BCL-2 inhibition targets oxidative phosphorylation and selectively eradicates quiescent human leukemia stem cells. Cell Stem Cell 12(3):329–341. https://doi.org/10.1016/j.stem.2012.12.013
Li C, Wu B, Li Y, Chen J, Ye Z, Tian X et al (2022) Amino acid catabolism regulates hematopoietic stem cell proteostasis via a GCN2-eIF2α axis. Cell Stem

Cell 29(7):1119–1134.e1117. https://doi.org/10.1016/j.stem.2022.06.004
Liang Y, Van Zant G, Szilvassy SJ (2005) Effects of aging on the homing and engraftment of murine hematopoietic stem and progenitor cells. Blood 106(4):1479–1487. https://doi.org/10.1182/blood-2004-11-4282
Lou Q, Jiang K, Xu Q, Yuan L, Xie S, Pan Y et al (2022) The RIG-I-NRF2 axis regulates the mesenchymal stromal niche for bone marrow transplantation. Blood 139(21):3204–3221. https://doi.org/10.1182/blood.2021013048
Lucas D, Scheiermann C, Chow A, Kunisaki Y, Bruns I, Barrick C et al (2013) Chemotherapy-induced bone marrow nerve injury impairs hematopoietic regeneration. Nat Med 19(6):695–703. https://doi.org/10.1038/nm.3155
Marchand T, Pinho S (2021) Leukemic stem cells: from leukemic niche biology to treatment opportunities. Front Immunol 12:775128. https://doi.org/10.3389/fimmu.2021.775128
Maryanovich M, Zahalka AH, Pierce H, Pinho S, Nakahara F, Asada N et al (2018) Adrenergic nerve degeneration in bone marrow drives aging of the hematopoietic stem cell niche. Nat Med 24(6): 782–791. https://doi.org/10.1038/s41591-018-0030-x
Meacham CE, Jeffery EC, Burgess RJ, Sivakumar CD, Arora MA, Stanley AM et al (2022) Adiponectin receptors sustain haematopoietic stem cells throughout adulthood by protecting them from inflammation. Nat Cell Biol 24(5):697–707. https://doi.org/10.1038/s41556-022-00909-9
Mendez-Ferrer S, Michurina TV, Ferraro F, Mazloom AR, Macarthur BD, Lira SA et al (2010) Mesenchymal and haematopoietic stem cells form a unique bone marrow niche. Nature 466(7308):829–834. https://doi.org/10.1038/nature09262
Mohrin M, Shin J, Liu Y, Brown K, Luo H, Xi Y et al (2015) Stem cell aging. A mitochondrial UPR-mediated metabolic checkpoint regulates hematopoietic stem cell aging. Science 347(6228): 1374–1377. https://doi.org/10.1126/science.aaa2361
Muller-Sieburg CE, Cho RH, Karlsson L, Huang JF, Sieburg HB (2004) Myeloid-biased hematopoietic stem cells have extensive self-renewal capacity but generate diminished lymphoid progeny with impaired IL-7 responsiveness. Blood 103(11):4111–4118. https://doi.org/10.1182/blood-2003-10-3448
Nakada D, Oguro H, Levi BP, Ryan N, Kitano A, Saitoh Y et al (2014) Oestrogen increases haematopoietic stem-cell self-renewal in females and during pregnancy. Nature 505(7484):555–558. https://doi.org/10.1038/nature12932
Nakahara F, Borger DK, Wei Q, Pinho S, Maryanovich M, Zahalka AH et al (2019) Engineering a haematopoietic stem cell niche by revitalizing mesenchymal stromal cells. Nat Cell Biol 21(5):560–567. https://doi.org/10.1038/s41556-019-0308-3
Nakamura-Ishizu A, Takubo K, Fujioka M, Suda T (2014) Megakaryocytes are essential for HSC quiescence through the production of thrombopoietin. Biochem Biophys Res Commun 454(2):353–357. https://doi.org/10.1016/j.bbrc.2014.10.095
Nishikawa K, Nakashima T, Takeda S, Isogai M, Hamada M, Kimura A et al (2010) Maf promotes osteoblast differentiation in mice by mediating the age-related switch in mesenchymal cell differentiation. J Clin Invest 120(10):3455–3465. https://doi.org/10.1172/JCI42528
Nombela-Arrieta C, Pivarnik G, Winkel B, Canty KJ, Harley B, Mahoney JE et al (2013) Quantitative imaging of haematopoietic stem and progenitor cell localization and hypoxic status in the bone marrow microenvironment. Nat Cell Biol 15(5):533–543. https://doi.org/10.1038/ncb2730
Oburoglu L, Tardito S, Fritz V, de Barros SC, Merida P, Craveiro M et al (2014) Glucose and glutamine metabolism regulate human hematopoietic stem cell lineage specification. Cell Stem Cell 15(2):169–184. https://doi.org/10.1016/j.stem.2014.06.002
Ossenkoppele GJ, Stussi G, Maertens J, van Montfort K, Biemond BJ, Breems D et al (2012) Addition of bevacizumab to chemotherapy in acute myeloid leukemia at older age: a randomized phase 2 trial of the Dutch-Belgian Cooperative Trial Group for Hemato-Oncology (HOVON) and the Swiss Group for Clinical Cancer Research (SAKK). Blood 120(24):4706–4711. https://doi.org/10.1182/blood-2012-04-420596
Padró T, Ruiz S, Bieker R, Bürger H, Steins M, Kienast J et al (2000) Increased angiogenesis in the bone marrow of patients with acute myeloid leukemia. Blood 95(8): 2637–2644
Pang WW, Price EA, Sahoo D, Beerman I, Maloney WJ, Rossi DJ et al (2011) Human bone marrow hematopoietic stem cells are increased in frequency and myeloid-biased with age. Proc Natl Acad Sci U S A 108(50):20012–20017. https://doi.org/10.1073/pnas.1116110108
Passaro D, Di Tullio A, Abarrategi A, Rouault-Pierre K, Foster K, Ariza-McNaughton L et al (2017) Increased vascular permeability in the bone marrow microenvironment contributes to disease progression and drug response in acute myeloid leukemia. Cancer Cell 32(3):324–341.e326. https://doi.org/10.1016/j.ccell.2017.08.001
Pierce H, Zhang D, Magnon C, Lucas D, Christin JR, Huggins M et al (2017) Cholinergic signals from the CNS regulate G-CSF-mediated HSC mobilization from bone marrow via a glucocorticoid signaling relay. Cell Stem Cell 20(5):648–658.e644. https://doi.org/10.1016/j.stem.2017.01.002
Pinho S, Frenette PS (2019) Haematopoietic stem cell activity and interactions with the niche. Nat Rev Mol Cell Biol 20(5):303–320. https://doi.org/10.1038/s41580-019-0103-9
Pinho S, Lacombe J, Hanoun M, Mizoguchi T, Bruns I, Kunisaki Y, Frenette PS (2013) PDGFRalpha and

CD51 mark human nestin+ sphere-forming mesenchymal stem cells capable of hematopoietic progenitor cell expansion. J Exp Med 210(7):1351–1367. https://doi.org/10.1084/jem.20122252

Pinho S, Marchand T, Yang E, Wei Q, Nerlov C, Frenette PS (2018) Lineage-biased hematopoietic stem cells are regulated by distinct niches. Dev Cell 44(5):634–641.e634. https://doi.org/10.1016/j.devcel.2018.01.016

Pioli PD, Casero D, Montecino-Rodriguez E, Morrison SL, Dorshkind K (2019) Plasma cells are obligate effectors of enhanced myelopoiesis in aging bone marrow. Immunity 51(2):351–366.e356. https://doi.org/10.1016/j.immuni.2019.06.006

Poulos MG, Ramalingam P, Gutkin MC, Llanos P, Gilleran K, Rabbany SY, Butler JM (2017) Endothelial transplantation rejuvenates aged hematopoietic stem cell function. J Clin Invest 127(11):4163–4178. https://doi.org/10.1172/jci93940

Qi L, Martin-Sandoval MS, Merchant S, Gu W, Eckhardt M, Mathews TP et al (2021) Aspartate availability limits hematopoietic stem cell function during hematopoietic regeneration. Cell Stem Cell 28(11):1982–1999.e1988. https://doi.org/10.1016/j.stem.2021.07.011

Raaijmakers MH, Mukherjee S, Guo S, Zhang S, Kobayashi T, Schoonmaker JA et al (2010) Bone progenitor dysfunction induces myelodysplasia and secondary leukaemia. Nature 464(7290):852–857. https://doi.org/10.1038/nature08851

Ranganath SH, Levy O, Inamdar MS, Karp JM (2012) Harnessing the mesenchymal stem cell secretome for the treatment of cardiovascular disease. Cell Stem Cell 10(3):244–258. https://doi.org/10.1016/j.stem.2012.02.005

Rossi DJ, Bryder D, Zahn JM, Ahlenius H, Sonu R, Wagers AJ, Weissman IL (2005) Cell intrinsic alterations underlie hematopoietic stem cell aging. Proc Natl Acad Sci U S A 102(26):9194–9199. https://doi.org/10.1073/pnas.0503280102

Rossi DJ, Bryder D, Seita J, Nussenzweig A, Hoeijmakers J, Weissman IL (2007) Deficiencies in DNA damage repair limit the function of haematopoietic stem cells with age. Nature 447(7145):725–729. https://doi.org/10.1038/nature05862

Sacma M, Pospiech J, Bogeska R, de Back W, Mallm JP, Sakk V et al (2019) Haematopoietic stem cells in perisinusoidal niches are protected from ageing. Nat Cell Biol 21(11):1309–1320. https://doi.org/10.1038/s41556-019-0418-y

Saito Y, Chapple RH, Lin A, Kitano A, Nakada D (2015) AMPK protects leukemia-initiating cells in myeloid leukemias from metabolic stress in the bone marrow. Cell Stem Cell 17(5):585–596. https://doi.org/10.1016/j.stem.2015.08.019

Sánchez-Aguilera A, Arranz L, Martín-Pérez D, García-García A, Stavropoulou V, Kubovcakova L et al (2014) Estrogen signaling selectively induces apoptosis of hematopoietic progenitors and myeloid neoplasms without harming steady-state hematopoiesis. Cell Stem Cell 15(6):791–804. https://doi.org/10.1016/j.stem.2014.11.002

Sanjuan-Pla A, Macaulay IC, Jensen CT, Woll PS, Luis TC, Mead A et al (2013) Platelet-biased stem cells reside at the apex of the haematopoietic stem-cell hierarchy. Nature 502(7470):232–236. https://doi.org/10.1038/nature12495

Schofield R (1978) The relationship between the spleen colony-forming cell and the haemopoietic stem cell. Blood Cells 4(1–2):7–25

Severe N, Karabacak NM, Gustafsson K, Baryawno N, Courties G, Kfoury Y et al (2019) Stress-induced changes in bone marrow stromal cell populations revealed through single-cell protein expression mapping. Cell Stem Cell 25(4):570–583.e577. https://doi.org/10.1016/j.stem.2019.06.003

Shen B, Tasdogan A, Ubellacker JM, Zhang J, Nosyreva ED, Du L et al (2021) A mechanosensitive periarteriolar niche for osteogenesis and lymphopoiesis. Nature 591(7850):438–444. https://doi.org/10.1038/s41586-021-03298-5

Shin JY, Hu W, Naramura M, Park CY (2014) High c-Kit expression identifies hematopoietic stem cells with impaired self-renewal and megakaryocytic bias. J Exp Med 211(2):217–231. https://doi.org/10.1084/jem.20131128

Sigurdsson V, Takei H, Soboleva S, Radulovic V, Galeev R, Siva K et al (2016) Bile acids protect expanding hematopoietic stem cells from unfolded protein stress in fetal liver. Cell Stem Cell 18(4):522–532. https://doi.org/10.1016/j.stem.2016.01.002

Stevens BM, Jones CL, Pollyea DA, Culp-Hill R, D'Alessandro A, Winters A et al (2020) Fatty acid metabolism underlies venetoclax resistance in acute myeloid leukemia stem cells. Nat Cancer 1(12):1176–1187. https://doi.org/10.1038/s43018-020-00126-z

Stier S, Ko Y, Forkert R, Lutz C, Neuhaus T, Grünewald E et al (2005) Osteopontin is a hematopoietic stem cell niche component that negatively regulates stem cell pool size. J Exp Med 201(11):1781–1791. https://doi.org/10.1084/jem.20041992

Stik G, Crequit S, Petit L, Durant J, Charbord P, Jaffredo T, Durand C (2017) Extracellular vesicles of stromal origin target and support hematopoietic stem and progenitor cells. J Cell Biol 216(7):2217–2230. https://doi.org/10.1083/jcb.201601109

Stoddart A, Wang J, Hu C, Fernald AA, Davis EM, Cheng JX, Le Beau MM (2017) Inhibition of WNT signaling in the bone marrow niche prevents the development of MDS in the Apc(del/+) MDS mouse model. Blood 129(22):2959–2970. https://doi.org/10.1182/blood-2016-08-736454

Straussman R, Morikawa T, Shee K, Barzily-Rokni M, Qian ZR, Du J et al (2012) Tumour micro-environment elicits innate resistance to RAF inhibitors through HGF secretion. Nature 487(7408):500–504. https://doi.org/10.1038/nature11183

Sugimura R, He XC, Venkatraman A, Arai F, Box A, Semerad C et al (2012) Noncanonical Wnt signaling maintains hematopoietic stem cells in the niche. Cell 150(2):351–365. https://doi.org/10.1016/j.cell.2012.05.041

Sugiyama T, Kohara H, Noda M, Nagasawa T (2006) Maintenance of the hematopoietic stem cell pool by CXCL12-CXCR4 chemokine signaling in bone marrow stromal cell niches. Immunity 25(6):977–988. https://doi.org/10.1016/j.immuni.2006.10.016

Taya Y, Ota Y, Wilkinson AC, Kanazawa A, Watarai H, Kasai M et al (2016) Depleting dietary valine permits nonmyeloablative mouse hematopoietic stem cell transplantation. Science 354(6316):1152–1155. https://doi.org/10.1126/science.aag3145

Tikhonova AN, Dolgalev I, Hu H, Sivaraj KK, Hoxha E, Cuesta-Dominguez A et al (2019) The bone marrow microenvironment at single-cell resolution. Nature 569(7755):222–228. https://doi.org/10.1038/s41586-019-1104-8

Velardi E, Tsai JJ, Radtke S, Cooper K, Argyropoulos KV, Jae-Hung S et al (2018) Suppression of luteinizing hormone enhances HSC recovery after hematopoietic injury. Nat Med 24(2):239–246. https://doi.org/10.1038/nm.4470

Wilson A, Laurenti E, Oser G, van der Wath RC, Blanco-Bose W, Jaworski M et al (2008) Hematopoietic stem cells reversibly switch from dormancy to self-renewal during homeostasis and repair. Cell 135(6):1118–1129. https://doi.org/10.1016/j.cell.2008.10.048

Wilson TR, Fridlyand J, Yan Y, Penuel E, Burton L, Chan E et al (2012) Widespread potential for growth-factor-driven resistance to anticancer kinase inhibitors. Nature 487(7408):505–509. https://doi.org/10.1038/nature11249

Xie T, Spradling AC (2000) A niche maintaining germ line stem cells in the Drosophila ovary. Science 290(5490):328–330. https://doi.org/10.1126/science.290.5490.328

Xu C, Gao X, Wei Q, Nakahara F, Zimmerman SE, Mar J, Frenette PS (2018) Stem cell factor is selectively secreted by arterial endothelial cells in bone marrow. Nat Commun 9(1):2449. https://doi.org/10.1038/s41467-018-04726-3

Yamamoto R, Morita Y, Ooehara J, Hamanaka S, Onodera M, Rudolph KL et al (2013) Clonal analysis unveils self-renewing lineage-restricted progenitors generated directly from hematopoietic stem cells. Cell 154(5):1112–1126. https://doi.org/10.1016/j.cell.2013.08.007

Yamashita M, Passegué E (2019) TNF-α coordinates hematopoietic stem cell survival and myeloid regeneration. Cell Stem Cell 25(3):357–372.e357. https://doi.org/10.1016/j.stem.2019.05.019

Yamazaki S, Ema H, Karlsson G, Yamaguchi T, Miyoshi H, Shioda S et al (2011) Nonmyelinating Schwann cells maintain hematopoietic stem cell hibernation in the bone marrow niche. Cell 147(5):1146–1158. https://doi.org/10.1016/j.cell.2011.09.053

Young K, Eudy E, Bell R, Loberg MA, Stearns T, Sharma D et al (2021) Decline in IGF1 in the bone marrow microenvironment initiates hematopoietic stem cell aging. Cell Stem Cell 28(8):1473–1482.e1477. https://doi.org/10.1016/j.stem.2021.03.017

Yu Z, Yang W, He X, Chen C, Li W, Zhao L et al (2022) Endothelial cell-derived angiopoietin-like protein 2 supports hematopoietic stem cell activities in bone marrow niches. Blood 139(10):1529–1540. https://doi.org/10.1182/blood.2021011644

Zahiragic L, Schliemann C, Bieker R, Thoennissen NH, Burow K, Kramer C et al (2007) Bevacizumab reduces VEGF expression in patients with relapsed and refractory acute myeloid leukemia without clinical antileukemic activity. Leukemia 21(6):1310–1312. https://doi.org/10.1038/sj.leu.2404632

Zhang J, Niu C, Ye L, Huang H, He X, Tong WG et al (2003) Identification of the haematopoietic stem cell niche and control of the niche size. Nature 425(6960):836–841. https://doi.org/10.1038/nature02041

Zhang B, Nguyen LXT, Li L, Zhao D, Kumar B, Wu H et al (2018) Bone marrow niche trafficking of miR-126 controls the self-renewal of leukemia stem cells in chronic myelogenous leukemia. Nat Med 24(4):450–462. https://doi.org/10.1038/nm.4499

Zhang J, Wu Q, Johnson CB, Pham G, Kinder JM, Olsson A et al (2021) In situ mapping identifies distinct vascular niches for myelopoiesis. Nature 590(7846):457–462. https://doi.org/10.1038/s41586-021-03201-2

Zhang D, Gao X, Li H, Borger DK, Wei Q, Yang E et al (2022) The microbiota regulates hematopoietic stem cell fate decisions by controlling iron availability in bone marrow. Cell Stem Cell 29(2):232–247.e237. https://doi.org/10.1016/j.stem.2021.12.009

Zhao M, Ross JT, Itkin T, Perry JM, Venkatraman A, Haug JS et al (2012) FGF signaling facilitates postinjury recovery of mouse hematopoietic system. Blood 120(9):1831–1842. https://doi.org/10.1182/blood-2011-11-393991

Zhao M, Perry JM, Marshall H, Venkatraman A, Qian P, He XC et al (2014) Megakaryocytes maintain homeostatic quiescence and promote post-injury regeneration of hematopoietic stem cells. Nat Med 20(11):1321–1326. https://doi.org/10.1038/nm.3706

Zhao M, Tao F, Venkatraman A, Li Z, Smith SE, Unruh J et al (2019) N-cadherin-expressing bone and marrow stromal progenitor cells maintain reserve hematopoietic stem cells. Cell Rep 26(3):652–669.e656. https://doi.org/10.1016/j.celrep.2018.12.093

Emerging Roles of Epigenetic Regulators in Maintaining Hematopoietic Stem Cell Homeostasis

3

Hui Wang, Yingli Han, and Pengxu Qian

Abstract

Hematopoietic stem cells (HSCs) are adult stem cells with the ability of self-renewal and multilineage differentiation into functional blood cells, thus playing important roles in the homeostasis of hematopoiesis and the immune response. Continuous self-renewal of HSCs offers fresh supplies for the HSC pool, which differentiate into all kinds of mature blood cells, supporting the normal functioning of the entire blood system. Nevertheless, dysregulation of the homeostasis of hematopoiesis is often the cause of many blood diseases. Excessive self-renewal of HSCs leads to hematopoietic malignancies (e.g., leukemia), while deficiency in HSC regeneration results in pancytopenia (e.g., anemia). The regulation of hematopoietic homeostasis is finely tuned, and the rapid development of high-throughput sequencing technologies has greatly boosted research in this field. In this chapter, we will summarize the recent understanding of epigenetic regulators including DNA methylation, histone modification, chromosome remodeling, noncoding RNAs, and RNA modification that are involved in hematopoietic homeostasis, which provides fundamental basis for the development of therapeutic strategies against hematopoietic diseases.

Keywords

Hematopoietic stem cell · Homeostasis · Epigenetics · Self-renewal

3.1 Introduction

Hematopoietic stem cells were first discovered in the early 1960s by two Canadian scientists, Ernest McCulloch and James Till. They discovered a kind of cell that could colonize the spleen while preserving the function of self-renewal and differentiation, and these cells are now called hematopoietic stem cells (HSCs) (Siminovitch et al. 1963). Then, Ernest McCulloch and James Till discovered the potential clinical applications of HSCs in treating various hematological disorders and established the theoretical foundations for bone marrow transplantation, which until now has benefited millions of patients with blood diseases by expanding their life span. The pioneering work of Ernest McCulloch and

H. Wang · Y. Han · P. Qian (✉)
Center for Stem Cell and Regenerative Medicine and Bone Marrow Transplantation Center of the First Affiliated Hospital, Zhejiang University School of Medicine, Hangzhou, China

Liangzhu Laboratory, Zhejiang University, Hangzhou, China

Institute of Hematology, Zhejiang University & Zhejiang Engineering Laboratory for Stem Cell and Immunotherapy, Hangzhou, China
e-mail: axu@zju.edu.cn

M. Zhao, P. Qian (eds.), *Hematopoietic Stem Cells*, Advances in Experimental Medicine and Biology 1442, https://doi.org/10.1007/978-981-99-7471-9_3

James Till was the foundation for the theory of HSC homeostasis, which holds that the characteristics of self-renewal and multilineage differentiation of HSCs are finely tuned and should be kept under precise balance and any disturbance would lead to severe blood diseases. If the self-renewal ability of HSCs is not properly regulated or the differentiation ability is blocked, there will be too many HSCs crowded in the bone marrow and no sufficient supply of mature blood cells, which is often the case in patients with leukemia. On the other hand, deficiency in HSC self-renewal leads to exhaustion of HSC pools, resulting in anemia or immunodeficiency (McCulloch and Till 2005). To date, the regulation of HSC homeostasis is still a research focus, and the development of high-throughput sequencing technologies has greatly promoted studies of HSC function as well as the underlying mechanisms. Recently, progress in epigenetics has revealed that epigenetic regulators are also involved in the regulation of HSC homeostasis, such as DNA methylation, histone modification, chromatin remodeling, noncoding RNAs (ncRNAs), RNA modification, etc., all of which affect HSC homeostasis through various mechanisms. In this chapter, we mainly summarize and discuss how epigenetic factors are involved in the regulation of HSC homeostasis (Fig. 3.1).

3.2 DNA Modification

Since the completion of the "Human Genome Project," researchers realized that in our genome, only a small portion of genes are being actively transcribed at different periods of life. DNA is packaged firmly and compactly into chromosomes with higher-order structures. While it is known that the expression of one gene requires RNA polymerase-based transcription, modifications to DNA can regulate this process by affecting chromosome structures and the accessibility of promoters to DNA. The most studied DNA modification, 5-methylcytosine (5mC), was first discovered by Razin and Cedar in 1977, during which research on gene expression regulation entered a new era (Razin and Cedar 1977; Pollack et al. 1980). In vertebrates, DNA methylation typically occurs at the CpG dinucleotide cytosine, which usually exists as clusters, namely, CpG islands. The modification is usually set up by DNA methyltransferases (DNMTs) and oxidized to 5-hydroxyl groups and then erased by TETs (10–11 translocation enzymes). In general, 5mC on *cis*-regulatory elements, promoters and enhancers, for example, can recruit methyl-binding proteins that prevent transcription factors from gaining access, finally resulting in transcription inhibition (Du et al. 2015; Zhu et al. 2016), and gene activities are regulated by methylation states. In mammals, most single CpGs are methylated, but CpG-rich regions near promoters (CpG islands) are often unmethylated, while regions near active enhancers are hypomethylated. In addition, gene activity in imprinting control regions (ICRs) is also regulated by allele-specific methylation states, resulting in expression in a parent-of-origin manner (Stadler et al. 2011; Skvortsova et al. 2018).

DNA methylation states are closely connected with HSC homeostasis, which undergoes extensive remodeling during the process of HSC differentiation. In recent years, advances in high-throughput sequencing technologies have greatly promoted research in epigenetics, and based on these technologies, genome-wide DNA methylation maps and the dynamic oscillations along the HSC differentiation trajectory were built at single-base resolution, offering us a more comprehensive landscape and deepening our understanding (Cabezas-Wallscheid et al. 2014; Qian et al. 2018; Farlik et al. 2016; Bock et al. 2012). During HSC differentiation, the genome methylation landscape changes in multiple regions simultaneously, collectively called differentially methylated regions (DMRs), and these regions are usually enhancers, promoters, and transcription factor-binding sites (Cabezas-Wallscheid et al. 2014).

DMRs frequently overlap with *cis*-regulating elements that are involved in HSC homeostasis regulation. For example, Hoxb genes are crucial for HSC homeostasis. During HSC

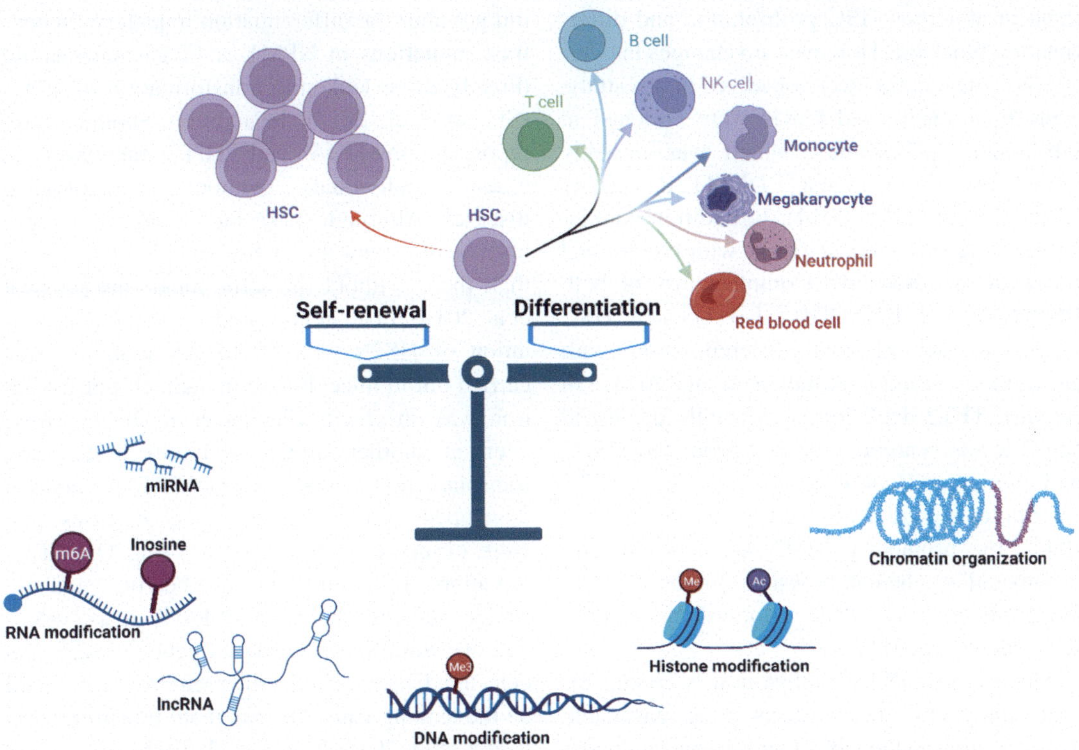

Fig. 3.1 Epigenetic factors that regulate HSC homeostasis. The balance of self-renewal and differentiation of HSCs is regulated by various epigenetic regulators, including DNA modification, histone modification, chromatin organization, RNA modification, and noncoding RNAs

differentiation, genes in the Hoxb cluster are hypermethylated, and thus, their expression is downregulated (Qian et al. 2018; Farlik et al. 2016). In addition, the imprinting control region H19-DMR is indispensable for HSC maintenance. Maternal-specific methylation leads to allele-specific expression of H19 and H19-derived miR-675 expression, and deletion of this region in HSCs results in rejuvenation of Igf2 expression and depletion of miR-675. This would cause the activation of HSCs and their entry into the cell cycle and eventually HSC exhaustion (Venkatraman et al. 2013). IG DMR is another imprinting control region located in the Gtl2-Dio3 region, and the methylation state controls the expression of a series of protein-coding and noncoding genes. It has been reported that the mega miRNA cluster in this region can extensively bind and suppress the translation of a series of genes of the PI3K-mTOR signaling pathway and inhibit mitochondrial metabolic activity, thus preserving HSCs in a quiescent state. Deletion of the IG-DMR region causes HSC activation and exhaustion (Qian et al. 2016).

DNA methylation also affects the process of HSC differentiation. During this process, stemness-related genes are suppressed by hypermethylation, while differentiation-related genes are activated by demethylation. Dysregulation of such coordination is the cause of many hematological disorders (Cabezas-Wallscheid et al. 2014). It has been reported that defects in DNMT1 (responsible for the maintenance of DNA methylation) activity result in HSC differentiation bias toward the myeloerythroid lineage rather than lymphoid progenies, demonstrating that constitutive DNA methylation activity is indispensable for HSC homeostasis (Broske et al. 2009; Trowbridge et al. 2009). Loss of DNMT3A was reported to result in abnormal HSC proliferation and differentiation blockage. However, no changes in overall DNA methylation were observed in this study, while hypermethylated CpGs were enriched in substantial CpG islands, which was possibly caused by complementary DNMT3B activity (Challen et al. 2011, 2014). In addition, similar results were observed in HSCs with conditional knockout of DNMT3B. Complete loss of both DNMT3A and DNMT3B also boosted HSC expansion and blocked differentiation, with much more severity (Challen et al. 2014). In contrast, TET2 depletion significantly decreased 5hmC levels genome-wide and promoted HSPC proliferation. Haploinsufficiency of TET2 contributes to the malignant transformation of HSCs by promoting HSC self-renewal and extramedullary hematopoiesis (Quivoron et al. 2011; Shide et al. 2012; Moran-Crusio et al. 2011; Ko et al. 2011).

Since proper DNA methylation is crucial for maintaining HSC homeostasis, it is reasonable that alterations in the DNA methylation landscape are related to the occurrence of hematological diseases. Indeed, DNA methylation-related factors are frequently mutated in leukemias, among which DNMT3A is the most frequently mutated in leukemias (17–34% in AML, 3–8% in MDS, 4% in CMML, 10% in MPN, and 4% in T-ALL). Similarly, a high mutation rate of the demethylase TET2 was also observed in leukemias (12% in AML, 42% in CMML, and 7.6% in MPN) (Roller et al. 2013; Abdel-Wahab et al. 2009). Intriguingly, the co-occurrence of DNMT3 and TET2 mutations could not counterbalance each other but instead exacerbated the situation and accelerated malignant transformation (Zhang et al. 2016). Currently, the exact mechanism by which DNMT3A or TET2 mutations drive leukemic transformation remains unclear. Findings from recent studies have led to the consensus that DNMT3A mutation is the early event of malignant transformation, priming normal HSCs into a preleukemic state, and other additional mutations are needed to finally drive disease initiation and progression. DNMT3A mutations were detected in T cells of AML patients, indicating that this mutation in HSCs did not alter the differentiation trajectory. In contrast, mutations in NPM1 or FLT3 genes could directly drive leukemic transformation of HSCs (Shlush et al. 2014). In addition, another study reported DNMT3A and TET2 mutations in elderly individuals without hematological diseases. Although these individuals can live a normal life, they are at higher risk of leukemia than those without the same mutations (Jaiswal et al. 2014). In an in vivo study, a serial transplantation of HSCs with DNMT3A mutation was carried out in mice, but no malignant transformation was observed (Challen et al. 2011). However, in another similar study, mice receiving transplantation of HSCs with DNMT3A mutation developed similar hematological diseases that were observed in patients harboring DNMT3A mutations. In addition, in mice bearing leukemic HSCs, several mutations (NPM1, Kras, c-Kit) that are common in leukemia patients were also detected. Changes in the microenvironment could be the leading cause of malignant transformation (Celik et al. 2015; Mayle et al. 2015).

DNA methylation in HSCs and leukemic cells is being extensively studied, and there is growing recognition of the importance of DNA methylation in disease progression. However, the specific roles and underlying mechanisms in HSC homeostasis and malignant transformation need further exploration. Advances in cancer-specific DMRs will further aid in the clinical diagnosis, treatment, and prognosis of hematological diseases.

3.3 Histone Modifications

DNA is packaged along with histones into nucleosomes, which are then further compacted into higher-order chromosome structures. Therefore, histones are also involved in gene expression regulation. In addition, histones are usually subjected to various posttranslational modifications, such as methylation, acetylation, and ubiquitination. Similar to the counterpart of the genetic code, these modifications are also called histone codes and can regulate gene expression by affecting chromosome opening states and accessibility. Depending on the kind of histone being modified and the number of methyl groups added, histone methylation can either promote or suppress RNA transcription. In contrast, the effects of histone acetylation and ubiquitination are usually definite and unequivocal, with the former opening chromatin and the latter being considered suppressive marks. Their roles in the regulation of HSC homeostasis have been reported extensively.

3.3.1 Histone Methylation

According to a previous report, the methylation number and location can both determine the function of histone methylation: methylation on lysine 4, 36, and 79 of histone 3 usually marks open chromosomes and active transcription, while H3K9, H3K27, and H4K20 are often enriched in transcriptionally inactive zones. PRC2 is mainly responsible for the inhibitory H3K27me3 modification, and several studies have revealed its role in HSC homeostasis regulation. It was reported that heterozygous mutation in *Ezh2*, *Eed*, or *Suz12*, constituents of the PRC2 complex, resulted in HSPC differentiation and enhanced engraftment of HSCs in transplantation. It is likely that the PRC2 complex suppresses the activation and differentiation of HSCs by introducing inhibitory histone methylation (Majewski et al. 2010). Controversially, it was also reported that overexpression of Ezh2 enhanced HSC reconstitution ability and maintained HSC stemness (Kamminga et al. 2006). In addition, Hidalgo and colleagues found that Ezh1, another component of the PRC2 complex, was crucial for HSC self-renewal and knockout of Ezh1 impaired HSC reconstitution ability and induced a senescence-like phenotype (Hidalgo et al. 2012).

Other regulators of histone methylation that induce active modifications are of equivalent importance for HSC homeostasis. For example, the H3K79 methylase Dot1l (disruptor of telomere silencing 1-like) was responsible for the regulation of GATA2 and PU.1 expression, which are both crucial transcription factors for HSC stemness and myeloid differentiation. Loss of Dot1l is embryonically lethal due to defects in erythroid lineage differentiation (Feng et al. 2010). Moreover, MLL1 (mixed-lineage leukemia 1), named due to its importance in mixed-lineage leukemia, is responsible for H3K4 methylation and is of crucial importance for HSC homeostasis and fetal hematopoiesis. It was reported that knockout of Mll1 caused severe loss of fetal liver cellularity and dramatic loss of HSC numbers (McMahon et al. 2007).

In summary, the functions of histone methylation in HSC homeostasis are context dependent, and different components may have divergent effects. Therefore, further studies are required to better understand their functions and underlying mechanisms.

3.3.2 Histone Acetylation

In general, histone acetylation causes more open chromosome states and thus higher accessibility. Genome-wide epigenetic profiling revealed that the acetylation states of genes crucial for HSC homeostasis are periodically changing and acetylation regulators play important roles during these processes. TRRAP (transformation/transcription domain-associated protein), the cofactor of histone acetyltransferase (HAT), is reported to be indispensable for HSC homeostasis. Knockout of TRRAP caused p53-dependent apoptosis in HSPCs and ultimately resulted in bone marrow failure (Loizou et al. 2009). Moreover, Moz (KAT6A), a type of HAT belonging to the MYST family, is also important for HSC homeostasis, and Moz fusion proteins are oncogenic and commonly observed in AML patients, indicating their important roles in HSC homeostasis regulation (Katsumoto et al. 2008). Genetic deletion of Moz resulted in HSC exhaustion and embryonic lethality but did not affect the lineage commitment of HSCs in mice. As a result, the reconstitution ability in transplantation was impaired, and the downregulation of c-Kit may be the leading cause (Katsumoto et al. 2006; Thomas et al. 2006).

Histone acetylation can also be removed by histone deacetylase (HDAC). Removement of histone acetylation is of equivalent importance for HSC homeostasis, usually by shutting down the expression of key factors in HSC self-renewal and/or differentiation. It was reported that knockout of HDAC1 and HDAC2 simultaneously resulted in severe HSC exhaustion, finally leading to anemia and cytopenia. However, single deletion of either HDAC1 or HDAC2 caused nothing abnormal in HSCs, indicating the functional redundancy of the HDAC family (Heideman et al. 2014). SIRT1, another HDAC, is also involved in the regulation of HSC homeostasis. Depletion of SIRT1 led to severe deficiency in HSC self-renewal and caused differentiation bias toward the myeloid lineage, a typical senescence-like phenotype. Furthermore, the authors also discovered that FOXO3, the crucial transcription factor related to longevity, was the key downstream factor mediating SIRT1's function in HSC homeostasis (Rimmelé et al. 2014).

3.3.3 Histone Ubiquitination

Histone ubiquitination mainly occurs at lysine 119 of H2A, labeled H2A-K119ub, and usually results in a closed chromosome state and decreased accessibility, thus suppressing gene expression. The ubiquitination states of chromosomes are also dynamically regulated by counteracting ubiquitination and deubiquitination enzymes. PRC1 is responsible for introducing ubiquitin modifications, and this is a heterogeneous multisubunit complex composed of five different subunits, which are all involved in HSC homeostasis regulation. For example, it was reported that ectopic expression of Bmi1, one component of the PRC1 complex, in embryonic stem cells (ESCs) promoted the hematopoiesis process by enhancing GATA2 expression, and the latter has been reported to be indispensable for primitive hematopoiesis. In addition, the authors also found that PRC1 suppressed the expression of $P16^{INK4A}/P19^{ARF}$, resulting in the proliferation of HSPCs derived from ESCs (Ding et al. 2012). Rae28, another component of the PRC1 complex, was found to be crucial for early hematopoietic development. Knockout of Rae28 led to embryonic lethality due to deficiency of fetal liver HSCs and impaired colony formation and reconstitution abilities of HSCs (Ohta et al. 2002). In contrast, knockout of Mel18, also one component of PRC1, led to G0 phase blockage and a quiescent state of HSCs, which was beneficial for the self-renewal ability of HSCs, possibly by preventing overactivation of HSCs. These results demonstrated that Mel18 mainly plays an inhibitory role in HSC self-renewal (Kajiume et al. 2004). However, whether the function of PRC1 in HSC homeostasis depends on its histone ubiquitination activity remains elusive, and further exploration is required (Valk-Lingbeek et al. 2004).

In the last decade, accumulating evidence has demonstrated that histone modifications play crucial roles in the regulation of HSC homeostasis, the dysfunction of which is the leading cause of many hematological diseases. These studies paved the way for explorations of targeting histone modifications in treating various hematological malignancies. However, how differentially modified histones in different genomic regions cooperate under the regulation of a few modifiers during hematopoiesis is still elusive, and further studies are needed.

3.4 Spatial Organization of Chromosomes

In recent years, with the development of second-generation sequencing technologies, an increasing number of studies have discovered the three-dimensional organization of chromosomes under patho-/physiological conditions. The crucial roles of the spatial organization of chromosomes in the regulation of gene expression and cell fate determination are receiving increasing attention. Tissue specificity often requires specific interactions between promoters and enhancers during cell fate determination. Spatial interactions of distant genomic regions rely on cooperation between cohesion and CTCF, which helps chromosomes form topologically associating domains (TADs) and chromosome loops. Different TADs are separated by TAD boundaries, which form under the control of CTCF and cohesion. Intra-TAD interactions are much more frequent than those among TADs. Mutations in cohesion and CTCF or sequence variation in their binding sites often disturb normal spatial interactions and lead to abnormal gene expression (Stadhouders et al. 2019).

Cohesin and CTCF are frequently mutated in AML, indicating their importance in HSC homeostasis and AML progression and demonstrating the fundamental role of spatial organization (Dolnik et al. 2012; Cuartero et al. 2018). It has been reported that depletion of Rad21, one component of cohesion, promotes HSC self-renewal by simultaneously upregulating a series of genes, including ERG, GATA2, RUNX1, and members of the Hox gene family. Knockout of this gene also led to blockage of differentiation induced by inflammatory stimuli (Cuartero et al. 2018; Mazumdar et al. 2015; Fisher et al. 2017). Moreover, many other factors were also found to affect chromosome organization, such as transcription factors and modifiers of DNA methylation and histone modifications, possibly through affecting nucleosome assembly.

Chromosome conformation capture (3C) and Hi-C derived from 3C are commonly used techniques to study chromosome organization and interaction. However, their application in the study of HSC homeostasis is limited due to the rarity of HSCs, and only a few studies have explored local interactions between specific regions. Javierre et al. reported genome-wide interactions among promoters in 17 kinds of human primary hematopoietic cells using the Hi-C technique. The study clearly showed obvious differences in promoter interactions depending on cell type, and unsupervised clustering based on these interactions reflected lineage differentiation trajectories of hematopoiesis. A higher mutation frequency was observed in interactions between intergenic regions, which could affect target gene expression (Javierre et al. 2016). Hu studied the chromosome conformation of every knot along the HSC differentiation cascade toward early T cells using multiple-enzyme Hi-C, which required low input cell numbers. They observed extensive dramatic genome-wide conformational changes in chromosomes during T-cell lineage commitment and accompanied gene expression shifts. It seems that this conformational change not only facilitates the shaping of cell-type specific gene expression signatures but also prevents cells from deviating, similar to railways for a train (Hu et al. 2018).

With the development of high-throughput sequencing techniques, genome-wide 3D maps of chromosome organization and interactions with higher resolution and lower input requirements can be expected. Such conformational information will certainly offer us a more concise and thorough understanding of HSC homeostasis regulation.

3.5 Noncoding RNAs (ncRNAs)

The Human Genome Project and ENCODE project offer us a landscape of genetic organizations of the human genome, which has helped us realize that only 5% of all genome transcripts are translated into proteins. Therefore, based on this ratio, noncoding transcripts constitute a large regulatory network, supervising gene expression during various physiological processes. HSC homeostasis is also under transcriptional and posttranscriptional regulation, and the roles of noncoding RNAs during this process are receiving increasing attention.

3.5.1 MiRNAs

MiRNAs are small regulatory ncRNAs of 19–23 nt and function by binding to the 3′UTR of target transcripts via base-pairing, which results in either translation inhibition or mRNA degradation. A few studies have reported that miRNAs regulate HSC homeostasis by targeting crucial mRNA transcripts for HSC self-renewal or differentiation. Based on high-throughput sequencing, miRNA expression signatures were profiled and compared among different hematological subpopulations, indicating their importance in HSC homeostasis. It was reported that the miRNA cluster, including miR-125a and miR-125b, is specifically upregulated in long-term HSCs. Ectopic expression of members of these clusters facilitated HSC reconstitution. MiR-125b could also promote the progression of myeloid leukemia. In addition, miR-196b was found to be overexpressed in patients with MLL (mixed-lineage leukemia) (O'Connell et al. 2010; Popovic et al. 2009).

Moreover, in another study, the author knocked out Dicer, the master processor of miRNA production, specifically in HSCs, and observed extensive downregulation of miRNAs, which in the end led to functional deficiency of HSCs. This study also revealed the crucial role of miR-125a in HSC homeostasis, and ectopic expression of this miRNA suppressed apoptosis and rescued HSC defects (Guo et al. 2010). Through comprehensive analysis of mRNA and miRNA expression in HSPC cells, Bissels et al. found that a series of miRNAs affect HSC self-renewal and differentiation by targeting crucial factors in hematopoiesis. For example, miR-29a could maintain HSC self-renewal by targeting the Wnt signaling pathway (Bissels et al. 2011). Another study also discovered that ectopic expression of miR-29a boosted HSC function in mice, and in AML patients, this miRNA was significantly upregulated (Han et al. 2010).

On the other hand, many miRNAs were reported to regulate HSC differentiation and lineage commitment by targeting various molecules. Tenedini et al. discovered that during the differentiation of human HSPCs (CD34+), miR-229 expression was specifically enhanced in megakaryoblasts, and ectopic expression of this miRNA in human HSCs induced megakaryocytic-granulocytic differentiation (Tenedini et al. 2010). In addition, miR-221 and miR-222 were reported to affect erythropoiesis. By targeting c-Kit, these miRNAs suppressed CD34+ progenitor cell proliferation and promoted erythropoietic differentiation (Felli et al. 2005). Fazi et al. reported that miR-223 facilitated HSC differentiation toward the myeloid lineage, partially by targeting NFI-a and CEBP-α (Fazi et al. 2005). It was reported that GATA1, a key factor for HSC differentiation, could induce the expression of miR-144 and miR-451, which contributed to erythroid differentiation of HSPCs (Dore et al. 2008; Papapetrou et al. 2010). MiR-150 was reported to be important for lymphoid differentiation, and c-Myb and NOTCH3 are both downstream targets mediating its function (Ghisi et al. 2011; Xiao et al. 2016).

Mounting evidence indicates that by targeting key molecules, miRNAs play pivotal roles in HSC homeostasis, and alterations in miRNA expression are hallmarks of many hematological malignancies, indicating their involvement in disease progression. miRNA-based studies and clinical applications are still promising for better understanding the mechanisms of HSC homeostasis and disease management.

3.5.2 LncRNAs

Different from miRNAs, the mechanisms of lncRNA function are more complicated. LncRNAs may take part in diverse patho-/physiological processes through interactions with proteins, RNA, and DNA. LncRNA H19 was reported to be upregulated specifically in long-term HSCs and decreased during differentiation to progenitor cells. Loss of H19 impaired HSC self-renewal function and reconstitution ability, resulting in HSC differentiation. Consequently, LT-HSC numbers decreased and ST-HSC increased (Venkatraman et al. 2013; Berg et al. 2011). LncHSC-1 and lncHSC-2 are two other lncRNAs that are specifically expressed in HSCs. With HSC differentiation, their expression decreased and could hardly be detected in progenitor cells. Luo et al. reported that depletion of either lncHSC-1 or lncHSC-2 impaired HSC functions and led to decreased HSC and progenitor numbers. Moreover, lineage differentiation increased after knocking down these two lncRNAs. Further exploration revealed that the transcription factor E2A was the key downstream factor (Luo et al. 2015).

LncRNAs are also involved in HSC differentiation via various mechanisms. Myeloid, lymphoid, and erythroid cells are the three lineages downstream of HSPC differentiation. Common myeloid progenitor cells (CMPs) are common progenitors of myeloid cells, and granulocytes, monocytes, and macrophages are all derived from CMPs. Accordingly, lymphocytes, such as T cells, B cells, and natural killer cells, are all differentiated from common lymphoid progenitors (CLPs). In addition, the millions of human red blood cells replenished every day are all derived from megakaryocytes and erythroid progenitor cells (MEPs). To date, many lncRNAs have been reported to take part in the process of lineage differentiation.

Wagner et al. first reported lncRNA EGO (eosinophil granule ontogeny) and found it to be conserved among many species. This lncRNA is involved in eosinophil formation, and its expression is restricted to mature eosinophils. Depletion of EGO in human HSCs (CD34+ progenitors) significantly prevented differentiation toward the eosinophil lineage, possibly through downregulation of neurotoxin (Wagner et al. 2007). Moreover, HOTAIRM1 (HOX antisense intergenic RNA myeloid 1), another lncRNA located at intron regions between HOXA1 and HOXA2, was also reported to be involved in myeloid differentiation. HOTAIRM1 is specifically expressed in myeloid cells and is dramatically induced accompanying myeloid differentiation. Depletion of HOTAIRM1 significantly decreased the expression of Mac-1 and CD18, blocking myeloid differentiation (Zhang et al. 2009). In addition, other lncRNAs, such as lnc-DC and lnc-MC, were also reported to affect myeloid differentiation via interactions with different molecules (Wang et al. 2014; Chen et al. 2015).

Similarly, many lncRNAs are involved in lymphoid differentiation. For example, lncRNA NRON was found to be crucial for T-cell differentiation. NRON was reported to bind to NFAT and suppress its nuclear translocation, thus inhibiting T-cell activation (Imam et al. 2015). Moreover, GAS5 (growth arrest-specific 5), another well-studied lncRNA, was found to suppress T-cell proliferation and eventually induce apoptosis. Further research demonstrated that GAS5 could interact with GREs and suppress target gene expression (Kino et al. 2010; Williams et al. 2011). LncRNA linc-MAF-4 is specifically expressed in Th1 cells and is indispensable for Th1-cell differentiation. This transcript could regulate MAF expression via interaction with its *cis*-regulating elements and recruit inhibitory PRC2 and EZH2. Depletion of this lncRNA enhanced MAF expression and caused the transition from Th1 to Th2 cells (Ranzani et al. 2015).

Erythropoiesis is also affected by many lncRNAs. Hu et al. reported that lncRNA-EPS is specifically expressed in erythroid lineages, which could affect HSC differentiation and erythropoiesis. Knockdown of EPS inhibited HSC differentiation and led to dysfunction of the erythropoiesis process. Mechanistically, EPS could transcriptionally inhibit Pycar, thus inhibiting apoptosis and safeguarding erythropoiesis (Hu et al. 2011). LncRNA EC7 was reported to be an enhancer RNA that could promote the expression of nearby *Band3*, thus regulating erythropoiesis. Further exploration revealed that lncRNA-EC7 could cooperate with CTCF and affect chromatin looping of the *Band3* region, thus activating transcription (Alvarez-Dominguez et al. 2014). Paralkar et al. systematically profiled lncRNAs that were specifically expressed in erythroid lineage cells. Most identified lncRNAs were poorly conserved and enriched in promoter regions, implying their role in transcription regulation. Among the identified lncRNAs, some candidates resided downstream of transcription factors that were crucial in erythropoiesis, such as GATA1 and TAL1, and their depletion significantly blocked erythroid differentiation (Paralkar et al. 2014).

In general, the functions of many noncoding RNAs in HSC homeostasis have been confirmed in various studies, but there are many more noncoding transcripts whose function and mechanism have not been fully elucidated. Thus, further investigation of noncoding transcripts is still needed and promising, which may shed new light on the regulation of HSC homeostasis.

3.6 RNA Modification

Epi-transcriptomic regulation, represented by RNA modifications, has received increasing attention in recent years. Compared with "epigenetic regulation," such as DNA methylation and histone modifications, RNA modifications are one step further in accessing the terminal readouts of genetic information and thus are more directly and efficiently exerting regulatory functions according to physiological clues. RNA modification has been reported to be essential for HSC homeostasis, and dysfunctional RNA modifications are often the cause of many kinds of hematological malignancies. Here, we summarize recent findings of *N*6-methyladenosine (m6A) and A-to-I editing regarding their roles in HSC homeostasis.

3.6.1 m6A Modification

Similar to epigenetic modifications, the *N*6-methyladenosine level is also dynamically modulated through cooperation between "writers" and "erasers" according to metabolic clues. M6A marks on mRNA are read by so-called "reader" proteins, such as proteins of the YTHD family and IGF2BP family. The crucial roles of m6A modification and corresponding modifiers in HSC homeostasis have been extensively reported, and their involvement in disease progression is also under extensive exploration.

Primitive hematopoiesis occurs mainly in hemogenic endothelial cells, and it was reported that METTL3, the methyltransferase of m6A, was only expressed in hematovascular regions. Depletion of METTL3 blocked endothelial-to-hemogenic transition (EHT), a crucial process of early hematopoiesis during embryonic development, and led to a drastic decrease in HSPCs. Further exploration identified Notch1a as the key downstream factor. M6A marks on its mRNA transcripts were recognized by YTHDF2 and thus mediated its degradation. Depletion of METTL3 prevented Notch1a degradation and accordingly inhibited EHT (Zhang et al. 2017). Similarly, Lv et al. reported that conditional knockout of METTL3 in mouse vascular cells blocked ETH, while definitive hematopoiesis was not affected (Lv et al. 2018). Overexpression of METTL3 in human cord blood HSCs (CD34+) enhanced proliferation, but differentiation toward the myeloid lineage was blocked (Vu et al. 2017). Accordingly, deletion of METTL3 resulted in the opposite phenomenon (Weng et al. 2018). In contrast, Lee et al. found that conditional knockout of METTL3 blocked myeloid differentiation of HSCs, but those differentiated cells were not affected (Lee et al. 2019).

Similarly, FTO, a m6A eraser, was also implicated in the regulation of HSC hematopoiesis. FTO was reported to be upregulated in AML patients, and overexpression of FTO also accelerated leukemogenesis by enhancing cell proliferation. Accordingly, deletion of FTO suppressed AML cell proliferation and colony formation. Further exploration revealed that ASB2 and RARα were the key downstream factors mediating the function of FTO (Xu et al. 2017). In addition, some studies indirectly confirmed the role of m6A in HSC homeostasis. In one study, Elkashef et al. found that D-2-hydroxyglutarate (D-2HG) abnormally accumulated in AML patients with IDH mutations, which inhibited FTO activity. In this case, enhanced m6A modification promoted the progression of leukemic cell expansion (Elkashef et al. 2017). However, in patients without IDH mutation, R-2HG, a product of IDH, also showed antileukemic activity by inhibiting FTO activity. Further exploration demonstrated that MYC was the key downstream factor (Su et al. 2018).

Except for m6A writers and erasers, those readers that recognize m6A marks were also reported to be crucial for HSC homeostasis. Our previous work reported that conditional knockout of YTHDF2 in mice promoted HSC proliferation and enhanced reconstitution ability. Depletion of YTHDF2 in human cord blood HSCs (CD34+) also resulted in HSC expansion (Li et al. 2018). Another study also reported that deletion of YTHDF2 resulted in mRNA decay and promoted HSC expansion, which was mediated by the Wnt signaling pathway (Wang et al. 2018). In addition, Paris et al. also found that functional inhibition of YTHDF2 specifically suppressed AML cells but enhanced the activity of normal HSCs (Paris et al. 2019). IGF2BP family proteins, containing the YTH domain, are also m6A readers that induce target mRNA degradation (Huang et al. 2018). It was reported that IGF2BP3 was upregulated in B-ALL patients with MLL rearrangement and ectopic expression of IGF2BP3 enhanced HSC proliferation and differentiation toward the lymphoid/myeloid lineage. Mechanistically, IGF2BP3 was found to bind to m6A at the 3′UTR of MYC and CDK6, stabilizing their mRNA transcripts (Palanichamy et al. 2016).

Based on these studies, the importance of m6A in HSC homeostasis has been revealed, but the specific role in HSC self-renewal and differentiation and the underlying mechanisms are still elusive. Specifically, how RNA methylation affects gene expression and cell fate commitment under various patho-/physiological conditions requires further investigation.

3.6.2 A-to-I Editing

A-to-I editing is another prevalent RNA modification detected in the mammalian transcriptome. The ADAR protein family is responsible for introducing A-to-I editing, during which adenosine is deaminated and turned into inosine. Inosine is recognized as guanosine during mRNA translation, thus resulting in amino acid substitution. There have been many studies concerning the effects of A-to-I editing on HSC self-renewal and differentiation. It was first reported that ADAR1 mutation resulted in defects in fetal liver hematopoiesis and erythroid lineage was the most affected (Hartner et al. 2004). The same group also reported that ADAR1 was indispensable for both fetal and adult HSCs, and loss of ADAR1 in HSCs resulted in increased type I and II interferon-inducible transcripts, finally inducing apoptosis and HSC loss (Wang et al. 2000). Wang et al. discovered that loss of one allele of ADAR1 caused embryonic lethality, and ADAR+/− embryos died from hematopoietic deficiency, implying the crucialness of correct A-to-I editing during embryonic hematopoiesis (Wang et al. 2000). Abrahamsson et al. reported that ADAR1 could cause splicing error of GSK3β in CML cells. This led to the accumulation of β-catenin, which could promote the proliferation of malignant cells (Abrahamsson et al. 2009).

These studies demonstrated the importance of A-to-I editing introduced by ADARs in HSC homeostasis. Nevertheless, the consequence of A-to-I editing is diverse, and the specific roles in HSC self-renewal and differentiation depend on the target mRNAs and the bases that are modified.

3.7 Conclusion

Recently, advances in high-throughput sequencing technologies have greatly deepened our understanding of epigenetic modulation in the regulation of HSC homeostasis. Nevertheless, applications of bulk sequencing were often limited when studying epigenetic regulation of HSC homeostasis due to the rarity of HSC numbers. For example, RIP-seq and miCLIP are techniques used to detect m6A, but neither of them can reach single-base resolution. Recently, Hu and colleagues reported the m6A-SAC-seq method, which can detect m6A marks in the whole transcriptome at single-base resolution. Using this technique, they profiled the dynamic landscape of m6A modification during HSPC differentiation, further deepening our knowledge of m6A modification and HSC homeostasis (Hu et al. 2022). In addition, Ai et al. reported the ChIP-seq method, which can detect chromatin modification states at the single-cell level (Ai et al. 2019). Therefore, in the future, high-throughput sequencing technologies with higher resolution and lower input demand would greatly facilitate the advancement of related research.

Acknowledgments This work was supported by grants from the National Key Research and Development Program of China (2022YFA1103500), the National Natural Science Foundation of China (82222003, 92268117, 81870080, 91949115, 82161138028), the Zhejiang Provincial Natural Science Foundation of China (Z24H080001), and the Leading Innovative and

Entrepreneur Team Introduction Program of Zhejiang (2020R01006).

Conflict of Interest The authors declare no competing interests.

References

Abdel-Wahab O, Mullally A, Hedvat C, Garcia-Manero G, Patel J, Wadleigh M, Malinge S, Yao J, Kilpivaara O, Bhat R, Huberman K, Thomas S, Dolgalev I, Heguy A, Paietta E, Le Beau MM, Beran M, Tallman MS, Ebert BL, Kantarjian HM, Stone RM, Gilliland DG, Crispino JD, Levine RL (2009) Genetic characterization of TET1, TET2, and TET3 alterations in myeloid malignancies. Blood 114(1):144–147

Abrahamsson AE, Geron I, Gotlib J, Dao KH, Barroga CF, Newton IG, Giles FJ, Durocher J, Creusot RS, Karimi M, Jones C, Zehnder JL, Keating A, Negrin RS, Weissman IL, Jamieson CH (2009) Glycogen synthase kinase 3beta missplicing contributes to leukemia stem cell generation. Proc Natl Acad Sci U S A 106(10):3925–3929

Ai S, Xiong H, Li CC, Luo Y, Shi Q, Liu Y, Yu X, Li C, He A (2019) Profiling chromatin states using single-cell itChIP-seq. Nat Cell Biol 21(9):1164–1172

Alvarez-Dominguez JR, Hu W, Yuan B, Shi J, Park SS, Gromatzky AA, van Oudenaarden A, Lodish HF (2014) Global discovery of erythroid long noncoding RNAs reveals novel regulators of red cell maturation. Blood 123(4):570–581

Berg JS, Lin KK, Sonnet C, Boles NC, Weksberg DC, Nguyen H, Holt LJ, Rickwood D, Daly RJ, Goodell MA (2011) Imprinted genes that regulate early mammalian growth are coexpressed in somatic stem cells. PLoS One 6(10):e26410

Bissels U, Wild S, Tomiuk S, Hafner M, Scheel H, Mihailovic A, Choi YH, Tuschl T, Bosio A (2011) Combined characterization of microRNA and mRNA profiles delineates early differentiation pathways of CD133+ and CD34+ hematopoietic stem and progenitor cells. Stem Cells (Dayton, Ohio) 29(5):847–857

Bock C, Beerman I, Lien WH, Smith ZD, Gu H, Boyle P, Gnirke A, Fuchs E, Rossi DJ, Meissner A (2012) DNA methylation dynamics during in vivo differentiation of blood and skin stem cells. Mol Cell 47(4):633–647

Broske AM, Vockentanz L, Kharazi S, Huska MR, Mancini E, Scheller M, Kuhl C, Enns A, Prinz M, Jaenisch R, Nerlov C, Leutz A, Andrade-Navarro MA, Jacobsen SE, Rosenbauer F (2009) DNA methylation protects hematopoietic stem cell multipotency from myeloerythroid restriction. Nat Genet 41(11): 1207–1215

Cabezas-Wallscheid N, Klimmeck D, Hansson J, Lipka DB, Reyes A, Wang Q, Weichenhan D, Lier A, von Paleske L, Renders S, Wunsche P, Zeisberger P, Brocks D, Gu L, Herrmann C, Haas S, Essers MAG, Brors B, Eils R, Huber W, Milsom MD, Plass C, Krijgsveld J, Trumpp A (2014) Identification of regulatory networks in HSCs and their immediate progeny via integrated proteome, transcriptome, and DNA methylome analysis. Cell Stem Cell 15(4):507–522

Celik H, Mallaney C, Kothari A, Ostrander EL, Eultgen E, Martens A, Miller CA, Hundal J, Klco JM, Challen GA (2015) Enforced differentiation of Dnmt3a-null bone marrow leads to failure with c-Kit mutations driving leukemic transformation. Blood 125(4):619–628

Challen GA, Sun D, Jeong M, Luo M, Jelinek J, Berg JS, Bock C, Vasanthakumar A, Gu H, Xi Y, Liang S, Lu Y, Darlington GJ, Meissner A, Issa JP, Godley LA, Li W, Goodell MA (2011) Dnmt3a is essential for hematopoietic stem cell differentiation. Nat Genet 44(1):23–31

Challen GA, Sun D, Mayle A, Jeong M, Luo M, Rodriguez B, Mallaney C, Celik H, Yang L, Xia Z, Cullen S, Berg J, Zheng Y, Darlington GJ, Li W, Goodell MA (2014) Dnmt3a and Dnmt3b have overlapping and distinct functions in hematopoietic stem cells. Cell Stem Cell 15(3):350–364

Chen MT, Lin HS, Shen C, Ma YN, Wang F, Zhao HL, Yu J, Zhang JW (2015) PU.1-regulated long noncoding RNA lnc-MC controls human monocyte/macrophage differentiation through interaction with MicroRNA 199a-5p. Mol Cell Biol 35(18):3212–3224

Cuartero S, Weiss FD, Dharmalingam G, Guo Y, Ing-Simmons E, Masella S, Robles-Rebollo I, Xiao X, Wang YF, Barozzi I, Djeghloul D, Amano MT, Niskanen H, Petretto E, Dowell RD, Tachibana K, Kaikkonen MU, Nasmyth KA, Lenhard B, Natoli G, Fisher AG, Merkenschlager M (2018) Control of inducible gene expression links cohesin to hematopoietic progenitor self-renewal and differentiation. Nat Immunol 19(9):932–941

Ding X, Lin Q, Ensenat-Waser R, Rose-John S, Zenke M (2012) Polycomb group protein Bmi1 promotes hematopoietic cell development from embryonic stem cells. Stem Cells Dev 21(1):121–132

Dolnik A, Engelmann JC, Scharfenberger-Schmeer M, Mauch J, Kelkenberg-Schade S, Haldemann B, Fries T, Kronke J, Kuhn MW, Paschka P, Kayser S, Wolf S, Gaidzik VI, Schlenk RF, Rucker FG, Dohner H, Lottaz C, Dohner K, Bullinger L (2012) Commonly altered genomic regions in acute myeloid leukemia are enriched for somatic mutations involved in chromatin remodeling and splicing. Blood 120(18): e83–e92

Dore LC, Amigo JD, Dos Santos CO, Zhang Z, Gai X, Tobias JW, Yu D, Klein AM, Dorman C, Wu W, Hardison RC, Paw BH, Weiss MJ (2008) A GATA-1-regulated microRNA locus essential for erythropoiesis. Proc Natl Acad Sci U S A 105(9):3333–3338

Du Q, Luu PL, Stirzaker C, Clark SJ (2015) Methyl-CpG-binding domain proteins: readers of the epigenome. Epigenomics 7(6):1051–1073

Elkashef SM, Lin AP, Myers J, Sill H, Jiang D, Dahia PLM, Aguiar RCT (2017) IDH mutation, competitive inhibition of FTO, and RNA methylation. Cancer Cell 31(5):619–620

Farlik M, Halbritter F, Muller F, Choudry FA, Ebert P, Klughammer J, Farrow S, Santoro A, Ciaurro V, Mathur A, Uppal R, Stunnenberg HG, Ouwehand WH, Laurenti E, Lengauer T, Frontini M, Bock C (2016) DNA methylation dynamics of human hematopoietic stem cell differentiation. Cell Stem Cell 19(6):808–822

Fazi F, Rosa A, Fatica A, Gelmetti V, De Marchis ML, Nervi C, Bozzoni I (2005) A minicircuitry comprised of microRNA-223 and transcription factors NFI-A and C/EBPalpha regulates human granulopoiesis. Cell 123(5):819–831

Felli N, Fontana L, Pelosi E, Botta R, Bonci D, Facchiano F, Liuzzi F, Lulli V, Morsilli O, Santoro S, Valtieri M, Calin GA, Liu CG, Sorrentino A, Croce CM, Peschle C (2005) MicroRNAs 221 and 222 inhibit normal erythropoiesis and erythroleukemic cell growth via kit receptor down-modulation. Proc Natl Acad Sci U S A 102(50):18081–18086

Feng Y, Yang Y, Ortega MM, Copeland JN, Zhang M, Jacob JB, Fields TA, Vivian JL, Fields PE (2010) Early mammalian erythropoiesis requires the Dot1L methyltransferase. Blood 116(22):4483–4491

Fisher JB, Peterson J, Reimer M, Stelloh C, Pulakanti K, Gerbec ZJ, Abel AM, Strouse JM, Strouse C, McNulty M, Malarkannan S, Crispino JD, Milanovich S, Rao S (2017) The cohesin subunit Rad21 is a negative regulator of hematopoietic self-renewal through epigenetic repression of Hoxa7 and Hoxa9. Leukemia 31(3):712–719

Ghisi M, Corradin A, Basso K, Frasson C, Serafin V, Mukherjee S, Mussolin L, Ruggero K, Bonanno L, Guffanti A, De Bellis G, Gerosa G, Stellin G, D'Agostino DM, Basso G, Bronte V, Indraccolo S, Amadori A, Zanovello P (2011) Modulation of microRNA expression in human T-cell development: targeting of NOTCH3 by miR-150. Blood 117(26): 7053–7062

Guo S, Lu J, Schlanger R, Zhang H, Wang JY, Fox MC, Purton LE, Fleming HH, Cobb B, Merkenschlager M, Golub TR, Scadden DT (2010) MicroRNA miR-125a controls hematopoietic stem cell number. Proc Natl Acad Sci U S A 107(32):14229–14234

Han YC, Park CY, Bhagat G, Zhang J, Wang Y, Fan JB, Liu M, Zou Y, Weissman IL, Gu H (2010) microRNA-29a induces aberrant self-renewal capacity in hematopoietic progenitors, biased myeloid development, and acute myeloid leukemia. J Exp Med 207(3):475–489

Hartner JC, Schmittwolf C, Kispert A, Müller AM, Higuchi M, Seeburg PH (2004) Liver disintegration in the mouse embryo caused by deficiency in the RNA-editing enzyme ADAR1. J Biol Chem 279(6): 4894–4902

Heideman MR, Lancini C, Proost N, Yanover E, Jacobs H, Dannenberg JH (2014) Sin3a-associated Hdac1 and Hdac2 are essential for hematopoietic stem cell homeostasis and contribute differentially to hematopoiesis. Haematologica 99(8):1292–1303

Hidalgo I, Herrera-Merchan A, Ligos JM, Carramolino L, Nuñez J, Martinez F, Dominguez O, Torres M, Gonzalez S (2012) Ezh1 is required for hematopoietic stem cell maintenance and prevents senescence-like cell cycle arrest. Cell Stem Cell 11(5):649–662

Hu W, Yuan B, Flygare J, Lodish HF (2011) Long non-coding RNA-mediated anti-apoptotic activity in murine erythroid terminal differentiation. Genes Dev 25(24):2573–2578

Hu G, Cui K, Fang D, Hirose S, Wang X, Wangsa D, Jin W, Ried T, Liu P, Zhu J, Rothenberg EV, Zhao K (2018) Transformation of accessible chromatin and 3D nucleome underlies lineage commitment of early T cells. Immunity 48(2):227–242.e228

Hu L, Liu S, Peng Y, Ge R, Su R, Senevirathne C, Harada BT, Dai Q, Wei J, Zhang L, Hao Z, Luo L, Wang H, Wang Y, Luo M, Chen M, Chen J, He C (2022) mA RNA modifications are measured at single-base resolution across the mammalian transcriptome. Nat Biotechnol 40:1210

Huang H, Weng H, Sun W, Qin X, Shi H, Wu H, Zhao BS, Mesquita A, Liu C, Yuan CL, Hu YC, Hüttelmaier S, Skibbe JR, Su R, Deng X, Dong L, Sun M, Li C, Nachtergaele S, Wang Y, Hu C, Ferchen K, Greis KD, Jiang X, Wei M, Qu L, Guan JL, He C, Yang J, Chen J (2018) Recognition of RNA N-methyladenosine by IGF2BP proteins enhances mRNA stability and translation. Nat Cell Biol 20(3): 285–295

Imam H, Bano AS, Patel P, Holla P, Jameel S (2015) The lncRNA NRON modulates HIV-1 replication in a NFAT-dependent manner and is differentially regulated by early and late viral proteins. Sci Rep 5: 8639

Jaiswal S, Fontanillas P, Flannick J, Manning A, Grauman PV, Mar BG, Lindsley RC, Mermel CH, Burtt N, Chavez A, Higgins JM, Moltchanov V, Kuo FC, Kluk MJ, Henderson B, Kinnunen L, Koistinen HA, Ladenvall C, Getz G, Correa A, Banahan BF, Gabriel S, Kathiresan S, Stringham HM, McCarthy MI, Boehnke M, Tuomilehto J, Haiman C, Groop L, Atzmon G, Wilson JG, Neuberg D, Altshuler D, Ebert BL (2014) Age-related clonal hematopoiesis associated with adverse outcomes. N Engl J Med 371(26):2488–2498

Javierre BM, Burren OS, Wilder SP, Kreuzhuber R, Hill SM, Sewitz S, Cairns J, Wingett SW, Varnai C, Thiecke MJ, Burden F, Farrow S, Cutler AJ, Rehnstrom K, Downes K, Grassi L, Kostadima M, Freire-Pritchett P, Wang F, BLUEPRINT Consortium, Stunnenberg HG, Todd JA, Zerbino DR, Stegle O, Ouwehand WH, Frontini M, Wallace C, Spivakov M, Fraser P (2016) Lineage-specific genome architecture links enhancers and non-coding disease variants to target gene promoters. Cell 167(5):1369–1384 e1319

Kajiume T, Ninomiya Y, Ishihara H, Kanno R, Kanno M (2004) Polycomb group gene mel-18 modulates the self-renewal activity and cell cycle status of hematopoietic stem cells. Exp Hematol 32(6):571–578

Kamminga LM, Bystrykh LV, de Boer A, Houwer S, Douma J, Weersing E, Dontje B, de Haan G (2006) The Polycomb group gene Ezh2 prevents hematopoietic stem cell exhaustion. Blood 107(5): 2170–2179

Katsumoto T, Aikawa Y, Iwama A, Ueda S, Ichikawa H, Ochiya T, Kitabayashi I (2006) MOZ is essential for maintenance of hematopoietic stem cells. Genes Dev 20(10):1321–1330

Katsumoto T, Yoshida N, Kitabayashi I (2008) Roles of the histone acetyltransferase monocytic leukemia zinc finger protein in normal and malignant hematopoiesis. Cancer Sci 99(8):1523–1527

Kino T, Hurt DE, Ichijo T, Nader N, Chrousos GP (2010) Noncoding RNA gas5 is a growth arrest- and starvation-associated repressor of the glucocorticoid receptor. Sci Signal 3(107):ra8

Ko M, Bandukwala HS, An J, Lamperti ED, Thompson EC, Hastie R, Tsangaratou A, Rajewsky K, Koralov SB, Rao A (2011) Ten-Eleven-Translocation 2 (TET2) negatively regulates homeostasis and differentiation of hematopoietic stem cells in mice. Proc Natl Acad Sci U S A 108(35):14566–14571

Lee H, Bao S, Qian Y, Geula S, Leslie J, Zhang C, Hanna JH, Ding L (2019) Stage-specific requirement for Mettl3-dependent mA mRNA methylation during haematopoietic stem cell differentiation. Nat Cell Biol 21(6):700–709

Li Z, Qian P, Shao W, Shi H, He XC, Gogol M, Yu Z, Wang Y, Qi M, Zhu Y, Perry JM, Zhang K, Tao F, Zhou K, Hu D, Han Y, Zhao C, Alexander R, Xu H, Chen S, Peak A, Hall K, Peterson M, Perera A, Haug JS, Parmely T, Li H, Shen B, Zeitlinger J, He C, Li L (2018) Suppression of mA reader Ythdf2 promotes hematopoietic stem cell expansion. Cell Res 28(9): 904–917

Loizou JI, Oser G, Shukla V, Sawan C, Murr R, Wang ZQ, Trumpp A, Herceg Z (2009) Histone acetyltransferase cofactor Trrap is essential for maintaining the hematopoietic stem/progenitor cell pool. J Immunol (Baltimore, Md : 1950) 183(10):6422–6431

Luo M, Jeong M, Sun D, Park HJ, Rodriguez BA, Xia Z, Yang L, Zhang X, Sheng K, Darlington GJ, Li W, Goodell MA (2015) Long non-coding RNAs control hematopoietic stem cell function. Cell Stem Cell 16(4): 426–438

Lv J, Zhang Y, Gao S, Zhang C, Chen Y, Li W, Yang YG, Zhou Q, Liu F (2018) Endothelial-specific mA modulates mouse hematopoietic stem and progenitor cell development via notch signaling. Cell Res 28(2): 249–252

Majewski IJ, Ritchie ME, Phipson B, Corbin J, Pakusch M, Ebert A, Busslinger M, Koseki H, Hu Y, Smyth GK, Alexander WS, Hilton DJ, Blewitt ME (2010) Opposing roles of polycomb repressive complexes in hematopoietic stem and progenitor cells. Blood 116(5):731–739

Mayle A, Yang L, Rodriguez B, Zhou T, Chang E, Curry CV, Challen GA, Li W, Wheeler D, Rebel VI, Goodell MA (2015) Dnmt3a loss predisposes murine hematopoietic stem cells to malignant transformation. Blood 125(4):629–638

Mazumdar C, Shen Y, Xavy S, Zhao F, Reinisch A, Li R, Corces MR, Flynn RA, Buenrostro JD, Chan SM, Thomas D, Koenig JL, Hong WJ, Chang HY, Majeti R (2015) Leukemia-associated cohesin mutants dominantly enforce stem cell programs and impair human hematopoietic progenitor differentiation. Cell Stem Cell 17(6):675–688

McCulloch EA, Till JE (2005) Perspectives on the properties of stem cells. Nat Med 11(10):1026–1028

McMahon KA, Hiew SY, Hadjur S, Veiga-Fernandes H, Menzel U, Price AJ, Kioussis D, Williams O, Brady HJ (2007) Mll has a critical role in fetal and adult hematopoietic stem cell self-renewal. Cell Stem Cell 1(3):338–345

Moran-Crusio K, Reavie L, Shih A, Abdel-Wahab O, Ndiaye-Lobry D, Lobry C, Figueroa ME, Vasanthakumar A, Patel J, Zhao X, Perna F, Pandey S, Madzo J, Song C, Dai Q, He C, Ibrahim S, Beran M, Zavadil J, Nimer SD, Melnick A, Godley LA, Aifantis I, Levine RL (2011) Tet2 loss leads to increased hematopoietic stem cell self-renewal and myeloid transformation. Cancer Cell 20(1):11–24

O'Connell RM, Chaudhuri AA, Rao DS, Gibson WS, Balazs AB, Baltimore D (2010) MicroRNAs enriched in hematopoietic stem cells differentially regulate long-term hematopoietic output. Proc Natl Acad Sci U S A 107(32):14235–14240

Ohta H, Sawada A, Kim JY, Tokimasa S, Nishiguchi S, Humphries RK, Hara J, Takihara Y (2002) Polycomb group gene rae28 is required for sustaining activity of hematopoietic stem cells. J Exp Med 195(6):759–770

Palanichamy JK, Tran TM, Howard JM, Contreras JR, Fernando TR, Sterne-Weiler T, Katzman S, Toloue M, Yan W, Basso G, Pigazzi M, Sanford JR, Rao DS (2016) RNA-binding protein IGF2BP3 targeting of oncogenic transcripts promotes hematopoietic progenitor proliferation. J Clin Invest 126(4):1495–1511

Papapetrou EP, Korkola JE, Sadelain M (2010) A genetic strategy for single and combinatorial analysis of miRNA function in mammalian hematopoietic stem cells. Stem Cells (Dayton, Ohio) 28(2):287–296

Paralkar VR, Mishra T, Luan J, Yao Y, Kossenkov AV, Anderson SM, Dunagin M, Pimkin M, Gore M, Sun D, Konuthula N, Raj A, An X, Mohandas N, Bodine DM, Hardison RC, Weiss MJ (2014) Lineage and species-specific long noncoding RNAs during erythro-megakaryocytic development. Blood 123(12): 1927–1937

Paris J, Morgan M, Campos J, Spencer GJ, Shmakova A, Ivanova I, Mapperley C, Lawson H, Wotherspoon DA, Sepulveda C, Vukovic M, Allen L, Sarapuu A, Tavosanis A, Guitart AV, Villacreces A, Much C, Choe J, Azar A, van de Lagemaat LN, Vernimmen D, Nehme A, Mazurier F, Somervaille TCP, Gregory RI, O'Carroll D, Kranc KR (2019) Targeting the RNA mA reader YTHDF2 selectively compromises cancer stem cells in acute myeloid leukemia. Cell Stem Cell 25(1):137–148.e136

Pollack Y, Stein R, Razin A, Cedar H (1980) Methylation of foreign DNA sequences in eukaryotic cells. Proc Natl Acad Sci U S A 77(11):6463–6467

Popovic R, Riesbeck LE, Velu CS, Chaubey A, Zhang J, Achille NJ, Erfurth FE, Eaton K, Lu J, Grimes HL, Chen J, Rowley JD, Zeleznik-Le NJ (2009) Regulation of mir-196b by MLL and its overexpression by MLL fusions contributes to immortalization. Blood 113(14): 3314–3322

Qian P, He Xi C, Paulson A, Li Z, Tao F, Perry John M, Guo F, Zhao M, Zhi L, Venkatraman A, Haug Jeffrey S, Parmely T, Li H, Dobrowsky Rick T, Ding W-X, Kono T, Ferguson-Smith Anne C, Li L (2016) The Dlk1-Gtl2 locus preserves LT-HSC function by inhibiting the PI3K-mTOR pathway to restrict mitochondrial metabolism. Cell Stem Cell 18(2):214–228

Qian P, De Kumar B, He XC, Nolte C, Gogol M, Ahn Y, Chen S, Li Z, Xu H, Perry JM, Hu D, Tao F, Zhao M, Han Y, Hall K, Peak A, Paulson A, Zhao C, Venkatraman A, Box A, Perera A, Haug JS, Parmely T, Li H, Krumlauf R, Li L (2018) Retinoid-sensitive epigenetic regulation of the Hoxb cluster maintains normal hematopoiesis and inhibits leukemogenesis. Cell Stem Cell 22(5):740–754.e747

Quivoron C, Couronne L, Della Valle V, Lopez CK, Plo I, Wagner-Ballon O, Do Cruzeiro M, Delhommeau F, Arnulf B, Stern MH, Godley L, Opolon P, Tilly H, Solary E, Duffourd Y, Dessen P, Merle-Beral H, Nguyen-Khac F, Fontenay M, Vainchenker W, Bastard C, Mercher T, Bernard OA (2011) TET2 inactivation results in pleiotropic hematopoietic abnormalities in mouse and is a recurrent event during human lymphomagenesis. Cancer Cell 20(1):25–38

Ranzani V, Rossetti G, Panzeri I, Arrigoni A, Bonnal RJ, Curti S, Gruarin P, Provasi E, Sugliano E, Marconi M, De Francesco R, Geginat J, Bodega B, Abrignani S, Pagani M (2015) The long intergenic noncoding RNA landscape of human lymphocytes highlights the regulation of T cell differentiation by linc-MAF-4. Nat Immunol 16(3):318–325

Razin A, Cedar H (1977) Distribution of 5-methylcytosine in chromatin. Proc Natl Acad Sci U S A 74(7): 2725–2728

Rimmelé P, Bigarella CL, Liang R, Izac B, Dieguez-Gonzalez R, Barbet G, Donovan M, Brugnara C, Blander JM, Sinclair DA, Ghaffari S (2014) Aging-like phenotype and defective lineage specification in SIRT1-deleted hematopoietic stem and progenitor cells. Stem Cell Rep 3(1):44–59

Roller A, Grossmann V, Bacher U, Poetzinger F, Weissmann S, Nadarajah N, Boeck L, Kern W, Haferlach C, Schnittger S, Haferlach T, Kohlmann A (2013) Landmark analysis of DNMT3A mutations in hematological malignancies. Leukemia 27(7): 1573–1578

Shide K, Kameda T, Shimoda H, Yamaji T, Abe H, Kamiunten A, Sekine M, Hidaka T, Katayose K, Kubuki Y, Yamamoto S, Miike T, Iwakiri H, Hasuike S, Nagata K, Marutsuka K, Iwama A, Matsuda T, Kitanaka A, Shimoda K (2012) TET2 is essential for survival and hematopoietic stem cell homeostasis. Leukemia 26(10):2216–2223

Shlush LI, Zandi S, Mitchell A, Chen WC, Brandwein JM, Gupta V, Kennedy JA, Schimmer AD, Schuh AC, Yee KW, JL ML, Doedens M, JJF M, Marke R, Kim HJ, Lee K, JD MP, Hudson TJ, HALT Pan-Leukemia Gene Panel Consortium, AMK B, Yousif F, Trinh QM, Stein LD, Minden MD, JCY W, Dick JE (2014) Identification of pre-leukaemic haematopoietic stem cells in acute leukaemia. Nature 506(7488):328–333

Siminovitch L, McCulloch EA, Till JE (1963) The distribution of colony-forming cells among spleen colonies. J Cell Comp Physiol 62:327–336

Skvortsova K, Iovino N, Bogdanovic O (2018) Functions and mechanisms of epigenetic inheritance in animals. Nat Rev Mol Cell Biol 19(12):774–790

Stadhouders R, Filion GJ, Graf T (2019) Transcription factors and 3D genome conformation in cell-fate decisions. Nature 569(7756):345–354

Stadler MB, Murr R, Burger L, Ivanek R, Lienert F, Schöler A, Wirbelauer C, Oakeley EJ, Gaidatzis D, Tiwari VK, Schübeler D (2011) DNA-binding factors shape the mouse methylome at distal regulatory regions. Nature 480(7378):490–495

Su R, Dong L, Li C, Nachtergaele S, Wunderlich M, Qing Y, Deng X, Wang Y, Weng X, Hu C, Yu M, Skibbe J, Dai Q, Zou D, Wu T, Yu K, Weng H, Huang H, Ferchen K, Qin X, Zhang B, Qi J, Sasaki AT, Plas DR, Bradner JE, Wei M, Marcucci G, Jiang X, Mulloy JC, Jin J, He C, Chen J (2018) R-2HG exhibits anti-tumor activity by targeting FTO/mA/MYC/CEBPA signaling. Cell 172(1–2): 90–105.e123

Tenedini E, Roncaglia E, Ferrari F, Orlandi C, Bianchi E, Bicciato S, Tagliafico E, Ferrari S (2010) Integrated analysis of microRNA and mRNA expression profiles in physiological myelopoiesis: role of hsa-mir-299-5p in CD34+ progenitor cells commitment. Cell Death Dis 1:e28

Thomas T, Corcoran LM, Gugasyan R, Dixon MP, Brodnicki T, Nutt SL, Metcalf D, Voss AK (2006) Monocytic leukemia zinc finger protein is essential for the development of long-term reconstituting hematopoietic stem cells. Genes Dev 20(9):1175–1186

Trowbridge JJ, Snow JW, Kim J, Orkin SH (2009) DNA methyltransferase 1 is essential for and uniquely regulates hematopoietic stem and progenitor cells. Cell Stem Cell 5(4):442–449

Valk-Lingbeek ME, Bruggeman SW, van Lohuizen M (2004) Stem cells and cancer; the polycomb connection. Cell 118(4):409–418

Venkatraman A, He XC, Thorvaldsen JL, Sugimura R, Perry JM, Tao F, Zhao M, Christenson MK, Sanchez R, Yu JY, Peng L, Haug JS, Paulson A, Li H, Zhong XB, Clemens TL, Bartolomei MS, Li L (2013) Maternal imprinting at the H19-Igf2 locus maintains adult haematopoietic stem cell quiescence. Nature 500(7462):345–349

Vu LP, Pickering BF, Cheng Y, Zaccara S, Nguyen D, Minuesa G, Chou T, Chow A, Saletore Y, MacKay M, Schulman J, Famulare C, Patel M, Klimek VM, Garrett-Bakelman FE, Melnick A, Carroll M, Mason CE, Jaffrey SR, Kharas MG (2017) The N-methyladenosine (mA)-forming enzyme METTL3 controls myeloid differentiation of normal hematopoietic and leukemia cells. Nat Med 23(11): 1369–1376

Wagner LA, Christensen CJ, Dunn DM, Spangrude GJ, Georgelas A, Kelley L, Esplin MS, Weiss RB, Gleich GJ (2007) EGO, a novel, noncoding RNA gene, regulates eosinophil granule protein transcript expression. Blood 109(12):5191–5198

Wang Q, Khillan J, Gadue P, Nishikura K (2000) Requirement of the RNA editing deaminase ADAR1 gene for embryonic erythropoiesis. Science (New York, NY) 290(5497):1765–1768

Wang P, Xue Y, Han Y, Lin L, Wu C, Xu S, Jiang Z, Xu J, Liu Q, Cao X (2014) The STAT3-binding long noncoding RNA lnc-DC controls human dendritic cell differentiation. Science (New York, NY) 344(6181): 310–313

Wang H, Zuo H, Liu J, Wen F, Gao Y, Zhu X, Liu B, Xiao F, Wang W, Huang G, Shen B, Ju Z (2018) Loss of YTHDF2-mediated mA-dependent mRNA clearance facilitates hematopoietic stem cell regeneration. Cell Res 28(10):1035–1038

Weng H, Huang H, Wu H, Qin X, Zhao BS, Dong L, Shi H, Skibbe J, Shen C, Hu C, Sheng Y, Wang Y, Wunderlich M, Zhang B, Dore LC, Su R, Deng X, Ferchen K, Li C, Sun M, Lu Z, Jiang X, Marcucci G, Mulloy JC, Yang J, Qian Z, Wei M, He C, Chen J (2018) METTL14 inhibits hematopoietic stem/progenitor differentiation and promotes leukemogenesis via mRNA mA modification. Cell Stem Cell 22(2): 191–205.e199

Williams GT, Mourtada-Maarabouni M, Farzaneh F (2011) A critical role for non-coding RNA GAS5 in growth arrest and rapamycin inhibition in human T-lymphocytes. Biochem Soc Trans 39(2):482–486

Xiao C, Calado DP, Galler G, Thai TH, Patterson HC, Wang J, Rajewsky N, Bender TP, Rajewsky K (2016) MiR-150 controls B cell differentiation by targeting the transcription factor c-Myb. Cell 165(4):1027

Xu K, Yang Y, Feng GH, Sun BF, Chen JQ, Li YF, Chen YS, Zhang XX, Wang CX, Jiang LY, Liu C, Zhang ZY, Wang XJ, Zhou Q, Yang YG, Li W (2017) Mettl3-mediated mA regulates spermatogonial differentiation and meiosis initiation. Cell Res 27(9):1100–1114

Zhang X, Lian Z, Padden C, Gerstein MB, Rozowsky J, Snyder M, Gingeras TR, Kapranov P, Weissman SM, Newburger PE (2009) A myelopoiesis-associated regulatory intergenic noncoding RNA transcript within the human HOXA cluster. Blood 113(11):2526–2534

Zhang X, Su J, Jeong M, Ko M, Huang Y, Park HJ, Guzman A, Lei Y, Huang Y-H, Rao A, Li W, Goodell MA (2016) DNMT3A and TET2 compete and cooperate to repress lineage-specific transcription factors in hematopoietic stem cells. Nat Genet 48(9):1014–1023

Zhang C, Chen Y, Sun B, Wang L, Yang Y, Ma D, Lv J, Heng J, Ding Y, Xue Y, Lu X, Xiao W, Yang YG, Liu F (2017) mA modulates haematopoietic stem and progenitor cell specification. Nature 549(7671):273–276

Zhu H, Wang G, Qian J (2016) Transcription factors as readers and effectors of DNA methylation. Nat Rev Genet 17(9):551–565

Metabolism in Hematopoiesis and Its Malignancy

4

Xiaoyuan Zeng, Yi-Ping Wang, and Cheuk-Him Man

Abstract

Hematopoietic stem cells (HSCs) are multipotent stem cells that can self-renew and generate all blood cells of different lineages. The system is under tight control in order to maintain a precise equilibrium of the HSC pool and the effective production of mature blood cells to support various biological activities. Cell metabolism can regulate different molecular activities, such as epigenetic modification and cell cycle regulation, and subsequently affects the function and maintenance of HSC. Upon malignant transformation, oncogenic drivers in malignant hematopoietic cells can remodel the metabolic pathways for supporting the oncogenic growth. The dysregulation of metabolism results in oncogene addiction, implying the development of malignancy-specific metabolism-targeted therapy. In this chapter, we will discuss the significance of different metabolic pathways in hematopoiesis, specifically, the distinctive metabolic dependency in hematopoietic malignancies and potential metabolic therapy.

Keywords

Hematopoiesis · Metabolism · Hematopoietic malignancy · Metabolism-targeted therapy · Nutrient metabolism

4.1 Introduction

Metabolism is tightly controlled in living organisms. The three categories of metabolism include catabolism, anabolism, and waste disposal (DeBerardinis and Thompson 2012). Catabolism breaks down complex organic or inorganic molecules, such as proteins, sugars, and fatty acids into smaller units for producing energy, neutralizing oxidative stress, and synthesizing biomacromolecules. Anabolism produces and assembles precursors into complex biomolecules. Besides, toxic metabolic wastes, including salts, phosphates, and nitrogenous wastes, generated after catabolism and anabolism are efficiently removed from the cells. All the metabolic reactions are catalyzed by enzymes so that the metabolic network can be fine-tuned to modulate various biological activities and support lives (Saxton and Sabatini 2017; Spinelli and Haigis 2018).

X. Zeng · C.-H. Man (✉)
Division of Haematology, Department of Medicine, Li Ka Shing Faculty of Medicine, The University of Hong Kong, Hong Kong SAR, China
e-mail: csman729@hku.hk

Y.-P. Wang (✉)
Precision Research Center for Refractory Diseases, Institute for Clinical Research, Shanghai General Hospital, Shanghai Jiao Tong University School of Medicine, Shanghai, China
e-mail: yiping.wang1@shgh.cn

M. Zhao, P. Qian (eds.), *Hematopoietic Stem Cells*, Advances in Experimental Medicine and Biology 1442, https://doi.org/10.1007/978-981-99-7471-9_4

Although most cell types require ATP and nucleotides to grow, different cells have specific preferences for nutrients to meet their metabolic addiction. For example, cardiac muscle and skeletal tissue make ATP by oxidizing fatty acids (Lopaschuk et al. 2010; Jensen 2002), while the cells in the central nervous system rely solely on glucose (Volkenhoff et al. 2015). However, many cells can shift their selection of fuel in different conditions. For example, skeletal muscle can shift glucose oxidation to fatty acid oxidation during intensive exercise (Hargreaves and Spriet 2020). The breakdown of glucose effectively generates building blocks for biosynthesis (nucleotides), energy production (ATP), and redox homeostasis (NAD/NADH or NADP/NADPH recycling). Therefore, glucose feeding into glycolysis and pentose phosphate pathway (PPP) is essential for active proliferating cells (Lin et al. 2012; West et al. 1990; Filosa et al. 2003). When stem cells commit to grow and proliferate from quiescence, the metabolic demands dramatically change (Puente et al. 2014; Cho et al. 2006; Wang et al. 2009). Doubling cell mass for growth requires increased energy and building blocks; therefore, dynamic metabolic regulation is essential to maintain normal cellular functions and respond to growth signals or stress, especially in hematopoietic cells.

Hematopoiesis produces all the cellular components of the blood. It occurs within the hematopoietic organs including the bone marrow, liver, and spleen. Hematopoietic stem cell (HSC) is the origin of the hematopoietic system that acquires the ability of self-renewal and gives rise to various progenies, known as multi-lineage differentiation (Weissman et al. 2001; Orkin and Zon 2008; Sawai et al. 2016; Busch et al. 2015; Sun et al. 2014). The processes are tightly regulated in response to physiological and disease conditions (Sawai et al. 2016; Hoggatt et al. 2018; Goncalves et al. 2016; Courties et al. 2015; Dutta et al. 2015), and therefore HSC transplantation, replacing abnormal HSCs with HSCs from a healthy donor, has been the major therapeutic strategy in hematological disorders (Appelbaum 2007; Copelan 2006). The cellular processes of HSC are regulated by multiple mediators such as the bone marrow microenvironment "niche" and cell-intrinsic factors including genetic and epigenetic regulations (Wilson et al. 2008; Chandel et al. 2016; Morrison and Scadden 2014; Cheng et al. 2000; Yu et al. 2017). Emerging evidence has suggested that metabolism is extremely important in mediating the maintenance and function of HSC (Ren et al. 2017; Nakamura-Ishizu et al. 2020; Ito et al. 2019). The long-term HSC (LT-HSC) is usually quiescent and resides in the hypoxic niches within the bone marrow, so that it can reduce cellular damage from oxidative stress, while the progenitors need to proliferate fast and are in huge demand of energy. Therefore, the determination of maintaining the stemness or committing to differentiate within HSC pool is highly associated with cellular metabolism. Apart from normal hematopoiesis, numerous oncogenes, such as MLL-AF9, NOTCH, PTEN, and MYC, have been shown to transform normal HSC into malignancies by remodeling various metabolic pathways (Weng et al. 2004; Yilmaz et al. 2006; Wilson et al. 2004; Krivtsov et al. 2006; Zhang et al. 2006). The metabolic addiction to a specific metabolic pathway in cancer suggests an effective therapeutic target that leaves normal blood cells unaffected. In this chapter, we will discuss the significance of cell metabolism in HSC and its malignancy, with an emphasis on emerging therapeutic strategies.

4.2 Glucose Flux Through Glycolysis and Pentose Phosphate Pathway

Glycolysis converts glucose to pyruvate and produces energy in the form of two ATP molecules. Ultimately, pyruvate either converts to lactate for efflux or enters the tricarboxylic acid (TCA) cycle in mitochondria fueling oxidative phosphorylation (Wang et al. 2021; Wang et al. 2022). Glucose can also flux into the PPP, branching at glucose-6-phosphate. PPP is crucial in producing five-carbon sugar for nucleotide biosynthesis and making reduced nicotinamide adenine dinucleotide phosphate (NADPH) for scavenging reactive oxygen species (ROS) (Wang et al. 2014a). Glycolysis or the PPP is tightly regulated at different stages by various enzymes such as hexokinase, pyruvate kinase, and glucose-6-phosphate dehydrogenase (Voet et al. 1981; Tanner et al. 2018).

4.2.1 Glucose Metabolism in Normal Hematopoiesis

Glucose metabolism specifically affects normal hematopoiesis. To ensure a sufficient supply of energy and building blocks for stem cell maintenance, HSC always upregulates the expression of glycolytic enzymes and activates glycolysis to fuel the anabolism (Vander Heiden et al. 2009). The enhanced glycolytic pathway and reduced mitochondrial activity are the signature of HSC. Instead, HSC is always quiescent at the hypoxic niche in bone marrow (Goncalves et al. 2016; Spencer et al. 2014); the low oxygen environment is not only well-tolerated by HSC but also appears to be essential for their survival and functions (Parmar et al. 2007; Eliasson and Jönsson 2010; Kubota et al. 2008). The preferential adaptation of glycolysis in HSC ensures lesser reliance on oxidative phosphorylation, which helps reduce ROS production from mitochondria. The low level of ROS prevents HSC from DNA or protein damage and facilitates the maintenance of HSC integrity and stemness (Suda et al. 2011; Zhang et al. 2016). At low oxygen, hypoxia-inducing factor 1a (HIF-1a) is stabilized, and the transcriptional expression of various glycolytic genes is induced, while the genes of TCA cycle are suppressed (Takubo et al. 2013; Takubo et al. 2010). FoxO3a, a downstream transcriptional effector of HIF-1a, is essential in HSC by repressing Myc function, mitochondrial mass, oxygen consumption, and ROS production and promoting cell survival in hypoxia (Miyamoto et al. 2007; Jensen et al. 2011). Deletion of FoxO family members in LT-HSC leads to a marked increase in ROS, increased cell cycling, and apoptosis, while the FoxO-deficient phenotype can be rescued by antioxidative agents (Tothova et al. 2007). Furthermore, another HIF-1a transcriptional activator, Meis1, is enriched in LT-HSC that regulates HSC metabolism through activating HIF-1a and favoring glucose utilization (Simsek et al. 2010; Kocabas et al. 2012). Besides, the deletion of PTPMT1 or PDK, the OXPHOS regulators, has been shown to suppress mitochondria respiration, increase glycolysis, maintain the quiescent state of HSC, and enhance HSC function (Takubo et al. 2013; Yu et al. 2013). Collectively, HSC function greatly relies on glycolysis rather than oxidative phosphorylation.

The differentiation of hematopoietic cells is also dependent on glycolysis. Lactate dehydrogenase A (LDHA) converts pyruvate and NADH to lactate and NAD. Loss of LDHA reduces glycolysis but induces oxidative phosphorylation, which in turn generates more ROS and suppresses the function of HSC and its progenitors (Wang et al. 2014b). Genetic knockdown of monocarboxylate transporter MCT1, a lactate exporter, also leads to a dramatic decrease in glucose metabolism and significant suppression of progenitors but not HSCs in murine model (Man et al. 2021). Further, the loss of pyruvate kinase isozyme M2 (PKM2), which converts phosphoenolpyruvate to pyruvate, impairs hematopoietic progenitor but not HSC activity (Wang et al. 2014b). Besides, glycolysis is also important in mature blood cells. During the activation of T lymphocytes, glucose metabolism is increased by upregulating the expression of glucose transporter and glycolytic enzymes which can support the TCR signal (Jones and Thompson 2007; Rathmell et al. 2003). Furthermore, TLR4-activated M1 macrophage and IL6-activated T-helper 17 cells (upon bacterial infection) depend on glycolysis by upregulation of glycolytic enzymes such as GLUT1 and PFKFB3 (Rodriguez-Prados et al. 2010). As for B lymphocytes, glucose metabolism markedly increases after B cell activation (Caro-Maldonado et al. 2014). Pharmacological inhibition of glycolysis by pyruvate dehydrogenase kinase (PDHK) inhibitor, dichloroacetate, significantly suppresses B cell proliferation and antibody secretion. The findings suggest that glycolysis is essential in the hematopoietic hierarchy.

Another glucose-utilizing metabolic pathway, the PPP, also plays a key role in normal hematopoiesis. G6PD, a key regulatory enzyme in the pathway, is important in red blood cells. Loss-of-function mutations in G6PD lead to clinical manifestations such as neonatal jaundice and acute hemolytic anemia (Mason et al. 2007). A tumor necrosis factor superfamily member, 4-1BBL, induces NADPH level, upregulates PPP genes, and promotes macrophage proliferation. The induced proliferation was completely abolished by 2-DG, an analog of glucose, indicating the essential role of glucose flux in supporting the normal function of macrophage (Tu et al. 2015). Furthermore, overactive PPP is pathogenic in patients with rheumatoid arthritis. Upregulation of G6PD and increased flux into the PPP result in NADPH accumulation and ROS clearance, which account for the hyperproliferation of CD4 T cells during clonal expansion and differentiation into cytokine-producing effector cells (Yang et al. 2016).

4.2.2 Glucose Addiction in Leukemias

Cancer cells preferentially use carbon for aerobic glycolysis over oxidative phosphorylation, despite the availability of oxygen and less efficient production of ATP molecules, known as the Warburg effect (Warburg 1956). Further, to support the higher demand for nucleotides and stronger REDOX capability, hyperactive PPP is also essential in supporting the survival of cancer cells.

Primary B cell acute lymphoid leukemia (B-ALL) shows upregulated glycolytic genes (GLUT1, SLC16A2, PFKL) and downregulated OXPHOS genes (PDHB, MDH1, MDH2) compared to normal HSPC (Boag et al. 2006). Therapeutic inhibition of glycolysis by 2-DG can induce apoptosis in B-ALL. Other studies further demonstrated that inhibition of glycolysis also suppressed the growth of B cell lymphoma survival (Liu et al. 2018). MYC oncogene has a broad effect on cancer cell biology that has been found to regulate cellular metabolism in various human cancers including hematopoietic malignancies (Stine et al. 2015). In chronic lymphoid leukemia (CLL), stromal niche activates the transcription of MYC in CLL and therefore leads to an increase in aerobic glycolysis in primary CLL cells by increasing the expression of glucose transporter and glycolytic enzymes (Jitschin et al. 2015). Further, lymphoma cell is dependent on MYC signaling which also upregulates glucose metabolism (including GLUT-1, HK2, and LDHA). Therapeutic inhibition of MYC reduces glucose uptake and sensitizes the growth-inhibitory effect of 2-DG in lymphoma cell lines (Broecker-Preuss et al. 2017). However, T-ALL, in which NOTCH is frequently mutated, exhibits less active aerobic glycolysis compared to normal T cells due to activated AMPK through inhibition of mTORC (Kishton et al. 2016). AMPK negatively regulates aerobic glycolysis, while AMPK deficiency or inhibition of oxidative phosphorylation leads to T-ALL cell death and reduces disease burden.

Besides lymphoid malignancy, glucose metabolism is also essential in myeloid malignancy. Acute myeloid leukemia (AML) exhibits high glucose uptake (Cunningham and Kohno 2016) and glycolytic metabolites are increased in patient serum (Chen et al. 2014). Active glycolysis in AML is associated with poor prognosis (Song et al. 2016). AML is addictive to mTORC1 pathway, which is essential to drive glycolysis and the PPP (Xu et al. 2003; Tamburini et al. 2009). mTOR increases the glycolytic pathway by activating the expression of glycolytic enzymes (including GLUT1, GLUT5, HK2, and PKM2) in AML (Wang et al. 2014b; Song et al. 2014; Chen et al. 2016). Further, AML-driving mutations, such as FLT3-ITD, RUNX1-EVI1, and HOXA9-Meis1, have been shown to promote aerobic glycolysis (Ju et al. 2017; Shen et al. 2015; Lynch et al. 2016; Man et al. 2022). A number of studies showed that genetic knockout of glycolytic genes, including PKM2, LDHA, and MCT4, significantly eliminated leukemic stem cells of AML without affecting normal HSPC function (Wang et al. 2014b; Man et al. 2021).

Unlike T-ALL, AMPK is activated in the leukemic stem cell of AML residing in bone marrow, while deletion of AMPK in AML disrupts glucose uptake, induces ROS, and eventually depletes leukemic stem cells (Saito et al. 2015). MCT4, a lactate transporter, also plays a role in sensing the intracellular pH and helps regulate glucose flux in AML. Genetic or therapeutic inhibition of MCT4 dramatically inhibits glucose metabolism and subsequently eliminates AML leukemic stem cells. Further, AML also induces insulin resistance in normal tissue and decreases glucose uptake by increasing insulin-like growth factor-binding protein 1 (IGFBP1), therefore facilitating the acquisition and utilization of glucose and conferring survival advantages to AML (Ye et al. 2018).

PPP is crucial in neutralizing oxidative stress via cycling NADP/NADPH. In B cells, the key enzyme G6PD is transcriptionally repressed by B cell transcription factors PAX5 and IKZF1, resulting in low PPP activity. In BCR-ABL and $NRAS^{G12D}$-transformed B-ALL, PP2A inhibits PFK2 and redirects glucose from glycolysis to the PPP, for protecting cells from oxidative stress. As a result, inhibition of PP2A and the PPP represents B-ALL-specific vulnerabilities. Therapeutic inhibition of PP2A abolished the shift in glucose flux to the PPP and resulted in the anti-B-ALL effect (Xiao et al. 2018). Further, elevated ROS level is observed in BCR-ABL-transformed chronic myeloid leukemia (CML), which can be inhibited by 2-DG or PI3K/mTOR inhibitors. 2-DG, in cooperation with imatinib (a potent BCR-ABL inhibitor), significantly suppresses CML growth. BCR-ABL upregulates the expression of NADPH oxidase Nox4 and increases ROS production in CML. The results suggest that both the active PPP and glucose supply are essential to CML growth. In AML, activation of SIRT2/mTORC supports leukemic growth and promotes glucose metabolism via PPP and results in glucose addiction (Man et al. 2021; Poulain et al. 2017; Xu et al. 2016). Significant dependence on oxidative PPP makes G6PD a potential therapeutic target in AML treatment.

Increased glycolytic flux also links to treatment resistance clinically. In ALL, glucocorticoid (prednisone and dexamethasone) has been the major treatment option (Inaba and Pui 2010). After binding to glucocorticoid, glucocorticoid receptors homodimerize, translocate to the nucleus, interact with glucocorticoid response elements, eventually inhibit cytokine production, and induce cell death. Glucocorticoid resistance is always associated with the upregulation of glycolytic genes and increased glucose consumption (Holleman et al. 2004), while the resistance can be overcome by the therapeutic inhibition of glycolysis (Hulleman et al. 2009). In CML, the emergence of imatinib resistance has been associated with enhanced expression of BCR-ABL and HIF-1a, accompanied by an increased rate of glycolysis. Inhibition of PDH and transketolase, a key enzyme in the non-oxidative phase of PPP, results in enhanced imatinib sensitivity of CML cells. In AML, increased expression of glycolytic enzymes and activation of glycolysis is associated with the resistance to sorafenib and chemotherapies (Song et al. 2016; Ryu et al. 2019; Huang et al. 2016). Further, FLT3-ITD, the commonest genetic mutation in AML, has been an attractive therapeutic target (Man et al. 2012; Mori et al. 2017). Inhibition of FLT3-ITD resulted in a dramatic increase in mitochondrial oxidative stress which caused apoptosis. Therefore, inhibiting G6PD can escalate mitochondrial ROS and potentiate the oxidative stress triggered by FLT3 inhibition, thus enhancing the efficacy of FLT3 inhibitor and helping eradicate AML (Gregory et al. 2016).

4.3 Amino Acid Metabolism

Amino acids are key nutrients that maintain normal HSC function and regulate its malignant transformation. Amino acids primarily serve as (DeBerardinis and Thompson 2012) precursors for the metabolic network to produce metabolites such as nucleotides, antioxidants or energy, and (Saxton and Sabatini 2017) building blocks for protein synthesis (van Galen et al. 2018). Besides, accumulating evidence has shown that dysregulation of amino acid metabolism is a key feature of leukemia during transformation, progression, and chemoresistance. Amino acid metabolism is emerging as a promising target in the treatment of hematopoietic cancers.

4.3.1 Amino Acids Regulate Normal Hematopoiesis

Amino acids are vital factors that modulate HSC function. In particular, valine and cysteine are indispensable for HSC survival and self-renewal (Taya et al. 2016). Valine-restricted diet leads to a dramatic decrease in HSC. The mechanism behind the specific reliance on valine remains unclear. HSC may have unique valine metabolism pathways, or valine may function as a signaling molecule to maintain HSC. The source of valine may be traced back to vascular endothelium or other stromal cells, highlighting a metabolic niche that supports HSC self-renewal. Further, methionine is a critical amino acid contributing to protein synthesis, methylation, glutathione synthesis, and generation of polyamines (Dutchak et al. 2015). Deletion of nitrogen permease regulator-like 2 (NPRL2) results in defects of cobalamin-dependent synthesis of methionine. The compromised hematopoiesis in NRPL2 null fetal liver is at least in part caused by insufficient methionine supply. As cobalamin is the cofactor of methionine synthase, the defects in NPRL2-deleted embryos can be restored by the addition of cyanocobalamin.

mTORC1 senses growth factors and amino acids and correspondingly promotes cell growth (Goberdhan et al. 2016). Unexpectedly, deletion of RagA, a key protein in mTORC1 pathway, selectively alters the population and output of hematopoietic progenitor cells. The function of RagA-deleted HSC remains competent in nutrient-rich or nutrient-scarce conditions. The indifference to nutrient availability, mediated by RagA, contributes to the stress resilience of HSC and organismal persistence during evolution (Kalaitzidis et al. 2017). Deletion of Raptor (a mTORC1 regulator) in mouse HSC abolishes mTORC1 activity and expands HSC population (Kalaitzidis et al. 2012). Suppression of PI3K-mTOR pathway by miRNA results in inhibition of mitochondrial metabolism and preserves LT-HSC function (Qian et al. 2016). The TSC-mTOR pathway also maintains HSC quiescence by repressing mitochondrial metabolism, therefore reducing ROS (Chen et al. 2008).

Amino acids have also been shown to contribute to cell fate decisions in hematopoiesis. In drosophila, the lymph gland is responsible for hematopoiesis as the progenitor cells share a close relationship to mammalian myeloid progenitors. Hematopoietic progenitors can sense the level of essential amino acids and activate WNT signaling. Consequently, blood progenitors link the availability of essential amino acids to their undifferentiated state (Shim et al. 2012). Functionally, activation of T cells is dependent on arginine modulated by BAZ1B, PSIP1, and TSN (Geiger et al. 2016).

4.3.2 Leukemia Cells Reprogram Amino Acid Metabolism

AML cells show enhanced uptake of glutamine to satisfy their biosynthetic needs (Fultang et al. 2021). Inhibition of glutaminase, an enzyme converting glutamine to glutamate and fueling mitochondrial oxidative phosphorylation, restricts leukemia cell growth with limited cytotoxic activity toward normal cells. Preclinical mouse models also support glutaminolysis as a therapeutic target for AML (Jacque et al. 2015; Gregory et al. 2019). The reliance of leukemia on specific amino acids opens a path of targeting these auxotrophic cancer cells by restricting corresponding amino acids. This can be achieved by either depleting circulating amino acids or pharmacologic inhibition of key enzymes or transporters (Vernieri et al. 2016). Deletion of SLC1A5 (also named ASCT2), the transporter of glutamine, impairs leukemogenesis and progression in MLL-AF9 or Pten deficiency-driven cancer model while causing mild defects in normal hematopoiesis. Additionally, loss of SLC1A5 suppresses leucine import and increased cell death (Ni et al. 2019).

Asparagine is an essential amino acid for the development of ALL (Müller and Boos 1998). A long-standing observation is that exposure of ALL cells to asparaginase, which reduced asparagine content in the extracellular environment, effectively eliminates cancer cells. Although some early investigations attributed the asparagine dependency of ALL cells to low expression of asparagine synthetase (ASNS) (Haskell and Canellos 1969), later studies suggest that basal expression of ASNS in ALL did not necessarily correlate with asparaginase sensitivity (Fine et al. 2005). Interestingly, ALL shows a similar pattern of transcriptomic changes upon asparaginase treatment, including decreased expression of tRNA synthetase and multiple transporters. This metabolic adaptation may be further targeted to enhance the effect of asparaginase treatment (Haskell and Canellos 1969). In another genome-wide study of single nucleotide polymorphism (SNP), aspartate metabolism is identified as the most relevant pathway to asparaginase sensitivity. Specifically, the genetic polymorphisms of two metabolic genes, adenylosuccinate lyase and aspartyl-tRNA synthetase, further contribute to asparaginase sensitivity (Haskell and Canellos 1969). Leukemia and solid tumor share a similar dependency on glutamine. A mechanistic study found that asparagine was able to suppress glutamine depletion-mediated cell death. Asparagine may have unknown fates in the leukemic metabolism network, as the addition of exogenous asparagine restores the levels of neither nonessential amino acids nor TCA cycle intermediates. As a result, asparagine is proposed to be an inhibitor of apoptosis in human cancer (Zhang et al. 2014). The mechanism by which asparagine suppresses cell death remains to be elucidated.

Branched-chain amino acids (leucine, isoleucine, and valine) are essential amino acids that serve as precursors in cellular metabolism and signaling metabolites in regulating protein production. For example, the supplement of leucine showed beneficial effects in an anemia mouse model. Exogenous leucine alleviates the deficiency in erythrocyte production caused by RPS19 deletion. Although BCAAs are activators of mTORC1, leucine does not affect mTOR activity in hematopoietic progenitors of the anemia model, indicating potential unknown downstream targets of leucine (Jaako et al. 2012). BCAAs are reversibly converted to branched-chain keto acids in the cytosol by BCAT1. Stable isotope tracing assays uncovered that BCAT1 was responsible for the production of BCAA in CML cells (Hattori et al. 2017). BCAT1 has been reported to collaborate with oncogenic NRAS mutations to maintain BCAA levels in leukemia (Gu et al. 2019). Chemical suppression of BCAT1 mediates cellular differentiation and delays myeloid leukemia progression (Hattori et al. 2017). Proteomic analysis of human AML cells uncovers stem cell-specific overexpression of proteins from BCAA pathway. BCAT1 is enriched in stem cell populations at both mRNA and protein levels. The overabundant BCAT1 in LSC limits intracellular alpha-ketoglutarate, a metabolite used for the demethylation of DNA and proteins (Wang et al. 2015). The consequent low alpha-ketoglutarate level in LSC mediates DNA hypermethylation by suppressing the TET2 DNA demethylase and maintaining HIF-1a protein level by reducing alpha-ketoglutarate-dependent demethylation and degradation of it. Genetic depletion of BCAT1 disrupts the tumor-initiating activity of LSCs, supporting BCAT1 as a potential target to compromise LSCs (Raffel et al. 2017). A more detailed proteomics analysis on $FLT3^{ITD}/NPM1^{mut}$ AML further demonstrates that LSC upregulates oxidative phosphorylation, validating the metabolic dependence in specific AML subtypes (Raffel et al. 2020).

Histidine metabolism was shown to regulate the sensitivity of leukemia cells to antifolates. Folate has been known for a long time to support leukemia proliferation and progression. The application of antifolates has seen success in the clinical treatment of hematopoietic malignancy. Methotrexate (MTX) is one of such compounds that is widely used as an anticancer drug. The cytotoxicity of MTX is achieved at least in part through inhibition of dihydrofolate reductase and consequent suppression of nucleotide synthesis. A genome-wide screening, aiming toward identifying the genes that regulate MTX sensitivity, pinpoints formimidoyltransferase cyclodeaminase (FTCD) as the determinant of

MTX efficacy. FTCD is responsible for the degradation of histidine, accompanied by the rapid exhaustion of intracellular tetrahydrofolate (THF), the product generated by dihydrofolate reductase to sustain nucleotide synthesis. Expression of FTCD is linked with MTX sensitivity of leukemic cell lines and with survival expectancy in cancer patients. Limiting dietary histidine holds promise to enhance the effect of MTX in the treatment of cancer patients (Kanarek et al. 2018).

Aberrant expression of amino acid transporter proteins also contributes to reprogrammed amino acid metabolism and thereby serves as potential targets for killing leukemia cells. T cell lymphoma expresses high levels of SLC77A5, also named system L amino acid transporter 1 (LAT1). Chemical inhibitors of SLC77A5 suppress the activity of mTORC1 and Akt and mediate an unfolded protein response. Importantly, SLC77A5 inhibitor is preferentially cytotoxic to leukemia (Rosilio et al. 2015).

LSCs are reported to utilize different sources of nutrients to fuel mitochondrial respiration (Wang et al. 2016; Wang and Lei 2018; Wang et al. 2020). Metabolomic analysis of LSC-like cells from primary AML revealed an elevation of 16 amino acids and 2 TCA cycle intermediates. LSCs rely on amino acids to maintain mitochondrial respiration and cell viability (Jones et al. 2018). Strikingly, relapsed LSCs that survived venetoclax and azacitidine treatment undergo rewiring of the metabolic network and no longer rely upon amino acids for respiration. These cells shift their preference to fatty acids to support mitochondrial metabolism. LSC has also been found to fuel glutamine to the biosynthesis of pyrimidine and glutathione rather than the TCA cycle after chemotherapeutic interventions. Interestingly, pyrimidine synthesis in malignant cells is dependent on bone marrow stromal cells-derived aspartate, providing another example of stromal regulation of leukemia metabolism. A timed inhibition of pyrimidine synthesis in residue cells allows for more efficient clearance of chemoresistant leukemia (van Gastel et al. 2020a).

4.4 Fatty Acid Metabolism

Normal HSCs are highly dependent on anaerobic respiration. Although HSCs rely heavily on glycolysis, lipid metabolism is also essential for HSC to maintain dormancy and to synthesize membrane lipid components under stress (Lee et al. 2018a). Fatty acid synthesis and oxidation can also contribute to restricting ROS levels by consuming NADP, or by indirectly increasing NADPH levels (Carracedo et al. 2013).

4.4.1 Fatty Acid in Normal Hematopoiesis

Fatty acid oxidation (FAO), also known as β-oxidation, is an important process for HSCs to keep quiescence and maintain self-renewal ability. Fatty acids are converted to activated form acyl-CoA, a step controlled by CD36, fatty-acid-binding proteins (FABP), and transporter proteins. Afterward, acyl-CoA is converted to carnitine by carnitine palmitoyltransferase 1A (CPT1A), the rate-limiting step of FAO. Once inside the mitochondrial matrix, acyl-CoA is decomposed into acetyl-CoA. Concurrently, NADH and FADH2 were generated to feed into the TCA cycle. Therapeutic inhibition of CPT1 by etomoxir exhausts HSC. Loss of peroxisome proliferator-activated receptor δ (PPARδ), a master regulator of the FAO pathway, controls the asymmetric division of HSC and therefore regulates the balance of stemness capacity and differentiating potential. The loss of PPARδ or FAO inhibition can induce the loss of HSC quiescent, promote HSC exhaustion, and induce differentiation (Ito et al. 2012).

Fatty acid synthesis (FAS) is essential for membrane synthesis, cell growth, and proliferation. Fatty acids are the building blocks for the synthesis of triacylglycerides, glycerophospholipids, cardiolipins, sphingolipids, and eicosanoids. Acetyl-CoA is catalyzed by acetyl-CoA carboxylase 1 and 2 (ACC1, ACC2) to malonyl-CoA, which is the limiting step of FAS. Subsequently, fatty acid synthase (FASN) elongates malonyl-CoA into fatty acids (Maier et al. 2008). FASN knockout causes neutropenia via inducing apoptosis and ER stress in mice (Lodhi et al. 2015). Further, lipoxygenase (LOX), which catalyzes the formation of unsaturated FA in the arachidonic acid pathway, is crucial to hematopoiesis. Loss of 12/15-LOX abrogates the repopulating capacity of HSC and reduces LT-HSC number (Kinder et al. 2010). Besides, the end product of the arachidonic acid pathway via cyclooxygenase (COX), PGE2, has been reported to enhance the number of HSPC and promote bone marrow recovery after stress, whereas nonsteroidal anti-inflammatory drugs (NSAIDs) or COX knockout decreased HSC number during embryogenesis in zebra fish model (North et al. 2007). PGE2 can also stabilize β-catenin in HSC, promote survival and proliferation during embryogenesis (Goessling et al. 2009), and increase anti-apoptotic protein survivin (Hoggatt et al. 2009; Fukuda et al. 2002). More recently, SPHK2 regulating S1P is highly expressed in HSC. Deletion of SPHK2 stabilizes HIF1-a, increases hypoxic response, and subsequently enhances anaerobic glycolysis that promotes HSC function (Li et al. 2022). These observations indicate the indispensable role of fatty acids in the maintenance of normal HSC function.

4.4.2 Fatty Acid in Leukemias

Compared to normal HSPC, AML cells upregulate FAO genes (PPARα, FABP4, CPT1A) and downregulate PHD3 in fatty acid metabolism (Tabe et al. 2017; Shi et al. 2016; German et al. 2016). FAO inhibitor or CPT1A inhibitor induces apoptosis of AML cells and synergizes with BCL2 and BCLXL inhibitor in inhibiting AML proliferation, while normal HSC is not affected (Ricciardi et al. 2015; Samudio et al. 2010). Survival was extended when FABP4 is knocked down in a Hoxa9/Meis1-driven murine leukemia model (Shafat et al. 2017). Furthermore, AMPK overactivation (Tabe et al. 2017) triggers the elevation of FAO by inhibiting ACC1/2 (German et al. 2016). Apart from high ROS and strong predominance of OXPHOS, the chemoresistant AML cells present high FAO and overexpression of CD36 (the FA transporter). The change of FAO in AML activates lipolysis and shows a close correlation with drug resistance (Farge et al. 2017; Ye et al. 2016).

In addition to myeloid malignancy, lipid metabolism also plays an important role in lymphoid malignancy. Lipids are stored in CLL cells and transformed to free fatty acids for energy production by OXPHOS, which is different from normal B lymphocytes (Rozovski et al. 2016). FAS can be inhibited by orlistat, a gastrointestinal lipase inhibitor, that kills primary CLL lymphocytes in vitro (Pallasch et al. 2008). Differently, the remarkably high expression of FAO-related enzymes strengthens FAO in CLL. CPT1 and CPT2, highly expressed primary CLL lymphocytes, are essential in importing fatty acids into mitochondria for FAO (Martinez Marignac et al. 2013). Perhexiline, a CPT1 inhibitor, results in the depletion of cardiolipin (an essential part of mitochondrial membranes), disrupts mitochondrial integrity, and induces programmed cell death (Liu et al. 2016).

Normal HSC survival remarkably depends by sphingosine 1-phosphate (S1P), sphingosine, ceramide, and other sphingolipids. In particular, S1P has a positive relationship with cell survival, while sphingosine and ceramide show negative correlations. Ceramide synthase (CerS), the enzyme catalyzing the formation of ceramide from sphingosine, is upregulated by BCL2 family inhibitor, ABT-263, upon apoptosis induction (Beverly et al. 2013). SACLAC, a ceramide analog, downregulates the level of MCL-1 (an anti-apoptotic protein) and induces apoptosis in AML (Pearson et al. 2020). In addition, FLT3 signaling is found to contribute to lower expression of CerS. FLT3 inhibition upregulates CerS, thus causing mitophagy-dependent death of AML cells (Dany et al. 2016). Inhibition of sphingosine kinase (SPHK), the enzyme that converts sphingosine to S1P, synergizes with vincristine chemotherapy in T-ALL (Evangelisti et al. 2014). Similarly, the inhibition of SPHK1 promotes apoptosis in AML blasts (Paugh et al. 2008).

Cholesterol is produced from the mevalonate pathway and contributes to the synthesis of ubiquinone, dolichol, vitamin D, and steroid hormones. Cholesterol metabolism is reported to show dysregulation in hematologic malignancies. Statin can inhibit HMG-CoA reductase (HMGCR), a rate-limiting enzyme in the mevalonate pathway, thus triggering cancer cell apoptosis (Xia et al. 2001; Wu et al. 2004). In a clinical study, CLL patients receiving hypolipidemic therapy had prolonged survival (Righolt et al. 2019). Further, a clinical evaluation of statins was conducted for examining the effect of chemotherapy in the treatment of AML. Statin treatment potentially improves the complete remission among favorable-risk AML groups (Kornblau et al. 2007; Advani et al. 2018). However, its role in the high-risk AML groups is less clear. In addition, statins increase the sensitivity of CLL and AML cells to ABT-199, a BCL2 inhibitor. This is because protein geranylgeranylation is inhibited, which increases the protein p53 upregulated modulator of apoptosis (PUMA) (Lee et al. 2018b).

The disorder of lipid metabolism has been considered as a major metabolic phenotype in hematopoietic malignancies. As demonstrated by substantial research, low FAS activity and high FAO rates appear in leukemic cells, which are associated with drug resistance (Samudio et al. 2010; Behan et al. 2009). The dependence of leukemic cells on FAO can be partially explained by the increased ATP production, which ensures higher survivability under metabolic stress.

4.5 Nucleotide Biosynthesis

Nucleotides are composed of three subunit molecules: a nitrogenous nucleobase, a five-carbon sugar (ribose or deoxyribose), and a phosphate group. Nucleotides play a central role in metabolism. Nucleic acid purines (adenosine and guanine) and pyrimidines (uracil, cytosine, and thymine) constitute deoxyribonucleic acid (DNA) and ribonucleic acid (RNA). Purines are the structural component of energy-carrying molecules such as ATP or signal transducers such as cAMP and cGMP. Purines can also be incorporated into cofactors of enzymatic reactions such as NADH and NADPH. Therefore, nucleotides are essential for cell metabolism and become a major vulnerability in proliferative cells. The synthesis of nucleotides is connected to anabolic pathways from multiple metabolic pathways such as amino acids, the PPP, and TCA cycle.

In mammalian cells, nucleotides can be synthesized de novo or recycled through salvage pathways. In de novo biosynthesis pathway, nucleotides are derived from the catabolism of sugar and amino acids or glutamine and CO_2. Purine synthesis starts from phosphoribosyl pyrophosphate (PRPP), which is produced from glucose via the PPP. PRPP is converted to inosine monophosphate (IMP) by IMDPH, the rate-determining step of the de novo biosynthesis of guanine. The reactions are incorporating some amino acids including aspartic acid, glycine, and glutamine and eventually produce GMP and AMP. Pyrimidine synthesis starts from carbamoyl phosphate from glutamine and CO_2. Upon the condensation and cyclization of aspartate and carbamoyl phosphate, dihydroorotate is produced and converted to orotate by dihydroorotate dehydrogenase (DHODH). Orotate is then covalently linked to a phosphorylated ribose unit and therefore orotidine monophosphate (OMP) is yielded. Further, UMP, UTP, and CTP are generated from OMP.

4.5.1 Nucleotide Biosynthesis in Hematopoiesis

Purine biosynthesis is important for HSC proliferation and lineage specification. Upon hematological stress, such as bone marrow transplantation and chemotherapy, p38-MAPK is activated to upregulate the expression of

IMPDH2, resulting in increased purine biosynthesis in HSPC. Eventually, quiescent HSC starts to proliferate, differentiate, and promote bone marrow recovery (Karigane et al. 2016). Further, IMPDH is essential to maintain the immune response of lymphocytes and the differentiation of erythrocytes, respectively (Gu et al. 2000). Besides, purine can be recycled from the degradation of RNA and DNA, known as the nucleotide salvage pathway. HSCs are highly dependent on hypoxanthine guanine phosphoribosyl transferase (HPRT)-associated purine salvaging. HSC with lower HPRT activity shows reduced engraftment potential, slower proliferation, and decreased mitochondrial activity (Vogel et al. 2019). It suggests that the HPRT-associated purine salvage pathway is important for HSC function.

Pyrimidine biosynthesis is also important after hematological stress on HSC. De novo pyrimidine biosynthesis is essential for HSC recovery after cyclophosphamide treatment in rats (Kudo et al. 1994). There has also been a link between glutamine, purine/pyrimidine biosynthesis, and HSC function (Oburoglu et al. 2014). Decreased purine and pyrimidine biosynthesis is associated with the inhibition of glutamine metabolism and subsequently affects the HSC function.

4.5.2 Nucleotide Biosynthesis in Leukemias

Nucleotide metabolism might be a therapeutic target for hematopoietic malignancy. Proof of principle for targeting nucleotide metabolism in clinics emerged early by the identification of aminopterin, a folate antagonist inhibitor of nucleotide biosynthesis in ALL (Farber and Diamond 1948). Nowadays, targeting nucleotide biosynthesis has been the mainstay of the treatment regime in hematopoietic malignancy. Commonly used agents included folic acid antagonists (methotrexate), purine analogs (fludarabine and clofarabine), and pyrimidine analogs (cytarabine).

In AML, PAICS (phosphoribosylaminoimidazole carboxylase) is responsible for the de novo purine biosynthesis, a process essential for AML survival (Yamauchi et al. 2019). A PAICS inhibitor (MRT252040) effectively suppressed AML growth via inducing cell cycle arrest and apoptosis. Further, knocking out Slc43a3 or Hprt by CRISPR-Cas9 significantly enhanced the antileukemic effect of PAICS inhibitor in AML cell lines. Besides, pyrimidine metabolism is also reported to be a possible target for AML (Christian et al. 2019). Dihydroorotate dehydrogenase (DHODH) functions in de novo pyrimidine synthesis by producing orotate from dihydroorotate. Treatment of AML with a DHODH inhibitor induces differentiation of leukemia into mature myeloid cells (Sykes et al. 2016). DHODH inhibition also reduces leukemic burden, decreases leukemia-initiating cells, and improves survival in animal models. Pyrimidine biosynthesis also confers chemoresistance in AML (van Gastel et al. 2020b). Upon chemotherapy, a small fraction of AML persists and exhibits a shift in metabolic dependence. Pyrimidine synthesis in chemoresistant AML cells is fueled by glutamine and niche-derived aspartate. Inhibiting pyrimidine synthesis by DHODH inhibitor overcomes chemoresistance. It suggested that inhibiting pyrimidine synthesis not only targets the differentiation block but also tackles the emergence of chemoresistance in AML.

In ALL, de novo purine synthesis is relatively more active compared to normal bone marrow lymphocytes (Piga et al. 1982; Scholar and Calabresi 1973). Therefore, 6-mercaptopurine (6-MP), a purine base analog commonly used in the treatment of ALL, inhibits de novo purine biosynthesis and induces apoptosis upon incorporation into DNA (De Abreu 1994). Genetic mutations in enzymes involved in purine biosynthesis have also been reported to regulate the sensitivity toward purine base analog in ALL. Cytosolic purine 5′-nucleotidase (NT5C2) is responsible for the dephosphorylation of 6-hydroxypurine nucleotide monophosphate such as IMP, dIM, GMP, dGMP, and XMP before export from the cells. NT5C2 can also antagonize 6-MP cytotoxic effect by facilitating the dephosphorylation and export of the active form of 6-MP derivative (TIMP). The emergence of NT5C2 mutant upon 6-MP treatment has been reported in relapsed T-ALL and B-ALL (Tzoneva et al. 2013; Meyer et al. 2013). Overexpressing NT5C2 mutant in T-ALL cells induced 6-MP resistance, confirming the mechanistic link between NT5C2 activity and 6-MP resistance. Besides, a relapse-specific mutation in phosphoribosyl pyrophosphate synthetase 1 (PRPS1), which converts ribose 5-phosphate into PRPP, is found in B-ALL. Mutant PRPS1 conferred resistance to 6-MP by abrogating the production of TIMP from 6-MP and deactivating the nucleotide feedback inhibition on 6-MP conversion (Li et al. 2015). However, certain ALL subtypes such as TEL-AML1 ALL have a lower rate of de novo purine synthesis compared to other childhood ALL subtypes (Zaza et al. 2004). This metabolic feature might be associated with the sensitivity toward purine base analogs such as methotrexate and mercaptopurine (Zaza et al. 2004; Shurtleff et al. 1995).

In addition, nucleotide degradation pathways are also important in the clearance of nucleotide metabolites. Purine nucleotide phosphorylase (PNP) catalyzes the breakdown of inosine, guanosine, and deoxyguanosine into the corresponding nitrogen base and sugar phosphate. Inhibition of PNP results in the accumulation of nucleosides and subsequently increases the conversion of deoxyguanosine into dGTP. This in turn inhibits ribonucleotide diphosphate reductase and blocks the synthesis of dCDP, dUDP, and DNA (Ullman et al. 1979). Therefore, forodesine, a transition state analog inhibitor of PNP, has shown an antiproliferative effect in T-ALL, T cell lymphoma, and B-ALL (Homminga et al. 2011; Makita et al. 2018; Balakrishnan et al. 2013).

Collectively, purine and pyrimidine metabolism is required for the proliferation and maintenance of an undifferentiated state in HSPCs. The distinct dependence of nucleotide metabolism is also the key signature of hematopoietic malignancies which serves as a possible therapeutic target.

4.6 Conclusion

The fundamental cellular activities have been closely associated with a complex network of metabolic pathways. Cell metabolism is crucial and indispensable in regulating HSC function and its malignant transformation (Table 4.1). However, better elucidating the function of respective metabolic pathways in the hematopoietic system is needed to resolve the pathophysiology and identify therapeutic strategies. Therefore, improving analytical sensitivity, coverage in metabolomics analysis, and cross-reference of multi-omics studies are critical for understanding the associated activities.

Besides, clonal evolution has been reported in leukemia which constantly adapts to environmental conditions. Leukemic cells reshape not only signaling or transcription but also cellular metabolism. Multi-omics analysis on genome and transcriptome has been applied to resolve the heterogeneity of a cell population at single-cell level (van Galen et al. 2019; Notta et al. 2011; Baryawno et al. 2019). Future technological advances in single-cell proteome and metabolome (Xiao et al. 2019; Hughes et al. 2014), in combination with genomic and functional data, might ultimately lead to a better understanding of metabolic regulation of hematopoiesis heterogeneity, oncogenic transformation, and clonal evolution during disease progression. More potent and precise treatment strategies targeting metabolism will aid in the combat against hematopoietic malignancies.

Table 4.1 The summary of essential metabolic enzymes in normal and malignant hematopoietic cells

Metabolic pathway	Normal	Leukemia		References
Glycolysis	HIF-1a LDHA PKM2 GLUT1 (Mac) PFKFB3 (T)	Gene (drug)		Takubo et al. (2010); Takubo et al. (2013); Wang et al. (2014a); Rodriguez-Prados et al. (2010); Ye et al. (2018); Chen et al. (2016); Song et al. (2014); Wang et al. (2014a); Jitschin et al. (2015); Saito et al. (2015); Tamburini et al. (2009); Xu et al. (2003); Boag et al. (2006); Man et al. (2021)
		Myeloid: IGFBP1 HIF-1a GLUT1 GLUT5 HK2 PKM2 (compound 3k) MCT4 (syrosingopine) mTORC1(everolimus, BEZ235)	*Lymphoid:* GLUT1 HK2 LDHA MCT2 PFKL PDHB MDH1/2	
PPP	G6PD	*Myeloid:* NOX4 (GLX351322) SIRT2/mTORC (veliparib) MCT4 (syrosingopine) PDH	*Lymphoid:* PP2A (LB-100)	Poulain et al. (2017); Xu et al. (2016); Xiao et al. (2018); Poulain et al. (2017); Xu et al. (2016); Man et al. (2021)
Amino acids	mTORC Valine and cysteine Arginine (T cell)	*Myeloid:* Asparagine (ASNase) Cysteine BCAA Methionine (pinometostat) Tryptophan (indoximod, epacadostat) Glutamine (CB839) Arginine (BCT-100) Ornithine (DFMO, AO476)	*Lymphoid:* Asparagine (ASNase) Glutamine (CB839) Arginine (CAT-1) Valine (BMS-906024)	Haskell and Canellos (1969); Fultang et al. (2021); Gregory et al. (2019); Rosilio et al. (2015); Vernieri et al. (2016)
Fatty acids	FABP CPT1A PPARg ACC1/2 LOX PEG2 SPHK2	*Myeloid:* CD36 CPT1A (etomoxir, ST1326) FABP4 PHD3 PPARα	*Lymphoid:* CPT1A (perhexiline) SPHK	German et al. (2016); Shi et al. (2016); Tabe et al. (2017) Farge et al. (2017); Ye et al. (2016); Evangelisti et al. (2014); Paugh et al. (2008)
Nucleotides	p38MAPK IMPDH HPRT	*Myeloid:* PNP (forodesine) PAICS (MRT252040) DHODH	*Lymphoid:* NT5C2 PRPS1 PNP	Ullman et al. (1979); Meyer et al. (2013); Tzoneva et al. (2013); Li et al. (2015); Yamauchi et al. (2019); Sykes et al. (2016)

References

Advani AS et al (2018) Report of the relapsed/refractory cohort of SWOG S0919: a phase 2 study of idarubicin and cytarabine in combination with pravastatin for acute myelogenous leukemia (AML). Leuk Res 67: 17–20. https://doi.org/10.1016/j.leukres.2018.01.021

Appelbaum FR (2007) Hematopoietic-cell transplantation at 50. N Engl J Med 357:1472–1475. https://doi.org/10.1056/NEJMp078166

Balakrishnan K, Ravandi F, Bantia S, Franklin A, Gandhi V (2013) Preclinical and clinical evaluation of forodesine in pediatric and adult B-cell acute lymphoblastic leukemia. Clin Lymphoma Myeloma Leuk 13: 458–466. https://doi.org/10.1016/j.clml.2013.04.009

Baryawno N et al (2019) A cellular taxonomy of the bone marrow stroma in homeostasis and leukemia. Cell 177: 1915–1932.e1916. https://doi.org/10.1016/j.cell.2019.04.040
Behan JW et al (2009) Adipocytes impair leukemia treatment in mice. Cancer Res 69:7867–7874. https://doi.org/10.1158/0008-5472.Can-09-0800
Beverly LJ et al (2013) BAK activation is necessary and sufficient to drive ceramide synthase-dependent ceramide accumulation following inhibition of BCL2-like proteins. Biochem J 452:111–119. https://doi.org/10.1042/bj20130147
Boag JM et al (2006) Altered glucose metabolism in childhood pre-B acute lymphoblastic leukaemia. Leukemia 20:1731–1737. https://doi.org/10.1038/sj.leu.2404365
Broecker-Preuss M, Becher-Boveleth N, Bockisch A, Dührsen U, Müller S (2017) Regulation of glucose uptake in lymphoma cell lines by c-MYC- and PI3K-dependent signaling pathways and impact of glycolytic pathways on cell viability. J Transl Med 15:158. https://doi.org/10.1186/s12967-017-1258-9
Busch K et al (2015) Fundamental properties of unperturbed haematopoiesis from stem cells in vivo. Nature 518:542–546. https://doi.org/10.1038/nature14242
Caro-Maldonado A et al (2014) Metabolic reprogramming is required for antibody production that is suppressed in anergic but exaggerated in chronically BAFF-exposed B cells. J Immunol 192:3626–3636. https://doi.org/10.4049/jimmunol.1302062
Carracedo A, Cantley LC, Pandolfi PP (2013) Cancer metabolism: fatty acid oxidation in the limelight. Nat Rev Cancer 13:227–232. https://doi.org/10.1038/nrc3483
Chandel NS, Jasper H, Ho TT, Passegué E (2016) Metabolic regulation of stem cell function in tissue homeostasis and organismal ageing. Nat Cell Biol 18:823–832. https://doi.org/10.1038/ncb3385
Chen C et al (2008) TSC-mTOR maintains quiescence and function of hematopoietic stem cells by repressing mitochondrial biogenesis and reactive oxygen species. J Exp Med 205:2397–2408. https://doi.org/10.1084/jem.20081297
Chen WL et al (2014) A distinct glucose metabolism signature of acute myeloid leukemia with prognostic value. Blood 124:1645–1654. https://doi.org/10.1182/blood-2014-02-554204
Chen WL et al (2016) Enhanced fructose utilization mediated by SLC2A5 is a unique metabolic feature of acute myeloid leukemia with therapeutic potential. Cancer Cell 30:779–791. https://doi.org/10.1016/j.ccell.2016.09.006
Cheng T et al (2000) Hematopoietic stem cell quiescence maintained by p21cip1/waf1. Science (New York, N. Y.) 287:1804–1808. https://doi.org/10.1126/science.287.5459.1804
Cho YM et al (2006) Dynamic changes in mitochondrial biogenesis and antioxidant enzymes during the spontaneous differentiation of human embryonic stem cells. Biochem Biophys Res Commun 348:1472–1478. https://doi.org/10.1016/j.bbrc.2006.08.020
Christian S et al (2019) The novel dihydroorotate dehydrogenase (DHODH) inhibitor BAY 2402234 triggers differentiation and is effective in the treatment of myeloid malignancies. Leukemia 33:2403–2415. https://doi.org/10.1038/s41375-019-0461-5
Copelan EA (2006) Hematopoietic stem-cell transplantation. N Engl J Med 354:1813–1826. https://doi.org/10.1056/NEJMra052638
Courties G et al (2015) Ischemic stroke activates hematopoietic bone marrow stem cells. Circ Res 116: 407–417. https://doi.org/10.1161/circresaha.116.305207
Cunningham I, Kohno B (2016) 18 FDG-PET/CT: 21st century approach to leukemic tumors in 124 cases. Am J Hematol 91:379–384. https://doi.org/10.1002/ajh.24287
Dany M et al (2016) Targeting FLT3-ITD signaling mediates ceramide-dependent mitophagy and attenuates drug resistance in AML. Blood 128:1944–1958. https://doi.org/10.1182/blood-2016-04-708750
De Abreu RA (1994) Nucleotide metabolism: mode of action of thiopurines in leukemia. Adv Exp Med Biol 370:195–200. https://doi.org/10.1007/978-1-4615-2584-4_42
DeBerardinis RJ, Thompson CB (2012) Cellular metabolism and disease: what do metabolic outliers teach us? Cell 148:1132–1144. https://doi.org/10.1016/j.cell.2012.02.032
Dutchak PA et al (2015) Regulation of hematopoiesis and methionine homeostasis by mTORC1 inhibitor NPRL2. Cell Rep 12:371–379. https://doi.org/10.1016/j.celrep.2015.06.042
Dutta P et al (2015) Myocardial infarction activates CCR2 (+) hematopoietic stem and progenitor cells. Cell Stem Cell 16:477–487. https://doi.org/10.1016/j.stem.2015.04.008
Eliasson P, Jönsson JI (2010) The hematopoietic stem cell niche: low in oxygen but a nice place to be. J Cell Physiol 222:17–22. https://doi.org/10.1002/jcp.21908
Evangelisti C et al (2014) Assessment of the effect of sphingosine kinase inhibitors on apoptosis, unfolded protein response and autophagy of T-cell acute lymphoblastic leukemia cells; indications for novel therapeutics. Oncotarget 5:7886–7901. https://doi.org/10.18632/oncotarget.2318
Farber S, Diamond LK (1948) Temporary remissions in acute leukemia in children produced by folic acid antagonist, 4-aminopteroyl-glutamic acid. N Engl J Med 238:787–793. https://doi.org/10.1056/NEJM194806032382301
Farge T et al (2017) Chemotherapy-resistant human acute myeloid leukemia cells are not enriched for leukemic stem cells but require oxidative metabolism. Cancer Discov 7:716–735. https://doi.org/10.1158/2159-8290.Cd-16-0441
Filosa S et al (2003) Failure to increase glucose consumption through the pentose-phosphate pathway results in

the death of glucose-6-phosphate dehydrogenase gene-deleted mouse embryonic stem cells subjected to oxidative stress. Biochem J 370:935–943. https://doi.org/10.1042/bj20021614

Fine BM, Kaspers GJ, Ho M, Loonen AH, Boxer LM (2005) A genome-wide view of the in vitro response to l-asparaginase in acute lymphoblastic leukemia. Cancer Res 65:291–299

Fukuda S, Foster RG, Porter SB, Pelus LM (2002) The antiapoptosis protein survivin is associated with cell cycle entry of normal cord blood CD34(+) cells and modulates cell cycle and proliferation of mouse hematopoietic progenitor cells. Blood 100:2463–2471. https://doi.org/10.1182/blood.V100.7.2463

Fultang L, Gneo L, De Santo C, Mussai FJ (2021) Targeting amino acid metabolic vulnerabilities in myeloid malignancies. Front Oncol 11:674720. https://doi.org/10.3389/fonc.2021.674720

van Galen P et al (2018) Integrated stress response activity Marks stem cells in Normal hematopoiesis and leukemia. Cell Rep 25:1109–1117.e1105. https://doi.org/10.1016/j.celrep.2018.10.021

van Galen P et al (2019) Single-cell RNA-Seq reveals AML hierarchies relevant to disease progression and immunity. Cell 176:1265–1281.e1224. https://doi.org/10.1016/j.cell.2019.01.031

van Gastel N et al (2020a) Induction of a timed metabolic collapse to overcome cancer Chemoresistance. Cell Metab 32:391–403.e396. https://doi.org/10.1016/j.cmet.2020.07.009

van Gastel N et al (2020b) Induction of a timed metabolic collapse to overcome cancer Chemoresistance. Cell Metab 32:391–403 e396. https://doi.org/10.1016/j.cmet.2020.07.009

Geiger R et al (2016) L-arginine modulates T cell metabolism and enhances survival and anti-tumor activity. Cell 167:829–842 e813. https://doi.org/10.1016/j.cell.2016.09.031

German NJ et al (2016) PHD3 loss in cancer enables metabolic reliance on fatty acid oxidation via deactivation of ACC2. Mol Cell 63:1006–1020. https://doi.org/10.1016/j.molcel.2016.08.014

Goberdhan DC, Wilson C, Harris AL (2016) Amino acid sensing by mTORC1: intracellular transporters mark the spot. Cell Metab 23:580–589. https://doi.org/10.1016/j.cmet.2016.03.013

Goessling W et al (2009) Genetic interaction of PGE2 and Wnt signaling regulates developmental specification of stem cells and regeneration. Cell 136:1136–1147. https://doi.org/10.1016/j.cell.2009.01.015

Goncalves KA et al (2016) Angiogenin promotes hematopoietic regeneration by dichotomously regulating quiescence of stem and progenitor cells. Cell 166:894–906. https://doi.org/10.1016/j.cell.2016.06.042

Gregory MA et al (2016) ATM/G6PD-driven redox metabolism promotes FLT3 inhibitor resistance in acute myeloid leukemia. Proc Natl Acad Sci U S A 113:E6669–e6678. https://doi.org/10.1073/pnas.1603876113

Gregory MA et al (2019) Targeting glutamine metabolism and redox state for leukemia therapy. Clinical cancer research : an official journal of the American Association for Cancer Research 25:4079–4090. https://doi.org/10.1158/1078-0432.Ccr-18-3223

Gu JJ et al (2000) Inhibition of T lymphocyte activation in mice heterozygous for loss of the IMPDH II gene. J Clin Invest 106:599–606. https://doi.org/10.1172/jci8669

Gu Z et al (2019) Loss of EZH2 reprograms BCAA metabolism to drive leukemic transformation. Cancer Discov 9:1228–1247. https://doi.org/10.1158/2159-8290.Cd-19-0152

Hargreaves M, Spriet LL (2020) Skeletal muscle energy metabolism during exercise. Nat Metab 2:817–828. https://doi.org/10.1038/s42255-020-0251-4

Haskell CM, Canellos GP (1969) l-asparaginase resistance in human leukemia--asparagine synthetase. Biochem Pharmacol 18:2578–2580. https://doi.org/10.1016/0006-2952(69)90375-x

Hattori A et al (2017) Cancer progression by reprogrammed BCAA metabolism in myeloid leukaemia. Nature 545:500–504. https://doi.org/10.1038/nature22314

Hoggatt J, Singh P, Sampath J, Pelus LM (2009) Prostaglandin E2 enhances hematopoietic stem cell homing, survival, and proliferation. Blood 113:5444–5455. https://doi.org/10.1182/blood-2009-01-201335

Hoggatt J et al (2018) Rapid mobilization reveals a highly Engraftable hematopoietic stem cell. *Cell* 172:191–204.e110. https://doi.org/10.1016/j.cell.2017.11.003

Holleman A et al (2004) Gene-expression patterns in drug-resistant acute lymphoblastic leukemia cells and response to treatment. N Engl J Med 351:533–542. https://doi.org/10.1056/NEJMoa033513

Homminga I et al (2011) In vitro efficacy of forodesine and nelarabine (ara-G) in pediatric leukemia. Blood 118:2184–2190. https://doi.org/10.1182/blood-2011-02-337840

Huang A et al (2016) Metabolic alterations and drug sensitivity of tyrosine kinase inhibitor resistant leukemia cells with a FLT3/ITD mutation. Cancer Lett 377:149–157. https://doi.org/10.1016/j.canlet.2016.04.040

Hughes CS et al (2014) Ultrasensitive proteome analysis using paramagnetic bead technology. Mol Syst Biol 10:757. https://doi.org/10.15252/msb.20145625

Hulleman E et al (2009) Inhibition of glycolysis modulates prednisolone resistance in acute lymphoblastic leukemia cells. Blood 113:2014–2021. https://doi.org/10.1182/blood-2008-05-157842

Inaba H, Pui CH (2010) Glucocorticoid use in acute lymphoblastic leukaemia. Lancet Oncol 11:1096–1106. https://doi.org/10.1016/S1470-2045(10)70114-5

Ito K, Bonora M, Ito K (2019) Metabolism as master of hematopoietic stem cell fate. Int J Hematol 109:18–27. https://doi.org/10.1007/s12185-018-2534-z

Ito K et al (2012) A PML–PPAR-δ pathway for fatty acid oxidation regulates hematopoietic stem cell maintenance. Nat Med 18:1350–1358. https://doi.org/10.1038/nm.2882

Jaako P et al (2012) Dietary L-leucine improves the anemia in a mouse model for Diamond-Blackfan anemia. Blood 120:2225–2228. https://doi.org/10.1182/blood-2012-05-431437

Jacque N et al (2015) Targeting glutaminolysis has antileukemic activity in acute myeloid leukemia and synergizes with BCL-2 inhibition. Blood 126:1346–1356. https://doi.org/10.1182/blood-2015-01-621870

Jensen KS et al (2011) FoxO3A promotes metabolic adaptation to hypoxia by antagonizing Myc function. EMBO J 30:4554–4570. https://doi.org/10.1038/emboj.2011.323

Jensen MD (2002) Fatty acid oxidation in human skeletal muscle. J Clin Invest 110:1607–1609. https://doi.org/10.1172/jci17303

Jitschin R et al (2015) Stromal cell-mediated glycolytic switch in CLL cells involves Notch-c-Myc signaling. Blood 125:3432–3436. https://doi.org/10.1182/blood-2014-10-607036

Jones CL et al (2018) Inhibition of amino acid metabolism selectively targets human leukemia stem cells. Cancer Cell 34:724–740 e724. https://doi.org/10.1016/j.ccell.2018.10.005

Jones RG, Thompson CB (2007) Revving the engine: signal transduction fuels T cell activation. Immunity 27:173–178. https://doi.org/10.1016/j.immuni.2007.07.008

Ju HQ et al (2017) ITD mutation in FLT3 tyrosine kinase promotes Warburg effect and renders therapeutic sensitivity to glycolytic inhibition. Leukemia 31:2143–2150. https://doi.org/10.1038/leu.2017.45

Kalaitzidis D et al (2012) mTOR complex 1 plays critical roles in hematopoiesis and Pten-loss-evoked leukemogenesis. Cell Stem Cell 11:429–439. https://doi.org/10.1016/j.stem.2012.06.009

Kalaitzidis D et al (2017) Amino acid-insensitive mTORC1 regulation enables nutritional stress resilience in hematopoietic stem cells. J Clin Invest 127:1405–1413. https://doi.org/10.1172/jci89452

Kanarek N et al (2018) Histidine catabolism is a major determinant of methotrexate sensitivity. Nature 559:632–636. https://doi.org/10.1038/s41586-018-0316-7

Karigane D et al (2016) p38α activates purine metabolism to initiate hematopoietic stem/progenitor cell cycling in response to stress. Cell Stem Cell 19:192–204. https://doi.org/10.1016/j.stem.2016.05.013

Kinder M et al (2010) Hematopoietic stem cell function requires 12/15-lipoxygenase-dependent fatty acid metabolism. Blood 115:5012–5022. https://doi.org/10.1182/blood-2009-09-243139

Kishton RJ et al (2016) AMPK is essential to balance glycolysis and mitochondrial metabolism to control T-ALL cell stress and survival. Cell Metab 23:649–662. https://doi.org/10.1016/j.cmet.2016.03.008

Kocabas F et al (2012) Meis1 regulates the metabolic phenotype and oxidant defense of hematopoietic stem cells. Blood 120:4963–4972. https://doi.org/10.1182/blood-2012-05-432260

Kornblau SM et al (2007) Blockade of adaptive defensive changes in cholesterol uptake and synthesis in AML by the addition of pravastatin to idarubicin + high-dose Ara-C: a phase 1 study. Blood 109:2999–3006. https://doi.org/10.1182/blood-2006-08-044446

Krivtsov AV et al (2006) Transformation from committed progenitor to leukaemia stem cell initiated by MLL-AF9. Nature 442:818–822. https://doi.org/10.1038/nature04980

Kubota Y, Takubo K, Suda T (2008) Bone marrow long label-retaining cells reside in the sinusoidal hypoxic niche. Biochem Biophys Res Commun 366:335–339. https://doi.org/10.1016/j.bbrc.2007.11.086

Kudo H et al (1994) Regulation of pyrimidine nucleotide synthesis in rat hematopoiesis. In Vivo 8:303–308

Lee JS et al (2018b) Statins enhance efficacy of venetoclax in blood cancers. Sci Transl Med 10. https://doi.org/10.1126/scitranslmed.aaq1240

Lee MKS, Al-Sharea A, Dragoljevic D, Murphy AJ (2018a) Hand of FATe: lipid metabolism in hematopoietic stem cells. Curr Opin Lipidol 29:240–245. https://doi.org/10.1097/mol.0000000000000500

Li B et al (2015) Negative feedback-defective PRPS1 mutants drive thiopurine resistance in relapsed childhood ALL. Nat Med 21:563–571. https://doi.org/10.1038/nm.3840

Li C et al (2022) Loss of sphingosine kinase 2 promotes the expansion of hematopoietic stem cells by improving their metabolic fitness. Blood 140:1686. https://doi.org/10.1182/blood.2022016112

Lin CJ, Lin CY, Chen CH, Zhou B, Chang CP (2012) Partitioning the heart: mechanisms of cardiac septation and valve development. Development 139:3277–3299. https://doi.org/10.1242/dev.063495

Liu PP et al (2016) Elimination of chronic lymphocytic leukemia cells in stromal microenvironment by targeting CPT with an antiangina drug perhexiline. Oncogene 35:5663–5673. https://doi.org/10.1038/onc.2016.103

Liu X et al (2018) B cell lymphoma with different metabolic characteristics show distinct sensitivities to metabolic inhibitors. J Cancer 9:1582–1591. https://doi.org/10.7150/jca.24331

Lodhi IJ, Link DC, Semenkovich CF (2015) Acute ether lipid deficiency affects neutrophil biology in mice. Cell Metab 21:652–653. https://doi.org/10.1016/j.cmet.2015.04.018

Lopaschuk GD, Ussher JR, Folmes CD, Jaswal JS, Stanley WC (2010) Myocardial fatty acid metabolism in health and disease. Physiol Rev 90:207–258. https://doi.org/10.1152/physrev.00015.2009

Lynch JR et al (2016) Gaq signaling is required for the maintenance of MLL-AF9-induced acute myeloid leukemia. Leukemia 30:1745–1748. https://doi.org/10.1038/leu.2016.24

Maier T, Leibundgut M, Ban N (2008) The crystal structure of a mammalian fatty acid synthase. Science (New York, N.Y.) 321:1315–1322. https://doi.org/10.1126/science.1161269

Makita S, Maeshima AM, Maruyama D, Izutsu K, Tobinai K (2018) Forodesine in the treatment of relapsed/refractory peripheral T-cell lymphoma: an evidence-based review. Onco Targets Ther 11:2287–2293. https://doi.org/10.2147/ott.S140756

Man CH et al (2012) Sorafenib treatment of FLT3-ITD(+) acute myeloid leukemia: favorable initial outcome and mechanisms of subsequent nonresponsiveness associated with the emergence of a D835 mutation. Blood 119:5133–5143. https://doi.org/10.1182/blood-2011-06-363960

Man CH et al (2021) Proton export alkalinizes intracellular pH and reprograms carbon metabolism to drive hematopoietic progenitor growth. Blood 139:502. https://doi.org/10.1182/blood.2021011563

Man CH et al (2022) Regulation of proton partitioning in kinase-activating acute myeloid leukemia and its therapeutic implication. Leukemia 36:1990–2001. https://doi.org/10.1038/s41375-022-01606-0

Martinez Marignac VL, Smith S, Toban N, Bazile M, Aloyz R (2013) Resistance to Dasatinib in primary chronic lymphocytic leukemia lymphocytes involves AMPK-mediated energetic re-programming. Oncotarget 4:2550–2566. https://doi.org/10.18632/oncotarget.1508

Mason PJ, Bautista JM, Gilsanz F (2007) G6PD deficiency: the genotype-phenotype association. Blood Rev 21:267–283. https://doi.org/10.1016/j.blre.2007.05.002

Meyer JA et al (2013) Relapse-specific mutations in NT5C2 in childhood acute lymphoblastic leukemia. Nat Genet 45:290–294. https://doi.org/10.1038/ng.2558

Miyamoto K et al (2007) Foxo3a is essential for maintenance of the hematopoietic stem cell pool. Cell Stem Cell 1:101–112. https://doi.org/10.1016/j.stem.2007.02.001

Mori M et al (2017) Gilteritinib, a FLT3/AXL inhibitor, shows antileukemic activity in mouse models of FLT3 mutated acute myeloid leukemia. Investig New Drugs 35:556–565. https://doi.org/10.1007/s10637-017-0470-z

Morrison SJ, Scadden DT (2014) The bone marrow niche for haematopoietic stem cells. Nature 505:327–334. https://doi.org/10.1038/nature12984

Müller HJ, Boos J (1998) Use of L-asparaginase in childhood ALL. Crit Rev Oncol Hematol 28:97–113. https://doi.org/10.1016/s1040-8428(98)00015-8

Nakamura-Ishizu A, Ito K, Suda T (2020) Hematopoietic stem cell metabolism during development and aging. Dev Cell 54:239–255. https://doi.org/10.1016/j.devcel.2020.06.029

Ni F et al (2019) Critical role of ASCT2-mediated amino acid metabolism in promoting leukaemia development and progression. Nat Metab 1:390–403. https://doi.org/10.1038/s42255-019-0039-6

North TE et al (2007) Prostaglandin E2 regulates vertebrate haematopoietic stem cell homeostasis. Nature 447:1007–1011. https://doi.org/10.1038/nature05883

Notta F et al (2011) Isolation of single human hematopoietic stem cells capable of long-term multilineage engraftment. Science (New York, N.Y.) 333:218–221. https://doi.org/10.1126/science.1201219

Oburoglu L et al (2014) Glucose and glutamine metabolism regulate human hematopoietic stem cell lineage specification. Cell Stem Cell 15:169–184. https://doi.org/10.1016/j.stem.2014.06.002

Orkin SH, Zon L (2008) I. SnapShot: hematopoiesis. Cell 132:712. https://doi.org/10.1016/j.cell.2008.02.013

Pallasch CP et al (2008) Targeting lipid metabolism by the lipoprotein lipase inhibitor orlistat results in apoptosis of B-cell chronic lymphocytic leukemia cells. Leukemia 22:585–592. https://doi.org/10.1038/sj.leu.2405058

Parmar K, Mauch P, Vergilio JA, Sackstein R, Down JD (2007) Distribution of hematopoietic stem cells in the bone marrow according to regional hypoxia. Proc Natl Acad Sci U S A 104:5431–5436. https://doi.org/10.1073/pnas.0701152104

Paugh SW et al (2008) A selective sphingosine kinase 1 inhibitor integrates multiple molecular therapeutic targets in human leukemia. Blood 112:1382–1391. https://doi.org/10.1182/blood-2008-02-138958

Pearson JM et al (2020) Ceramide analogue SACLAC modulates sphingolipid levels and MCL-1 splicing to induce apoptosis in acute myeloid leukemia. Molecular cancer research : MCR 18:352–363. https://doi.org/10.1158/1541-7786.Mcr-19-0619

Piga A et al (1982) Nucleoside incorporation into DNA and RNA in acute leukaemia: differences between the various leukaemia sub-types. Br J Haematol 52:195–204. https://doi.org/10.1111/j.1365-2141.1982.tb03881.x

Poulain L et al (2017) High mTORC1 activity drives glycolysis addiction and sensitivity to G6PD inhibition in acute myeloid leukemia cells. Leukemia 31:2326–2335. https://doi.org/10.1038/leu.2017.81

Puente BN et al (2014) The oxygen-rich postnatal environment induces cardiomyocyte cell-cycle arrest through DNA damage response. Cell 157:565–579. https://doi.org/10.1016/j.cell.2014.03.032

Qian P et al (2016) The Dlk1-Gtl2 locus preserves LT-HSC function by inhibiting the PI3K-mTOR pathway to restrict mitochondrial metabolism. Cell Stem Cell 18:214–228. https://doi.org/10.1016/j.stem.2015.11.001

Raffel S et al (2017) BCAT1 restricts αKG levels in AML stem cells leading to IDHmut-like DNA hypermethylation. Nature 551:384–388. https://doi.org/10.1038/nature24294

Raffel S et al (2020) Quantitative proteomics reveals specific metabolic features of acute myeloid leukemia stem cells. Blood 136:1507–1519. https://doi.org/10.1182/blood.2019003654

Rathmell JC, Elstrom RL, Cinalli RM, Thompson CB (2003) Activated Akt promotes increased resting T cell size, CD28-independent T cell growth, and development of autoimmunity and lymphoma. Eur J Immunol 33:2223–2232. https://doi.org/10.1002/eji.200324048

Ren R, Ocampo A, Liu GH, Izpisua Belmonte JC (2017) Regulation of stem cell aging by metabolism and epigenetics. Cell Metab 26:460–474. https://doi.org/10.1016/j.cmet.2017.07.019

Ricciardi MR et al (2015) Targeting the leukemia cell metabolism by the CPT1a inhibition: functional preclinical effects in leukemias. Blood 126:1925–1929. https://doi.org/10.1182/blood-2014-12-617498

Righolt CH et al (2019) Statin use and chronic lymphocytic leukemia incidence: a nested case-control study in Manitoba, Canada. Cancer epidemiology, biomarkers & prevention : a publication of the American Association for Cancer Research, cosponsored by the American Society of Preventive Oncology 28: 1495–1501. https://doi.org/10.1158/1055-9965.Epi-19-0107

Rodriguez-Prados JC et al (2010) Substrate fate in activated macrophages: a comparison between innate, classic, and alternative activation. J Immunol 185:605–614. https://doi.org/10.4049/jimmunol.0901698

Rosilio C et al (2015) L-type amino-acid transporter 1 (LAT1): a therapeutic target supporting growth and survival of T-cell lymphoblastic lymphoma/T-cell acute lymphoblastic leukemia. Leukemia 29:1253–1266. https://doi.org/10.1038/leu.2014.338

Rozovski U, Hazan-Halevy I, Barzilai M, Keating MJ, Estrov Z (2016) Metabolism pathways in chronic lymphocytic leukemia. Leuk Lymphoma 57:758–765. https://doi.org/10.3109/10428194.2015.1106533

Ryu MJ et al (2019) PTEN/AKT signaling mediates chemoresistance in refractory acute myeloid leukemia through enhanced glycolysis. Oncol Rep 42:2149–2158. https://doi.org/10.3892/or.2019.7308

Saito Y, Chapple RH, Lin A, Kitano A, Nakada D (2015) AMPK protects leukemia-initiating cells in myeloid Leukemias from metabolic stress in the bone marrow. Cell Stem Cell 17:585–596. https://doi.org/10.1016/j.stem.2015.08.019

Samudio I et al (2010) Pharmacologic inhibition of fatty acid oxidation sensitizes human leukemia cells to apoptosis induction. J Clin Invest 120:142–156. https://doi.org/10.1172/jci38942

Sawai CM et al (2016) Hematopoietic stem cells are the major source of multilineage hematopoiesis in adult animals. Immunity 45:597–609. https://doi.org/10.1016/j.immuni.2016.08.007

Saxton RA, Sabatini DM (2017) mTOR signaling in growth, metabolism, and disease. Cell 168:960–976. https://doi.org/10.1016/j.cell.2017.02.004

Scholar EM, Calabresi P (1973) Identification of the enzymatic pathways of nucleotide metabolism in human lymphocytes and leukemia cells. Cancer Res 33:94–103

Shafat MS et al (2017) Leukemic blasts program bone marrow adipocytes to generate a protumoral microenvironment. Blood 129:1320–1332. https://doi.org/10.1182/blood-2016-08-734798

Shen L et al (2015) RUNX1-Evi-1 fusion gene inhibited differentiation and apoptosis in myelopoiesis: an in vivo study. BMC Cancer 15:970. https://doi.org/10.1186/s12885-015-1961-y

Shi J et al (2016) High expression of CPT1A predicts adverse outcomes: a potential therapeutic target for acute myeloid leukemia. EBioMedicine 14:55–64. https://doi.org/10.1016/j.ebiom.2016.11.025

Shim J, Mukherjee T, Banerjee U (2012) Direct sensing of systemic and nutritional signals by haematopoietic progenitors in drosophila. Nat Cell Biol 14:394–400. https://doi.org/10.1038/ncb2453

Shurtleff SA et al (1995) TEL/AML1 fusion resulting from a cryptic t(12;21) is the most common genetic lesion in pediatric ALL and defines a subgroup of patients with an excellent prognosis. Leukemia 9:1985–1989

Simsek T et al (2010) The distinct metabolic profile of hematopoietic stem cells reflects their location in a hypoxic niche. Cell Stem Cell 7:380–390. https://doi.org/10.1016/j.stem.2010.07.011

Song K et al (2014) HIF-1α and GLUT1 gene expression is associated with chemoresistance of acute myeloid leukemia. Asian Pac J Cancer Prev 15:1823–1829. https://doi.org/10.7314/apjcp.2014.15.4.1823

Song K et al (2016) Resistance to chemotherapy is associated with altered glucose metabolism in acute myeloid leukemia. Oncol Lett 12:334–342. https://doi.org/10.3892/ol.2016.4600

Spencer JA et al (2014) Direct measurement of local oxygen concentration in the bone marrow of live animals. Nature 508:269–273. https://doi.org/10.1038/nature13034

Spinelli JB, Haigis MC (2018) The multifaceted contributions of mitochondria to cellular metabolism. Nat Cell Biol 20:745–754. https://doi.org/10.1038/s41556-018-0124-1

Stine ZE, Walton ZE, Altman BJ, Hsieh AL, Dang CV (2015) MYC, metabolism, and cancer. Cancer Discov 5:1024–1039. https://doi.org/10.1158/2159-8290.Cd-15-0507

Suda T, Takubo K, Semenza GL (2011) Metabolic regulation of hematopoietic stem cells in the hypoxic niche. Cell Stem Cell 9:298–310. https://doi.org/10.1016/j.stem.2011.09.010

Sun J et al (2014) Clonal dynamics of native haematopoiesis. Nature 514:322–327. https://doi.org/10.1038/nature13824

Sykes DB et al (2016) Inhibition of dihydroorotate dehydrogenase overcomes differentiation blockade in acute myeloid leukemia. Cell 167:171–186 e115. https://doi.org/10.1016/j.cell.2016.08.057

Tabe Y et al (2017) Bone marrow adipocytes facilitate fatty acid oxidation activating AMPK and a transcriptional network supporting survival of acute Monocytic leukemia cells. Cancer Res 77:1453–1464. https://doi.org/10.1158/0008-5472.Can-16-1645

Takubo K et al (2010) Regulation of the HIF-1alpha level is essential for hematopoietic stem cells. Cell Stem Cell 7:391–402. https://doi.org/10.1016/j.stem.2010.06.020

Takubo K et al (2013) Regulation of glycolysis by Pdk functions as a metabolic checkpoint for cell cycle quiescence in hematopoietic stem cells. Cell Stem Cell 12:49–61. https://doi.org/10.1016/j.stem.2012.10.011

Tamburini J et al (2009) Targeting translation in acute myeloid leukemia: a new paradigm for therapy? Cell cycle (Georgetown Tex) 8:3893–3899. https://doi.org/10.4161/cc.8.23.10091

Tanner LB et al (2018) Four key steps control glycolytic flux in mammalian cells. Cell systems 7:49–62.e48. https://doi.org/10.1016/j.cels.2018.06.003

Taya Y et al (2016) Depleting dietary valine permits nonmyeloablative mouse hematopoietic stem cell transplantation. Science (New York, N.Y.) 354:1152–1155. https://doi.org/10.1126/science.aag3145

Tothova Z et al (2007) FoxOs are critical mediators of hematopoietic stem cell resistance to physiologic oxidative stress. Cell 128:325–339. https://doi.org/10.1016/j.cell.2007.01.003

Tu TH et al (2015) 4-1BBL signaling promotes cell proliferation through reprogramming of glucose metabolism in monocytes/macrophages. FEBS J 282:1468–1480. https://doi.org/10.1111/febs.13236

Tzoneva G et al (2013) Activating mutations in the NT5C2 nucleotidase gene drive chemotherapy resistance in relapsed ALL. Nat Med 19:368–371. https://doi.org/10.1038/nm.3078

Ullman B, Gudas LJ, Clift SM, Martin DW Jr (1979) Isolation and characterization of purine-nucleoside phosphorylase-deficient T-lymphoma cells and secondary mutants with altered ribonucleotide reductase: genetic model for immunodeficiency disease. Proc Natl Acad Sci U S A 76:1074–1078. https://doi.org/10.1073/pnas.76.3.1074

Vander Heiden MG, Cantley LC, Thompson CB (2009) Understanding the Warburg effect: the metabolic requirements of cell proliferation. Science 324:1029–1033. https://doi.org/10.1126/science.1160809

Vernieri C et al (2016) Targeting cancer metabolism: dietary and pharmacologic interventions. Cancer Discov 6:1315–1333. https://doi.org/10.1158/2159-8290.Cd-16-0615

Voet JG, Coe J, Epstein J, Matossian V, Shipley T (1981) Electrostatic control of enzyme reactions: effect of ionic strength on the pKa of an essential acidic group on glucose oxidase. Biochemistry 20:7182–7185. https://doi.org/10.1021/bi00528a020

Vogel M et al (2019) HPRT and purine salvaging are critical for hematopoietic stem cell function. Stem cells (Dayton, Ohio) 37:1606–1614. https://doi.org/10.1002/stem.3087

Volkenhoff A et al (2015) Glial glycolysis is essential for neuronal survival in drosophila. Cell Metab 22:437–447. https://doi.org/10.1016/j.cmet.2015.07.006

Wang J et al (2009) Dependence of mouse embryonic stem cells on threonine catabolism. Science (New York, N.Y.) 325:435–439. https://doi.org/10.1126/science.1173288

Wang Y et al (2015) WT1 recruits TET2 to regulate its target gene expression and suppress leukemia cell proliferation. Mol Cell 57:662–673. https://doi.org/10.1016/j.molcel.2014.12.023

Wang YH et al (2014b) Cell-state-specific metabolic dependency in hematopoiesis and leukemogenesis. Cell 158:1309–1323. https://doi.org/10.1016/j.cell.2014.07.048

Wang YP, Lei QY (2018) Metabolite sensing and signaling in cell metabolism. Signal Transduct Target Ther 3:30. https://doi.org/10.1038/s41392-018-0024-7

Wang YP, Li JT, Qu J, Yin M, Lei QY (2020) Metabolite sensing and signaling in cancer. J Biol Chem 295:11938–11946. https://doi.org/10.1074/jbc.REV119.007624

Wang YP et al (2014a) Regulation of G6PD acetylation by SIRT2 and KAT9 modulates NADPH homeostasis and cell survival during oxidative stress. EMBO J 33:1304–1320. https://doi.org/10.1002/embj.201387224

Wang YP et al (2016) Arginine methylation of MDH1 by CARM1 inhibits glutamine metabolism and suppresses pancreatic cancer. Mol Cell 64:673–687. https://doi.org/10.1016/j.molcel.2016.09.028

Wang YP et al (2021) Malic enzyme 2 connects the Krebs cycle intermediate fumarate to mitochondrial biogenesis. Cell Metab 33:1027–1041 e1028. https://doi.org/10.1016/j.cmet.2021.03.003

Wang ZH, Chen L, Li W, Chen L, Wang YP (2022) Mitochondria transfer and transplantation in human health and diseases. Mitochondrion 65:80–87. https://doi.org/10.1016/j.mito.2022.05.002

Warburg O (1956) On the origin of cancer cells. Science (New York, N.Y.) 123:309–314. https://doi.org/10.1126/science.123.3191.309

Weissman IL, Anderson DJ, Gage F (2001) Stem and progenitor cells: origins, phenotypes, lineage commitments, and transdifferentiations. Annu Rev Cell Dev Biol 17:387–403. https://doi.org/10.1146/annurev.cellbio.17.1.387

Weng AP et al (2004) Activating mutations of NOTCH1 in human T cell acute lymphoblastic leukemia. Science (New York, N.Y.) 306:269–271. https://doi.org/10.1126/science.1102160

West JD, Flockhart JH, Peters J, Ball ST (1990) Death of mouse embryos that lack a functional gene for glucose phosphate isomerase. Genet Res 56:223–236. https://doi.org/10.1017/s0016672300035321

Wilson A et al (2004) C-Myc controls the balance between hematopoietic stem cell self-renewal and differentiation. Genes Dev 18:2747–2763. https://doi.org/10.1101/gad.313104

Wilson A et al (2008) Hematopoietic stem cells reversibly switch from dormancy to self-renewal during homeostasis and repair. Cell 135:1118–1129. https://doi.org/10.1016/j.cell.2008.10.048
Wu J, Wong WW, Khosravi F, Minden MD, Penn LZ (2004) Blocking the Raf/MEK/ERK pathway sensitizes acute myelogenous leukemia cells to lovastatin-induced apoptosis. Cancer Res 64:6461–6468. https://doi.org/10.1158/0008-5472.Can-04-0866
Xia Z et al (2001) Blocking protein geranylgeranylation is essential for lovastatin-induced apoptosis of human acute myeloid leukemia cells. Leukemia 15:1398–1407. https://doi.org/10.1038/sj.leu.2402196
Xiao G et al (2018) B-cell-specific diversion of glucose carbon utilization reveals a unique vulnerability in B cell malignancies. Cell 173:470–484.e418. https://doi.org/10.1016/j.cell.2018.02.048
Xiao Z, Dai Z, Locasale JW (2019) Metabolic landscape of the tumor microenvironment at single cell resolution. Nat Commun 10:3763. https://doi.org/10.1038/s41467-019-11738-0
Xu Q, Simpson SE, Scialla TJ, Bagg A, Carroll M (2003) Survival of acute myeloid leukemia cells requires PI3 kinase activation. Blood 102:972–980. https://doi.org/10.1182/blood-2002-11-3429
Xu SN, Wang TS, Li X, Wang YP (2016) SIRT2 activates G6PD to enhance NADPH production and promote leukaemia cell proliferation. Sci Rep 6:32734. https://doi.org/10.1038/srep32734
Yamauchi T et al (2019) Paics, a De novo purine synthetic enzyme, is a novel target for AML therapy. Blood 134: 1390–1390. https://doi.org/10.1182/blood-2019-127047
Yang Z et al (2016) Restoring oxidant signaling suppresses proarthritogenic T cell effector functions in rheumatoid arthritis. Sci Transl Med 8:331ra338. https://doi.org/10.1126/scitranslmed.aad7151
Ye H et al (2016) Leukemic stem cells evade chemotherapy by metabolic adaptation to an adipose tissue niche. Cell Stem Cell 19:23–37. https://doi.org/10.1016/j.stem.2016.06.001
Ye H et al (2018) Subversion of systemic glucose metabolism as a mechanism to support the growth of leukemia cells. Cancer Cell 34:659–673.e656. https://doi.org/10.1016/j.ccell.2018.08.016
Yilmaz OH et al (2006) Pten dependence distinguishes haematopoietic stem cells from leukaemia-initiating cells. Nature 441:475–482. https://doi.org/10.1038/nature04703
Yu VWC et al (2017) Epigenetic memory underlies cell-autonomous heterogeneous behavior of hematopoietic stem cells. Cell 168:944–945. https://doi.org/10.1016/j.cell.2017.02.010
Yu WM et al (2013) Metabolic regulation by the mitochondrial phosphatase PTPMT1 is required for hematopoietic stem cell differentiation. Cell Stem Cell 12:62–74. https://doi.org/10.1016/j.stem.2012.11.022
Zaza G et al (2004) Acute lymphoblastic leukemia with TEL-AML1 fusion has lower expression of genes involved in purine metabolism and lower de novo purine synthesis. Blood 104:1435–1441. https://doi.org/10.1182/blood-2003-12-4306
Zhang J et al (2006) PTEN maintains haematopoietic stem cells and acts in lineage choice and leukaemia prevention. Nature 441:518–522. https://doi.org/10.1038/nature04747
Zhang J et al (2014) Asparagine plays a critical role in regulating cellular adaptation to glutamine depletion. Mol Cell 56:205–218. https://doi.org/10.1016/j.molcel.2014.08.018
Zhang Y et al (2016) CXCR4/CXCL12 axis counteracts hematopoietic stem cell exhaustion through selective protection against oxidative stress. Sci Rep 6:37827. https://doi.org/10.1038/srep37827

The Origin of Clonal Hematopoiesis and Its Implication in Human Diseases

5

Zhen Zhang and Jianlong Sun

Abstract

Clonal expansion of hematopoietic cells is first observed in hematological malignancies where all the leukemic cells can be traced back to a single cell carrying oncogenic alterations. Interestingly, expansion of hematopoietic clones with defined genomic alterations, including single nucleotide variants (SNVs), small insertions and deletions (indels), and large structural chromosomal alterations (CAs), is also found in the healthy population. These genomic changes often affect leukemia driver genes. As a result, healthy individuals bearing such clonal hematopoiesis (CH) are at a higher risk of hematological malignancies. In addition to blood cancers, SNV/indel-related CH has been found associated with elevated cardiovascular and all-cause mortality, indicating adverse impacts of abnormalities in the blood on the normal functions of non-hematological tissues. In the past decade, much effort has been invested in understanding the origins of CH and its causal relationship with diseases in hematological and non-hematological tissues. Here, we review recent progress in these areas and discuss future directions that can be pursued to translate the acquired knowledge into better management of CH-related diseases.

Keywords

Hematopoietic stem cells · Clonal hematopoiesis · Hematological malignancy · Inflammation · Chronic inflammatory diseases

5.1 Introduction

Hematopoiesis generates trillions of blood cells every day. Compared with the large cell numbers of the mature cells, the hematopoietic stem cells (HSCs), which are common origins of all the downstream cells, are exceedingly rare. This discrepancy is overcome by successive amplifications of HSC progeny along a cascade of differentiation steps, which ultimately generates clones of cells with a common origin from the initiating HSCs. Therefore, mature cells in the blood are inherently clonal (Kondo et al. 2003).

Assessment of clonal relationships among individual blood cells relies on the presence of common inheritable markers that should be unique to each HSC. For human subjects, this is achieved mainly by examining the naturally occurring SNVs/indels (Genovese et al. 2014; Jaiswal et al. 2014; Xie et al. 2014) and structural alterations (Jacobs et al. 2012; Laurie et al. 2012) on chromosomes, which are unique to each HSC

Z. Zhang · J. Sun (✉)
School of Life Science and Technology, ShanghaiTech University, Shanghai, China
e-mail: sunjl@shanghaitech.edu.cn

M. Zhao, P. Qian (eds.), *Hematopoietic Stem Cells*, Advances in Experimental Medicine and Biology 1442, https://doi.org/10.1007/978-981-99-7471-9_5

as it is exceedingly rare that two different cells will initially acquire the same mutations. With these cell-specific "barcodes," a positive call for the presence of CH will be made if multiple single blood cells are found to carry the same genetic markers. However, if hematopoiesis is supported by equal contributions of many small HSC clones, the unique markers of the individual HSCs will have a high chance of being present at a frequency lower than the detection limit. In such a scenario, the blood sample, although clonal by nature, may appear "non-clonal" in an experimental measurement.

While hematopoietic clones of limited size are challenging to detect, a large body of literature indicates that under aging (Fey et al. 1994) or leukemia (Fialkow et al. 1967) conditions, selected clones can expand to such an extent that even methods with limited sensitivity will be sufficient for their detection, leading to the observation of CH. In fact, the term "clonal hematopoiesis," widely used in the literature, refers to the presence of large hematopoietic clones detectable under certain technical constraints. Apparently, the call of CH for a blood sample depends heavily on the detection methods and the somewhat arbitrarily determined cell fraction cutoff for CH.

The phenomenon of CH has raised widespread interest since its discovery. Although first identified mainly in blood disorders, such as MPN, MDS, AML, and CLL, aberrantly expanded hematopoietic clones, with SNVs, indels, and CAs, are also found with a high frequency in the healthy population, especially in the elderly (Genovese et al. 2014; Jaiswal et al. 2014; Xie et al. 2014; Jacobs et al. 2012; Laurie et al. 2012). These aberrant clones are less likely caused by spontaneous expansion of wild-type HSC clones. Instead, genetic alterations affecting blood cancer genes are highly enriched in these clones, suggesting that acquired growth advantages similar to that of the corresponding neoplastic clones are the potential cause of their excess growth. On the other hand, most individuals, although having a higher risk of hematological malignancies, do not succumb to such diseases in their lifetime (Genovese et al. 2014; Jaiswal et al. 2014; Jacobs et al. 2012; Laurie et al. 2012), which raises the question of their physiological relevance. In this book chapter, we review the studies on the detection of CH, especially in populations without a history of blood cancers. We describe the characteristics of the genetic alterations associated with CH development and summarize our current understandings of the mechanisms that drive aberrant clonal growth. Finally, we discuss the clinical significance of CH and the potential translational application of knowledge obtained in this area of research.

5.2 Examination of CH by XCI

Hematologists have a long-standing interest in understanding the hematopoiesis process from a clonal perspective. Historically, locus-specific assays developed to determine the X chromosomal inactivation (XCI) patterns have been used to estimate the number of clones in the peripheral blood of females (Gale et al. 1991; Fialkow 1973). XCI occurs randomly in one of the two X chromosomes in early development, and the established inactivation pattern is stably inherited by the progeny of these embryonic cells. Due to the random nature of XCI, the frequency of cells expressing either the paternal or the maternal alleles will be similar. When many cell clones are involved in blood cell production, the two alleles will be expressed in peripheral blood at an equal frequency. However, if the blood contains a very large clone, its XCI pattern will dominate the blood cells, which will lead to a deviation from the 1:1 ratio between the two alleles.

A skewing from a balanced XCI pattern is first observed in a female CML patient who carries two alleles for the X-linked *G-6-PD* gene (Fialkow et al. 1967). In healthy individuals, the two *G-6-PD* gene alleles are expressed in mature blood cells with a similar frequency due to random inactivation of one of the alleles during XCI. However, in this CML patient, only a single type of isoform is detected in granulocytes and erythrocytes, which suggests that the blood is dominated by the progeny of a single multipotent progenitor cell.

Cancer development is known to have a latent phase before a cell carrying the initial cancer-causing mutation evolves into a full-blown disease. It is therefore not surprising that skewed XCI patterns, or CH, are also found in blood samples of healthy individuals (Fey et al. 1994), as they may have preleukemic clones still undergoing clonal evolution. However, what truly makes this observation unusual is the high prevalence of CH in this cohort. The frequency of CH far exceeds that of leukemia in the population, which means that the vast majority of these expanded clones do not eventually develop into blood cancers. A second intriguing finding of this earlier work is the discovery of age as the most prominent risk factor for CH. At the time, it was unclear whether this age-dependency is due to increased susceptibility of the elderly to CH or that the affected clones will take time to grow into a detectable size, hence more frequently detected in the elderly.

The three individuals with skewed XCI patterns were later found to contain mutations in the *TET2* gene in their neutrophils. Further validation indicated that *TET2* mutations were detected in 5.6% (10/182) of older women with XCI skewing but not in those without such a phenomenon (0/105) or young women (0/95) (Busque et al. 2012). This work is seminal in that it demonstrates for the first time that a leukemia driver mutation can cause CH without developing devastating leukemia. The obvious follow-up question is whether other leukemia driver mutations can also be found in the elderly with no blood cancers.

5.3 Detection of CH in the Era of Next-Generation Sequencing

The above example illustrated that a skewed XCI pattern is more likely a sign than a cause of CH. Therefore, the phenomenon of XCI skewing itself does not offer insights into the molecular mechanisms that cause excess clonal growth. In this regard, SNV/indels (Genovese et al. 2014; Jaiswal et al. 2014; Xie et al. 2014) and larger structural CA (Jacobs et al. 2012; Laurie et al. 2012) have become useful markers for clonal identification that can provide direct clues to the mechanisms driving CH.

The HSC genome accumulates genetic mutations, mostly SNV/indels, in both a cell cycle-dependent and a cell cycle-independent manner (Osorio et al. 2018). It is estimated that every HSC will carry approximately 40 mutations at the end of embryonic development (Osorio et al. 2018) and further accumulates mutations at an annual rate of 14.2 (Osorio et al. 2018) to 17 (Lee-Six et al. 2018) mutations across the genome and 0.13 mutations (Welch et al. 2012) in exome per year. Therefore, the individual HSCs are expected to have distinct combinations of mutations, which can serve as unique markers of their clonal identities. When these HSCs generate clones of cells through cell proliferation and differentiation, the progeny of the same clone will carry identical combinations of mutations as their ancestral HSCs. These mutations can be detected by whole-genome sequencing or whole-exome sequencing, and their abundance, correlated with the sizes of the clones, can be measured by variant allele frequency (VAF).

VAF estimated from a sequencing dataset refers to the percentage of sequencing reads carrying the SNV among all the reads from the same genomic locus. If we assume each read corresponds to an allele in a separate cell, the VAF will tell the percentage of cells carrying that SNV in the blood cell population under examination. Therefore, sufficient coverage of the specific genomic locus, as reflected by the total number of reads obtained for that locus, will be essential to detect the small clones carrying the corresponding SNV. The higher the coverage is, the smaller the detectable clones will be.

Due to inherent sequencing error rates and the financial constraint on the sequencing coverage one can achieve, a typical sequencing experiment will have a lower bound for the detectable VAF or clone size. Those clones that contain fewer cells than this lower bound will become undetectable. Therefore, the observed prevalence of CH in any population is heavily influenced by this technical constraint, and conclusions on the frequency of CH in different studies shall be interpreted and compared by considering the various VAF lower bounds in these studies.

5.4 CH Associated with Leukemia Driver Mutations

In 2014, three groups reported analyses of SNV and indels by whole-exome sequencing on blood samples from large-scale cohorts of individuals with no history of hematologic malignancies (Genovese et al. 2014; Jaiswal et al. 2014; Xie et al. 2014). With a lower bound of VAF at around 2%, all three studies commonly identified frequent mutations in leukemia driver genes, among which *DNMT3A*, *TET2*, and *ASXL1* are the most affected. Therefore, these mutations are associated with a clonal excess state and do not necessarily transform the affected clones into a full-blown malignancy. Consistent with age-related XCI skewing, the incidence of CH with leukemia driver mutations rises with age. While mutations of these genes are rarely detected at the age of 40 or younger, their frequencies increase with each decade after that and reach 10% of the individuals 70 years or older (Genovese et al. 2014; Jaiswal et al. 2014). After removing the influence of sex and the status of blood-unrelated diseases, age appears to be the most significant contributor to the risk of CH, reinforcing the concept that clonal expansion occurs with time.

It is perhaps not surprising that individuals carrying expanded clones with leukemia driver mutations are at a higher risk of developing hematologic cancers in the subsequent longitudinal follow-up when compared with individuals without such clones (Genovese et al. 2014; Jaiswal et al. 2014). Unexpectedly, individuals with CH are also associated with increased all-cause mortality and a higher risk of death from cardiovascular causes (Jaiswal et al. 2014). However, the overall risks of blood cancers or mortality are still very low for these affected individuals, which led the authors to propose the term "clonal hematopoiesis of indeterminate potential" or "CHIP" for describing this physiological phenomenon (Steensma et al. 2015). Correspondingly, screening the elderly for CH is considered premature as it is still unclear when and under what conditions an expanded clone will develop into cancer.

These initial works by WES likely underestimate the frequencies of CH in the healthy population due to limitations in sequencing depth and systemic errors introduced during PCR amplification and sequencing. Several aspects of the analysis protocols have been improved to overcome these technical hurdles. First, targeted sequencing of leukemia driver genes has been used to focus the sequencing power on selected regions of the genome so that coverage of these loci can be dramatically increased (McKerrell et al. 2015). Second, unique oligonucleotide indexes have been introduced into PCR primers to label the individual DNA molecules with unique molecular identifiers before PCR amplification (Young et al. 2016). This error-corrected sequencing significantly reduces errors caused by base substitution errors during PCR and sequencing. Together, these modifications reduce the lower bound of VAF below 0.1%, considerably increasing detection power for small hematopoietic clones. With these improvements, CH associated with leukemia driver genes is found in 19.5% of individuals older than 90 years of age (McKerrell et al. 2015), which is much more prevalent than previously thought. Moreover, CHIP is also found in 95% of middle-aged individuals (50–70 years old), and the average VAF of these mutations is tenfold lower than the detection limit in early work (Young et al. 2016). Although the study's sample size is relatively small, the finding suggests that clonal expansion is much more widespread in the population than previously estimated.

While most studies on CHIP examine cancer mutations associated with myeloid malignancies, mutations in lymphoid leukemia driver genes are also found in expanded peripheral blood clones of healthy individuals (Niroula et al. 2021). Although the frequency of lymphoid CHIP is lower than myeloid CHIP, its prevalence similarly increases with age. The lymphoid CHIP mutations are more evenly distributed across many genes. This contrasts with myeloid CHIP mutations, primarily enriched in the top three genes—*DNMT3A*, *TET2*, and *ASXL1*. Individuals with lymphoid CHIP have a higher risk of lymphoid but not myeloid malignancies. The reciprocal is also true for individuals with myeloid CHIP, who have an equal chance of developing lymphoid malignancies as people with no CHIP. This observation suggests that the promoting effects of leukemia driver mutations on clonal expansion and leukemic transformation both occur in a cell type-specific manner.

5.5 CH with Non-leukemia Driver Mutations

The above studies indicate a strong effect of leukemia driver mutations in aberrant clonal growth of hematopoietic cells. However, not all the expanded blood cell clones from individuals with CHIP have mutations in cancer driver genes. A whole-genome sequencing study of a cohort with no previous history of hematologic malignancies observed much broader spectrums of mutations at the expense of sequencing depth (Zink et al. 2017). While more than 50% of the subjects older than 85 have CH, 85% of the cases do not have mutations in known leukemia driver genes. One of the earlier WES works made a similar observation (Genovese et al. 2014), although clones with leukemia driver mutations in that study do have a higher average VAF value and are associated with a higher risk of blood cancers. Whether these non-leukemia driver genes affected by CH-associated mutations can regulate clonal expansion is currently unknown.

5.6 CH Associated with Mosaic Chromosomal Alterations

The expression and functions of driver genes in hematologic malignancies are disrupted by short variants and large structural chromosomal alterations. Although the short variants, such as somatic SNVs and indels, are readily detected in next-generation sequencing as discussed above, structural chromosomal alterations (CAs), including loss, gain, and copy number neutral loss of heterozygosity (CN-LOH), are much more challenging to quantify. While the reports on CA-related CH appeared as early as 2012, many more CAs were discovered when high-quality genomic data, large population cohorts, and sensitive methods for CA quantification became available in recent years (Loh et al. 2020; Terao et al. 2020).

The most notable CA in peripheral blood cells is the loss of Y chromosome (LOY) in a fraction of peripheral blood cells from males. This phenomenon, a typical example of mosaic chromosomal alteration (mCA), was first identified over five decades ago (Pierre and Hoagland 1972). Similar to CHIP, its frequency increases with age (Pierre and Hoagland 1972). In addition, smoking status is another major risk factor for LOY (Zhou et al. 2016). The association of LOY with human disease is complex. While it is reported that individuals with LOY in their blood have a risk of acute myeloid leukemia and other hematological diseases (Herens et al. 1992), other studies suggest that LOY in leukocytes may be a biomarker of genome instability in non-hematological tissues (Thompson et al. 2019) and hence is associated with a shorter life span and risk of non-hematological malignancies (Forsberg et al. 2014). This notion, however, is challenged by studies of independent populations with different ethnicities (Terao et al. 2019). This investigation fails to identify the association between LOY and risks for solid tumors or all-cause and cancer-related mortality, which underscores our still limited understanding of the physiological consequences of LOY.

In addition to LOY, large numbers of mCAs affecting autosomes are discovered from blood cell-derived DNA using small nucleotide polymorphism (SNP) genotyping arrays (Jacobs et al. 2012; Laurie et al. 2012). Publically available datasets, such as those generated by the UK Biobank study (Loh et al. 2018, 2020) and the BioBank Japan study (Terao et al. 2020) that provide comprehensive phenotypic and genotypic details of its participants, together with robust computational analysis pipelines, have facilitated mCA identification and the discovery of inherited genetic variants that promote their occurrence. These studies reveal that approximately 3.5–5% of the participants aged between 40 and 70 years carry mCA at a cell fraction as low as 0.75–1% (Terao et al. 2020). Similar to SNV-associated CHIP, the frequency of mCA increases significantly with age in the UK Biobank cohort and reaches almost inevitability (35%) for individuals over the age of 90 in the Japanese population (Terao et al. 2020). This age factor explains the much more frequent incidence of mCA in the Japanese cohort, which contains older participants than the UK group.

Around two-thirds of the mCAs can be classified with high confidence. These alterations include mosaic deletion, duplications, and CN-LOH in which either the reference (wild-type) allele or the variant allele is replaced by their homologous alleles. Most deletion events are interstitial, spanning neither telomere nor centromere (Jacobs et al. 2012). Duplication events often involve the entire chromosomes, which leads to complete trisomies (Jacobs et al. 2012; Loh et al. 2020). CN-LOH events most frequently begin at the telomeres and extend across a portion of the chromosomes (Jacobs et al. 2012). Pointing to a potential role of CA in promoting clonal expansion, CN-LOH events on selected chromosome regions (13q, 2p, 4q) span the same commonly deleted regions (CDRs) (Laurie et al. 2012; Loh et al. 2018). Moreover, the genetic loci of DNMT3A, TET2, ETV6, NF1, and CHEK2, which are frequently mutated in cancers, are also hotspots of focal deletions (Terao et al. 2020; Loh et al. 2018). These findings collectively suggest that haploinsufficiency at these genetic loci contributes to clonal expansion and predisposes affected individuals to a high risk of hematological malignancies. Consistent with the notion that CA results in clonal outgrowth by modulating the gene dosage of cancer-related genes, a more focused analysis of CAs found in both myeloid and lymphoid leukemia also reveals age-dependent CH among individuals with no prior history of blood cancers (Niroula et al. 2021), which directly proves the effect of cancer-related CA in promoting clonal outgrowth before triggering the leukemic transformation.

5.7 Co-occurrence of CHIP Mutations and mCAs in CH

Both SNV/indels and mCAs-related CHs are associated with the risk of hematological malignancies. This raises an interesting question of whether the two types of genomic alterations of CH are mutually exclusive or occur within the same leukocyte clones as frequently found in blood cancers. In this regard, joint analyses of the same subjects for SNV/small indels and mCA in their blood cells provide critical insights into the cooperation between these two types of genomic alterations in CH development (Loh et al. 2018; Saiki et al. 2021; Gao et al. 2021; Ljungström et al. 2022). First, individuals with LOY in their blood cells significantly enriched those carrying pathogenic variants. In-depth examination suggests both alterations occur within the same lineage, and larger clones with LOY are closely associated with variants of high VAFs (Ljungström et al. 2022). Second, CN-LOH or focal deletion, two common types of mCA, frequently co-localizes with mutations in cancer-related genes such as *TET2*, *JAK2*, and *DNMT3A*, leading to the biallelic alterations when the two types of alterations occur in the same cells (Gao et al. 2021). The cellular fractions of mCA and SNV at these gene loci are similar in a subset of individuals with CH. This strongly argues that these *cis* multi-hit events occur very early in aberrant clonal growth (Saiki et al. 2021). Third, *trans* interaction between mCA and SNV that affect separate gene and genomic locus is also observed in CH. Interestingly, these pairs of co-mutations, such as 4q24 CN-LOH/*SRSF2*, 7q CH-LOH/*ASXL1*, and 20q del/*U2AF1*, have been implicated in myeloid diseases. Similarly, co-mutation pairs trisomy 12/*NOTCH1*, *MYD88*, or *FBXW7* and 13q del/*ATM* are characteristic recurrent mutations in CLL (Gao et al. 2021). Last but not least, *TP53* is found in both *cis* and *trans* interactions between SNV/small indels and mCA. While some *TP53* clones frequently have LOH of 17p where *TP53* localizes, the others carry deletion at 5q, a chromosomal alteration that strongly interacts with *TP53* mutations in myelodysplastic syndromes (Saiki et al. 2021; Gao et al. 2021). In addition, *TP53* clones often are associated with many other chromosomal alterations, which highlight the critical role of TP53 in safeguarding the integrity of human genomes (Gao et al. 2021).

Only a single SNV or CA can be detected in most individuals with CH. However, some do carry multiple somatic variations of the same or different types, and their co-occurrence within the same clones can be verified by single-cell analysis in a few cases (Saiki et al. 2021). These co-occurrence events, either between different kinds of variations or within the same types, are detected at a frequency higher than expected by chance, indicating a potential synergistic effect among these genetic alterations. Moreover, there is a strong correlation between clone size and the number of genetic alterations, further validating the selective advantages of these aberrant clones.

5.8 Inherited Risk Factors for SNV/Indel- and mCA-Associated CH

The development of CH appears inevitable if one lives sufficiently long. However, population studies indicate that the onset of CH and the extent to which hematopoietic clones expand vary dramatically among individuals. Is this heterogeneity a natural outcome of clonal evolution? Alternatively, are certain groups of individuals predisposed to CH development? Answers to these questions will likely help identify the susceptible population for close monitoring of CH onset.

One of the best-known examples of genetic predisposition to somatically acquired mutations comes from the quest for inherited variants associated with the acquisition of *JAK2*^{V617F}, the most frequently observed gain-of-function mutation in MPN. Interestingly, such variants turn out to be on the same locus of *JAK2*, where a small number of haplotypes, defined by a combination of SNP genotypes spanning *JAK2*, have a much higher chance of acquiring the *JAK2*^{V617F} mutations than the rest of *JAK2* haplotypes (Jones et al. 2009; Olcaydu et al. 2009; Kilpivaara et al. 2009). The precise genomic locations of the corresponding variants in these highly susceptible haplotypes and the underlying mechanisms for their predisposition to *JAK2*^{V617F} are yet to be defined.

Unlike the *cis*-association between *JAK2*^{V617F} mutation and specific *JAK2* haplotypes, many inherited variants on autosomes have been found to associate with mLOY (Thompson et al. 2019; Terao et al. 2019; Wright et al. 2017). These variants, often affecting expression or protein functions, are enriched in genes involved in the regulation of cell cycle checkpoint, mitosis, DNA damage response, and apoptosis, which suggests that insufficient maintenance of genome integrity and escaping of the apoptosis pathway are major causes of LOY (Thompson et al. 2019; Terao et al. 2019; Wright et al. 2017). Moreover, these variants overlap significantly with susceptible loci associated with non-hematological cancers and predispose women to the risk of breast cancer (Thompson et al. 2019). Therefore, LOY-associated CH may be a biomarker observed in blood cells for an overall increase in genomic instability.

The patterns of genetic predisposition to *JAK2*^{V617F} and LOY are observed for inherited variants associated with autosomal mCA (Loh et al. 2018, 2020; Terao et al. 2020). For instance, variants at loci of multiple genes, such as those involved in HSC homeostasis control (*MPL*, *SH2B3*) and DNA double-strand break repair (*ATM*, *TM2D3*, *NBN*, and *MRE11*), are associated with mCA occurring in nearby regions (Loh et al. 2020; Terao et al. 2020). The most frequent type of mCA predisposed by these variants is CN-LOH. The result of these CN-LOH events often leads to homozygosity by retaining the variant or reference alleles that give a growth advantage to the cells in the individuals carrying such variants (Loh et al. 2020). On the other hand, variants at the loci of *TERT*, *CHEK2*, *TP53*, and *MAD1L1* are associated with mCAs on other chromosomes (Terao et al. 2020; Loh et al. 2018). While the intronic variants in *TERT* and the frameshift variant in DNA damage response gene *CHEK2* predispose the subjects to multiple types of CAs (Loh et al. 2018), the *3-UTR* variants in *TP53* (Loh et al. 2018) and the missense variant in spindle assembly checkpoint gene *MAD1L1* (Terao et al. 2020) are associated with only chromosome losses and gains, respectively. The mechanistic basis of these disparities is currently unknown.

By performing a single-variant genome-wide association analysis, the germline variants associated with SNV/indel found in CHIP were identified. (Bick et al. 2020). It appears that three types of germline variants are linked with an increased propensity to develop mutations conferring growth advantage. The first set of variants are near the *TET2* gene and most likely suppress its expression. These inherited variants are associated with the most frequent somatic mutations in *DNMT3A*, *ASXL1*, and *TET2* in CHIP-carrying individuals of African ancestry, indicating a widespread effect of these germline *TET2* variants on the acquisition of somatic SNV/small indels. Moreover, *TET2* has been implicated in the regulation of HSC self-renewal. Thus, an accelerated proliferation of HSCs with inherited *TET2* mutations may increase the chance of accumulating other somatic mutations that can reinforce the growth advantage. The second set of variants is associated with acquiring somatic CHIP mutations in specific genes. For instance, inherited variants at the promoter region of *TCL1A* are specifically associated with an increased risk of *DNMT3A* mutations, and the *JAK2* 46/1 haplotype is associated with the aforementioned $JAK2^{V617F}$ mutations. The third set of variants occurs in genes (*TERT* and *CHEK2*) that generally affect the maintenance of genome integrity (Bick et al. 2020). Variants in these genes are associated with a broad spectrum of genomic alterations, such as LOY, mCA, and SNV/small indel, which strongly suggests the intimate connection between the maintenance of genome integrity and the development of CH.

Genome-wide association analysis on large cohorts demonstrates the role of genetic predisposition in acquiring somatic alterations that drive the clonal growth of hematopoietic cells. These inherited risk factors are mainly genetic variants affecting the expression or functions of genes involved in maintaining genome integrity and stem cell homeostasis. However, future studies will be required to test whether these variants can help identify individuals at a high risk of CHIP development.

5.9 Impact of CH-Related SNVs and mCAs on Clonal Outgrowth

Many SNVs/indels and mCAs found in CH affect driver genes of blood cancers. Therefore, investigating the molecular mechanisms by which the CH-related alterations cause aberrant clonal growth helps understand the initial biological events in the leukemic transformation process. In this regard, much effort has been made in the past few years to elucidate the functions of the most frequent CHIP mutations in genes such as *DNMT3A*, *TET2*, *ASXL1*, and those involved in mRNA splicing. These studies reveal deregulation in HSPC fate choices as a common mechanism through which CHIP mutations promote clonal outgrowth.

5.10 Epigenetic Regulators

DNMT3A and *TET2* are the two most frequently mutated genes in CHIP (Bick et al. 2020). DNMT3A is a DNA methyltransferase responsible for de novo methylation (Okano et al. 1999), whereas TET2 catalyzes the first step of DNA demethylation (Koh et al. 2011). Although these two proteins have opposite functions in DNA methylation, their depletion in mice stimulates HSC functions in bone marrow transplantation (Li et al. 2011; Moran-Crusio et al. 2011; Challen et al. 2012). DNMT3A-mediated DNA methylation appears to suppress the expression of HSC self-renewal genes so that its knockout in mice led to enhanced self-renewal in HSCs. Interestingly, *Dnmt3a*-KO results in promiscuous expression of self-renewal genes in differentiated cells, a phenomenon that may stimulate self-renewal expansion of lineage-restricted progenitors (Challen et al. 2012).

In comparison with DNMT3A, TET2 promotes HSC self-renewal via different mechanisms. It has been shown that TET2-mediated demethylation sustains the expression of tumor suppressor genes (Rasmussen et al. 2015) and is required for appropriate transcriptional activation of genes involved in erythroid differentiation (Madzo et al. 2014). Moreover, TET2 may play a role in suppressing the expression of repetitive elements such as LINE, SINE, and LTR, whose expression is high in pluripotent stem cells and cancer cells (Lopez-Moyado et al. 2019). Therefore, loss of function mutations in TET2 may interfere with normal cell differentiation and tumor suppression, leading to deregulated clonal growth of hematopoietic cells. Counterintuitively, Tet2-KO in murine HSCs upregulates transcription of key erythroid differentiation regulators Klf1 and Epor and regulators of the myeloid and B cell lineages (Zhang et al. 2016). The expression of these factors does not drive erythroid differentiation as one would have expected (Zhang et al. 2016). Instead, they activate the JAK-STAT signaling pathway, essential to HSC self-renewal expansion (Zhang et al. 2016). This observation indicates that loss of TET2 can stimulate HSC clonal outgrowth by modulating the expression of specific self-renewal regulators.

ASXL1 is a putative scaffold protein that interacts with multiple histone-modifying enzymes (Bick et al. 2020). Most *ASXL1* mutations generate truncation in its C-terminus and result in a global reduction in the levels of H3K27me3 (Abdel-Wahab et al. 2012; Inoue et al. 2013), H2AK119Ub (Asada et al. 2018; Balasubramani et al. 2015), and H3K4me3 (Nagase et al. 2018; Inoue et al. 2018). Alterations of these histone modifications interfere with the expression of *Sox6*, *Id3*, *Tjp*, *Hba*, *p16Ink4a*, and the Hoxa family genes and impair myeloid differentiation (Nagase et al. 2018; Uni et al. 2019), potentially causing clonal outgrowth. However, unlike the *Tet2*-KO or *Dnmt3a*-KO HSCs, murine HSCs carrying *Asxl1* mutations mimicking those found in CH individuals display engraftment deficiency when transplanted in lethally irradiated recipient mice (Inoue et al. 2018; Hsu et al. 2017; Abdel-Wahab et al. 2013), indicating that Asxl1-mutant HSCs use alternative mechanisms for clonal expansion during aging. In addition to epigenetic modulation, mutated ASXL1 has been found to deubiquitinate and stabilize AKT in cooperation with BAP1. The enhanced AKT/mTOR signaling activates cell proliferation, mitochondria activities, ROS production, and DNA damages, which collectively result in deregulated HSC proliferation (Fujino et al. 2021). Moreover, truncation of the carboxyl terminus of ASXL1 loses the motifs mediating phase separation, disrupts paraspeckle formation, alters subcellular localization of NONO, and interferes with RNA splicing and HSC stem cell functions (Yamamoto et al. 2021). Therefore, epigenetic-dependent and epigenetic-independent mechanisms can potentially drive aberrant HSC clonal growth with *ASXL1* mutations.

5.11 Splicing Factors

The second-largest family of genes frequently mutated in CHIP is the splicing factors (Genovese et al. 2014; Jaiswal et al. 2014; Xie et al. 2014), among which *SF3B1*, *SRSF2*, and *U2AF1* are the most commonly mutated. CHIP-related mutations in these factors often cause errors in splicing. The resulted mis-spliced transcripts then trigger nonsense-mediated decay (NMD) that further reduces the expression of the affected genes. A representative example of this mechanism is from mutations in *SF3B1* that promote mis-splicing and downregulation of *MAP3K7* transcript through NMD. Reduced *MAP3K7* expression further activates the NFkB signaling (Lee et al. 2018), which drives myeloid-biased hematopoiesis. Deregulation in splicing reactions also generates isoforms with alternative functions. For instance, *SRSF2* and *U2AF1* mutations induce the generation of aberrant isoforms of *Caspase 8* and *IRAK4*, both causing hyperactivation of the NFkB signaling and myeloid-biased hematopoiesis (Lee et al. 2018;

Smith et al. 2019). In addition to the skewing of lineage choice, splicing factor mutations can alter the expression and function of leukemia gene products such as *GNAS* (Cara et al. 2015), *PICALM* (Cara et al. 2015), and *EZH2* (Kim et al. 2015), which results in aberrant clonal growth and CHIP development.

5.12 DNA Damage Response Factors

PPM1D and *TP53* are the two most recurrently mutated genes in CHIP that interfere with normal functions of DNA damage responses (Genovese et al. 2014; Xie et al. 2014; Bick et al. 2020). In response to DNA damage, the p53 protein is activated to arrest the cell cycle progression for DNA repair, or to induce cell apoptosis should the damage be too extensive to repair. PPM1D is part of a feedback loop that limits p53 activity. Activated p53 induced the expression of PPM1D, which dephosphorylates p53 and related proteins in DNA damage response through its phosphatase activity. Given the critical roles of these two proteins in DNA damage responses, it is not entirely unexpected to observe an enrichment of mutations in these two genes in chemotherapy-induced CHIP (Hsu et al. 2018; Kahn et al. 2018; Marusyk et al. 2010). In contrast to the apparent effect of TP53 and PPM1D mutations in clonal outgrowth after DNA damage stress, it is still unclear how these mutations would confer a growth advantage to hematopoietic cells under steady-state conditions. Mutant p53 has been found to interact with EZH2 and enhance its chromatin association, thereby increasing the levels of H3K27me3 on genes regulating HSC self-renewal and differentiation (Chen et al. 2019). Notably, *PPM1D* mutation does not increase the engraftment capacities of HSCs in transplantation (Hsu et al. 2018; Wong et al. 2018), indicating alternative mechanisms underlying CHIP development driven by *PPM1D* mutations.

5.13 CHIP Mutations in Non-cancer Driver Genes

As described in previous sections, many of the CHIP mutations are not found in leukemia driver genes (Genovese et al. 2014; Zink et al. 2017). How the affected clones carrying these non-driver mutations gain a growth advantage has remained elusive. It has been suggested that HSCs randomly choose their cell fates between self-renewal, differentiation, and death (Abkowitz et al. 1996). Hence, some of the HSCs will be depleted from the HSC pool through symmetric differentiation or cell death, which will inevitably result in a loss of clonal complexity. As a result of this neutral drift process, it is conceivable that the number of clones involved in ongoing hematopoiesis will drop over time (Zink et al. 2017), and some of them will become so large that they will be detected in studies of CH through the detection of neutral mutations they carry. These mutations may not necessarily be regulating HSC clonal growth. Instead, they may have been detected as CH-related mutations merely because they happen to exist in the expanded clones that are naturally evolved.

That said, mathematical modeling has been used to estimate how likely it is for such a neutral drift process to create detectable clones under the current detection limitations with an inferred 25,000 to 1.3 million of HSCs involved in ongoing hematopoiesis (Watson et al. 2020; Ashcroft et al. 2017). It turns out that clone size does increase over time. However, the speed of this increase is so slow that it would take >2000 years to generate the detectable clones should neutral drift be the only driving force for clonal outgrowth (Watson et al. 2020). Therefore, although these non-driver mutations may never be involved in the leukemogenesis process, they can still affect the cell fate regulation mechanisms in a way that would result in clonal outgrowth, as found in people with CH.

5.14 mCA

The mechanisms of clonal dominance driven by mCAs are less studied than CHIP mutations. CH-associated focal deletions frequently target regions of leukemia driver genes such as *DNMT3A* and *TET2*, which are also frequently mutated in CHIPs (Laurie et al. 2012; Terao et al. 2020; Loh et al. 2018). CN-LOH, on the other hand, often results in homozygosity by retaining either the inherited variants or the reference wild-type alleles, depending on which allele will confer the growth advantage to the affected cell (Loh et al. 2020). Both DNA damage response genes (*MRE11*, *MBN*, *ATM*) and HSC self-renewal regulators (*MPL*, *SH2B3*) are affected by CN-LOH in CH (Loh et al. 2020), suggesting that escaping from the quality control or homeostasis maintenance mechanisms drives clonal outgrowth. Mosaic LOY enhances the reconstitution capacity of HSCs and gives rise to CH in the mouse model by losing the chromosome Y-linked tumor suppressor gene *Kdm5d* (Zhang et al. 2022a). KDM5D is an H3K4 demethylase, and its loss of function leads to increased DNA damage and elevated expression of MYC target genes (Zhang et al. 2022a). Outside of the Y chromosome, LOY also regulates the expression of autosomal genes involved in inflammation, immune response (Dumanski et al. 2021), and HSPC regulation (GABRR1 (Zhu et al. 2019), PRLR (Abdelbaset-Ismail et al. 2016), and LEP (Bennett et al. 1996)). However, whether these gene expression changes occur in mLOY HSPCs and how these changes contribute to clonal dominance remain to be validated.

5.15 Cellular Origins of Expanded Clones in CHIP

It is generally believed that aberrant clonal growth observed in the blood reflects a deregulated clonal growth in HSCs. Moreover, the CHIP mutations, such as those in *DNMT3A* and *TET2*, do promote HSC self-renewal, as shown by investigations in model organisms (Li et al. 2011; Moran-Crusio et al. 2011; Challen et al. 2012). However, whether the clonal expansion in downstream mature cells is always associated with concurrent HSC expansion has not been rigorously tested. The presence of CHIP mutations in multiple blood cell lineages (Young et al. 2016; Arends et al. 2018; Buscarlet et al. 2018), the recurrent detection of the same CHIP mutations in short-lived myeloid cells collected years apart, and the identification of the same driver mutations in both HSCs and mature blood cells all suggest that the self-renewing multipotent cells are the clonal origins of peripheral blood clonal expansion (Young et al. 2016). On the other hand, the observation of high concordance between mature myeloid cells and progenitors, but not HSCs, in their VAFs of CHIP mutations argues that the growth advantage conferred by the CHIP mutations may occur at the progenitor stage (Arends et al. 2018). In support of this notion, *Tet2* or *Dnmt3a* mutated progenitors have been found to possess stem cell properties (Kunimoto et al. 2014; Ostrander et al. 2020). The inherited genetic variants associated with mLOY affect the genes mainly expressed in multipotent progenitor cells (MPPs) and HSCs, suggesting that the initial LOY events can happen in both MPPs and HSCs (Thompson et al. 2019; Terao et al. 2019). Last but not least, clonal analysis and lineage tracing studies in unperturbed hematopoiesis indicate extensive self-renewal abilities of MPPs that are devoid of stem cell properties in a transplantation setting (Sun et al. 2014; Busch et al. 2015; Pei et al. 2017). These observations suggest that CH-associated mutations may affect multiple stages of the hematopoietic differentiation hierarchy. The expression patterns and the mutations' impact on protein functions may ultimately determine the exact cellular origins of aberrant clones in peripheral blood.

5.16 Impact of Extrinsic Factors on Clonal Outgrowth

While CHIP-related mutations and mCAs confer a growth advantage to the affected cells, the dynamic environment in which these cells reside may impose additional pressure that helps select the fittest clones. Among all the extrinsic factors, pro-inflammatory cytokines, the critical mediators of inflammaging, have been suggested to impair HSC self-renewal and survival (Pietras et al. 2016; Yamashita and Passegue 2019; Matatall et al. 2016). However, depletion of genes affected by genetic alterations found in CH appears to help HSCs better survive such an adverse selection, potentially contributing to the increased frequency of CH in the elderly. For instance, *Tet2* depletion in HSCs allows better survival and self-renewal in the presence of IL-6 (Cai et al. 2018; Meisel et al. 2018) and TNFα (Abegunde et al. 2018), which are cytokines known to promote cell differentiation and apoptosis in wild-type HSCs. Similarly, *Dnmt3a*-mutant murine HSCs are resistant to IFNγ (Hormaechea-Agulla et al. 2021) or TNFα (Liao et al. 2022)-induced cell differentiation and cell death, and deletion of *Asxl1* in zebra fish HSCs increases their expression of anti-inflammation genes, which in turn boosts their survival and clonal dominance in an inflammatory environment (Avagyan et al. 2021). These observations collectively highlight the critical role of inflammation in selecting clones carrying CHIP-related mutations.

5.17 Association of CH with Hematological Malignancies

A large body of literature indicates that individuals with SNV/small indel or mCA-related CHIP have a high risk of hematological malignancies (Genovese et al. 2014; Jaiswal et al. 2014; Laurie et al. 2012). However, the absolute risk of succumbing to such diseases remains low (Jaiswal et al. 2014; Laurie et al. 2012), which raises the question of what additional factors contribute to a malignant transformation in these aberrant clones.

In this regard, association studies find that the number of driver mutations concurrently present in an aberrant clone positively correlates with the risk of hematologic malignancies. Among the many CHIP mutations, those in *TP53* by themselves are sufficient for predicting a high risk of immediate AML. In contrast, mutations in *DNMT3A* have to be combined with concurrent mutations in splicing factor genes to serve as a significant risk factor for the early-onset AML (Desai et al. 2018). Similarly, mutations in splicing factors alone can predict the risk of MDS, but the variants in *TET2*, *DNMT3A*, and *ASXL1* only confer a high risk when combined with other mutations (Malcovati et al. 2017). Not only does the number of mutations distinguish different risk groups, but the nature of the mutations and affected genes would affect their association with disease risk. For instance, while mutations in *TP53*, splicing factors, *IDH1*, and *IDH2* are much rarer in CHIP than those in *DNMT3A* and *TET2*, they are often associated with a higher risk of AML (Desai et al. 2018; Abelson et al. 2018). These differences in the prediction power likely reflect the different functions of these genes in developing myeloid malignancies and the various combinations of mutations required for malignant transformation.

Compared to the number and type of mutations, the size of the aberrant clones and the overall blood cell counts are more robust predictors of future malignancies. For instance, a comparison between preleukemic patients and healthy individuals with CHIP identifies the size of mutated clones and subtle differences in normal blood count (especially a higher red blood cell distribution width, RDW) as the major factors that can identify those subjects with a high risk of progression to AML (Abelson et al. 2018). Additionally, individuals carrying mCA-related CH have a substantially increased risk of malignant leukemia if they display abnormal blood cell count parameters (Niroula et al. 2021) or the mCAs have high fitness effects (Watson and Blundell 2022). Therefore, the size and the impact of aberrant clones on normal hematopoiesis, as reflected by abnormalities in blood cell counts, can better reveal the status of the mutant clones in their progression to full-blown malignancies.

5.18 Functions of CH-Related Mutations in Immune Cells

While many of the CH-related genetic variations have a limited effect on leukemogenesis in HSCs, the downstream cells inheriting these mutations often display abnormal cell functions, among which the deregulation of the inflammation response pathway is one of the most prominent outcomes. The best-understood example of such an effect is the role of TET2 in regulating cytokine production in innate immune cells. It has been shown that *Tet2* deletion in macrophages augments the production of multiple inflammatory cytokines, including IL-6 (Zhang et al. 2015; Fuster et al. 2017; Jaiswal et al. 2017), IL-1β (Fuster et al. 2017; Jaiswal et al. 2017), and TNFα (Cull et al. 2017), in response to LPS stimulation. A similar effect is observed for *DNMT3A* (Sano et al. 2018a; Leoni et al. 2017), *ASXL1* (Avagyan et al. 2021), *PPM1D* (Yura et al. 2021), *JAK2* (Kleppe et al. 2015), and genes involved in spliceosomes (Pollyea et al. 2019) in different innate immune cell types in mice and zebra fishes. Together, these observations suggest that CH mutations alter the inflammation milieu, which may subsequently impose additional pressure on the selection of the corresponding aberrant clones themselves.

Although CHIP-related mutations are more frequently observed in myeloid cells, some of them do regulate immune responses in lymphoid cells of the adaptive immune system. For instance, loss of *Dnmt3a* or *Tet2* promotes the development of $CD8^+$ T memory cells and increases the expression of IFN-γ in $CD8^+$ T effector cells (Carty et al. 2018; Ladle et al. 2016). However, a similar perturbation of the two genes shows opposite effects on IFN-γ production in $CD4^+$ T helper cells (Gamper et al. 2009; Thomas et al. 2012; Ichiyama et al. 2015), highlighting the complexity in the functions of these epigenetic regulators.

Compared with CHIP-related mutations, the effect of mCA on the immune response is less clear. It is known that germline copy number variation of chromosome Y-linked genes, such as *Sly* and *Rbmy*, modulates gene expression in $CD4^+$ T cells and macrophages and shows an inverse correlation with IL-6 and IFN-γ production in activated $CD4^+$ T cells (Case et al. 2013). Likewise, LOY in immune cells interferes with the expression of genes with critical roles in inflammatory responses and immune functions (Dumanski et al. 2021). Autosomal mCA has also been found to increase the risk of infection (Zekavat et al. 2021), although a mechanistic understanding of such a correlation is still lacking.

5.19 Association of CH with Nonmalignant Diseases

Given the impact of CH mutations on the inflammatory response in immune cells, it is not surprising that individuals with CHIP have a high risk of cardiovascular diseases (Jaiswal et al. 2014), in which the dysregulated immune system plays an essential role in disease onset. A causal effect of CH mutations on cardiovascular diseases is established when the transplantation of *Tet2*-deficient HSPCs accelerates symptoms of atherosclerosis in recipient mice fed with a high-fat diet. As expected, this effect is mediated by elevated production of IL-6 and IL-1β in *Tet2*-deficient macrophages (Fuster et al. 2017; Jaiswal et al. 2017). In line with their effects on cytokine production, mutations in genes such as *Dnmt3a* (Sano et al. 2018a), *JAK2* (Sano et al. 2019), and *Ppm1d* (Yura et al. 2021) are all causal risk factors of coronary heart disease, and the *JAK2*^{V617F} mutation, in particular, increases the risk by more than 12-fold (Jaiswal et al. 2017), which is consistent with its central role in the control of inflammation responses.

Unlike CHIP mutations, autosomal mCAs are not associated with cardiovascular disease risk, except for *JAK2* CN-LOH (Loh et al. 2020). Although mLOY has been shown to increase the incidence of cardiovascular disease (Loftfield et al. 2018; Haitjema et al. 2017), it is challenging to ascertain the role of mLOY in disease progression as it is often associated with SNVs/small indels found in CHIP (Ljungström et al. 2022). A clear understanding of this association hence requires further investigations.

Besides cardiovascular diseases, CHIP mutations are associated with many other age-related illnesses. For instance, CHIP is a causal risk factor for the development and progression of chronic obstructive pulmonary disease (Zink et al. 2017; Miller et al. 2022), and studies in mouse models demonstrate an enhanced immune response as the underlying cause of smoking-induced emphysema in *Tet2*-knockout animals (Miller et al. 2022). *Tet2* knockout also aggravates insulin resistance and hyperglycemia in obese or aged mice. Mechanistically, *Tet2* deletion stimulates IL-1β production in macrophages, which suppresses the adipocytic expression of *Irs1*, a pivotal mediator of insulin signaling (Fuster et al. 2020). These observations establish *Tet2* mutation as a causal risk factor for type 2 diabetes.

5.20 Perspective

Our genome is under constant assault from intrinsic and extrinsic hazards over the entire life span. The resulting genomic errors create an extensive repertoire of somatic clones distinct in their genetic composition or gene expression patterns. Under selection pressures such as those encountered in a chronic inflammatory environment, the clones better surviving the selection will grow out, leading to the appearance of dominant clones in the tissues. Such an evolution process is best manifested in the hematopoietic system, in which CH becomes so prevalent over time that it almost reaches the inevitability for people in their advanced ages. However, the acquisition of CH is not a sentence for blood cancers or non-hematological diseases. In fact, many of the carriers remain healthy, and only a minority of the affected individuals eventually progress to malignant or other disease stages. Hence, we suggest that future studies of CH should help determine the group of mutations, their assembly, the inherited risk factors, and the environmental cues that, in combination, can predict future malignancies, cardiovascular disease, and other related illnesses with high confidence. In this regard, large-scale association studies suggest that specific mutations, clone size, and associated abnormalities in blood cell counts more accurately identify individuals with a high risk of blood cancers. Similar work on risk stratifications and related factors for diseases affecting the non-hematological tissues is still lacking and shall be the focus of future association studies. On the other hand, mechanistic studies on the malignant transformation of preleukemic diseases and the development of non-hematological diseases in individuals with CH have generated important insights into the cooperating events required for disease progression (Velten et al. 2021; Williams et al. 2022). These understandings of the disease etiology will provide additional clues to define better the factors that could be examined for risk assessment.

Given the causal relationships between CHIP and the risk of hematological malignancies and chronic inflammatory diseases, early intervention in the evolution of aberrant clones may reduce the likelihood of disease onset. Based on insights from the mechanistic studies, strategies that directly dampen the growth-promoting effects of specific mutations or inflammatory milieu have demonstrated potential impact in disease prevention.

For several high-risk CHIP mutations, such as those in *TP53* (Zhang et al. 2022b; Hsiue et al. 2021), *IDH* (Liu and Gong 2019), and splicing factors (Seiler et al. 2018), targeted therapies have been developed to inhibit the growth advantage of affected cells specifically. Likewise, vitamin C can restore TET2 deficiency and suppress leukemogenesis (Agathocleous et al. 2017; Cimmino et al. 2017). Rapamycin can inhibit the Akt/mTOR signaling induced by mutant ASXL1 and ameliorate the aberrant expansion of mutant HSCs (Fujino et al. 2021). While many of these targeted therapies are in clinical trials, the timing of initiating such treatment for individuals with CH is still challenging to determine. Although a delay in treatment may miss the window for the best therapeutic effects of these agents, an earlier than necessary intervention may create unwanted selection pressures that can result in the outgrowth of additional clones with unknown risks for future malignancies. Therefore, the benefits of CH treatments require further examination and shall be balanced with undesirable consequences they may accrue.

Since inflammation is both an extrinsic selective pressure for CH development and a central mediator linking CH and non-hematological diseases, prevention of inflammation is likely a more practical approach to alleviate the adverse impacts of CH. It has been demonstrated in mouse models that inhibition of inflammation hinders the expansion of mutant clones (Meisel et al. 2018) and the progression of chronic inflammatory diseases, including cardiovascular diseases (Forsberg et al. 2014; Sano et al. 2018b) and diabetes (Fuster et al. 2020). In addition, a restraint on the growth of mutant clones may decrease the probability of acquiring secondary mutations, reducing the chance of malignant transformation. Although inflammation suppression seems beneficial from the perspective of CH inhibition, an overall reduction in immune functions may dampen tumor surveillance and increase the risk of infection. Hence, novel approaches that target CH with high specificity need to be developed to address such a problem.

The past decade witnessed a dramatic increase in our understanding of the cellular and molecular mechanisms that drive aberrant clonal growth in the hematopoietic system. The revealing of its association with blood cancers and non-hematological diseases has stimulated widespread interest, as the early detection of this clonal outgrowth phenomenon may offer rare opportunities to prevent the onset of many diseases that still lack a cure. Research in the next decade will test if this promise can be realized and if the management of related diseases can be improved by reacting to the alarming signals from our blood cell clones.

References

Abdelbaset-Ismail A et al (2016) Human haematopoietic stem/progenitor cells express several functional sex hormone receptors. J Cell Mol Med 20:134–146. https://doi.org/10.1111/jcmm.12712

Abdel-Wahab O et al (2012) ASXL1 mutations promote myeloid transformation through loss of PRC2-mediated gene repression. Cancer Cell 22:180–193. https://doi.org/10.1016/j.ccr.2012.06.032

Abdel-Wahab O et al (2013) Deletion of Asxl1 results in myelodysplasia and severe developmental defects in vivo. J Exp Med 210:2641–2659. https://doi.org/10.1084/jem.20131141

Abegunde SO, Buckstein R, Wells RA, Rauh MJ (2018) An inflammatory environment containing TNFalpha favors Tet2-mutant clonal hematopoiesis. Exp Hematol 59:60–65. https://doi.org/10.1016/j.exphem.2017.11.002

Abelson S et al (2018) Prediction of acute myeloid leukaemia risk in healthy individuals. Nature 559:400–404. https://doi.org/10.1038/s41586-018-0317-6

Abkowitz JL, Catlin SN, Guttorp P (1996) Evidence that hematopoiesis may be a stochastic process in vivo. Nat Med 2:190–197. https://doi.org/10.1038/nm0296-190

Agathocleous M et al (2017) Ascorbate regulates haematopoietic stem cell function and leukaemogenesis. Nature 549:476–481. https://doi.org/10.1038/nature23876

Arends CM et al (2018) Hematopoietic lineage distribution and evolutionary dynamics of clonal hematopoiesis. Leukemia 32:1908–1919. https://doi.org/10.1038/s41375-018-0047-7

Asada S et al (2018) Mutant ASXL1 cooperates with BAP1 to promote myeloid leukaemogenesis. Nat Commun 9:2733. https://doi.org/10.1038/s41467-018-05085-9

Ashcroft P, Manz MG, Bonhoeffer S (2017) Clonal dominance and transplantation dynamics in hematopoietic stem cell compartments. PLoS Comput Biol 13:e1005803. https://doi.org/10.1371/journal.pcbi.1005803

Avagyan S et al (2021) Resistance to inflammation underlies enhanced fitness in clonal hematopoiesis. Science 374:768–772. https://doi.org/10.1126/science.aba9304

Balasubramani A et al (2015) Cancer-associated ASXL1 mutations may act as gain-of-function mutations of the ASXL1–BAP1 complex. Nat Commun 6:7307. https://doi.org/10.1038/ncomms8307

Bennett BD et al (1996) A role for leptin and its cognate receptor in hematopoiesis. Curr Biol 6:1170–1180. https://doi.org/10.1016/s0960-9822(02)70684-2

Bick AG et al (2020) Inherited causes of clonal haematopoiesis in 97,691 whole genomes. Nature 586:763–768. https://doi.org/10.1038/s41586-020-2819-2

Buscarlet M et al (2018) Lineage restriction analyses in CHIP indicate myeloid bias for TET2 and multipotent stem cell origin for DNMT3A. Blood 132:277–280. https://doi.org/10.1182/blood-2018-01-829937

Busch K et al (2015) Fundamental properties of unperturbed haematopoiesis from stem cells in vivo. Nature 518:542–546. https://doi.org/10.1038/nature14242

Busque L et al (2012) Recurrent somatic TET2 mutations in normal elderly individuals with clonal

hematopoiesis. Nat Genet 44:1179–1181. https://doi.org/10.1038/ng.2413
Cai Z et al (2018) Inhibition of inflammatory signaling in Tet2 mutant preleukemic cells mitigates stress-induced abnormalities and clonal hematopoiesis. Cell Stem Cell 23:833–849. https://doi.org/10.1016/j.stem.2018.10.013
Cara et al (2015) Mutant U2AF1 expression alters hematopoiesis and Pre-mRNA splicing in vivo. Cancer Cell 27:631–643. https://doi.org/10.1016/j.ccell.2015.04.008
Carty SA et al (2018) The loss of TET2 promotes CD8^{+} T cell memory differentiation. J Immunol 200:82–91. https://doi.org/10.4049/jimmunol.1700559
Case LK et al (2013) The Y chromosome as a regulatory element shaping immune cell transcriptomes and susceptibility to autoimmune disease. Genome Res 23: 1474–1485. https://doi.org/10.1101/gr.156703.113
Challen GA et al (2012) Dnmt3a is essential for hematopoietic stem cell differentiation. Nat Genet 44: 23–31. https://doi.org/10.1038/ng.1009
Chen S et al (2019) Mutant p53 drives clonal hematopoiesis through modulating epigenetic pathway. Nat Commun 10:5649. https://doi.org/10.1038/s41467-019-13542-2
Cimmino L et al (2017) Restoration of TET2 function blocks aberrant self-renewal and leukemia progression. Cell 170:1079–1095. https://doi.org/10.1016/j.cell.2017.07.032
Cull AH, Snetsinger B, Buckstein R, Wells RA, Rauh MJ (2017) Tet2 restrains inflammatory gene expression in macrophages. Exp Hematol 55:56–70. https://doi.org/10.1016/j.exphem.2017.08.001
Desai P et al (2018) Somatic mutations precede acute myeloid leukemia years before diagnosis. Nat Med 24:1015–1023. https://doi.org/10.1038/s41591-018-0081-z
Dumanski JP et al (2021) Immune cells lacking Y chromosome show dysregulation of autosomal gene expression. Cell Mol Life Sci 78:4019–4033. https://doi.org/10.1007/s00018-021-03822-w
Fey M et al (1994) Clonality and X-inactivation patterns in hematopoietic cell populations detected by the highly informative M27 beta DNA probe. Blood 83:931–938. https://doi.org/10.1182/blood.v83.4.931.931
Fialkow PJ (1973) Primordial cell pool size and lineage relationships of five human cell types. Ann Hum Genet 37:39–48. https://doi.org/10.1111/j.1469-1809.1973.tb01813.x
Fialkow PJ, Gartler SM, Yoshida A (1967) Clonal origin of chronic myelocytic leukemia in man. Proc Natl Acad Sci U S A 58:1468–1471. https://doi.org/10.1073/pnas.58.4.1468
Forsberg LA et al (2014) Mosaic loss of chromosome Y in peripheral blood is associated with shorter survival and higher risk of cancer. Nat Genet 46:624–628. https://doi.org/10.1038/ng.2966
Fujino T et al (2021) Mutant ASXL1 induces age-related expansion of phenotypic hematopoietic stem cells through activation of Akt/mTOR pathway. Nat Commun 12:1826. https://doi.org/10.1038/s41467-021-22053-y
Fuster JJ et al (2017) Clonal hematopoiesis associated with TET2 deficiency accelerates atherosclerosis development in mice. Science 355:842–847. https://doi.org/10.1126/science.aag1381
Fuster JJ et al (2020) TET2-loss-of-function-driven clonal hematopoiesis exacerbates experimental insulin resistance in aging and obesity. Cell Rep 33:108326. https://doi.org/10.1016/j.celrep.2020.108326
Gale RE, Wheadon H, Linch DC (1991) X-chromosome inactivation patterns using HPRT and PGK polymorphisms in haematologically normal and post-chemotherapy females. Br J Haematol 79:193–197. https://doi.org/10.1111/j.1365-2141.1991.tb04521.x
Gamper CJ, Agoston AT, Nelson WG, Powell JD (2009) Identification of DNA methyltransferase 3a as a T cell receptor-induced regulator of Th1 and Th2 differentiation. J Immunol 183:2267–2276. https://doi.org/10.4049/jimmunol.0802960
Gao T et al (2021) Interplay between chromosomal alterations and gene mutations shapes the evolutionary trajectory of clonal hematopoiesis. Nat Commun 12: 338. https://doi.org/10.1038/s41467-020-20565-7
Genovese G et al (2014) Clonal hematopoiesis and blood-cancer risk inferred from blood DNA sequence. N Engl J Med 371:2477–2487. https://doi.org/10.1056/nejmoa1409405
Haitjema S et al (2017) Loss of Y chromosome in blood is associated with major cardiovascular events during follow-up in men after carotid endarterectomy. Circ Cardiovasc Genet 10:e001544. https://doi.org/10.1161/circgenetics.116.001544
Herens C, Brasseur E, Jamar M, Vierset L, Schoenen I, Koulischer L (1992) Loss of the Y chromosome from normal and neoplastic bone marrows. Genes Chromosom Cancer 5:83–88. https://doi.org/10.1002/gcc.2870050112
Hormaechea-Agulla D et al (2021) Chronic infection drives Dnmt3a-loss-of-function clonal hematopoiesis via IFNγ signaling. Cell Stem Cell 28:1428–1442. https://doi.org/10.1016/j.stem.2021.03.002
Hsiue EH et al (2021) Targeting a neoantigen derived from a common TP53 mutation. Science 371:8697. https://doi.org/10.1126/science.abc8697
Hsu YC et al (2017) The distinct biological implications of Asxl1 mutation and its roles in leukemogenesis revealed by a knock-in mouse model. J Hematol Oncol 10:139. https://doi.org/10.1186/s13045-017-0508-x
Hsu JI et al (2018) PPM1D mutations drive clonal hematopoiesis in response to cytotoxic chemotherapy. Cell Stem Cell 23:700–713. https://doi.org/10.1016/j.stem.2018.10.004
Ichiyama K et al (2015) The methylcytosine dioxygenase Tet2 promotes DNA demethylation and activation of cytokine gene expression in T cells. Immunity 42:613–626. https://doi.org/10.1016/j.immuni.2015.03.005

Inoue D et al (2013) Myelodysplastic syndromes are induced by histone methylation "altering ASXL1 mutations". J Clin Investig 123:4627–4640. https://doi.org/10.1172/jci70739

Inoue D et al (2018) A novel ASXL1-OGT axis plays roles in H3K4 methylation and tumor suppression in myeloid malignancies. Leukemia 32:1327–1337. https://doi.org/10.1038/s41375-018-0083-3

Jacobs KB et al (2012) Detectable clonal mosaicism and its relationship to aging and cancer. Nat Genet 44:651–658. https://doi.org/10.1038/ng.2270

Jaiswal S et al (2014) Age-related clonal hematopoiesis associated with adverse outcomes. N Engl J Med 371: 2488–2498. https://doi.org/10.1056/nejmoa1408617

Jaiswal S et al (2017) Clonal hematopoiesis and risk of atherosclerotic cardiovascular disease. N Engl J Med 377:111–121. https://doi.org/10.1056/nejmoa1701719

Jones AV et al (2009) JAK2 haplotype is a major risk factor for the development of myeloproliferative neoplasms. Nat Genet 41:446–449. https://doi.org/10.1038/ng.334

Kahn JD et al (2018) PPM1D-truncating mutations confer resistance to chemotherapy and sensitivity to PPM1D inhibition in hematopoietic cells. Blood 132:1095–1105. https://doi.org/10.1182/blood-2018-05-850339

Kilpivaara O et al (2009) A germline JAK2 SNP is associated with predisposition to the development of JAK2(V617F)-positive myeloproliferative neoplasms. Nat Genet 41:455–459. https://doi.org/10.1038/ng.342

Kim E et al (2015) SRSF2 mutations contribute to myelodysplasia by mutant-specific effects on exon recognition. Cancer Cell 27:617–630. https://doi.org/10.1016/j.ccell.2015.04.006

Kleppe M et al (2015) JAK–STAT pathway activation in malignant and nonmalignant cells contributes to MPN pathogenesis and therapeutic response. Cancer Discov 5:316–331. https://doi.org/10.1158/2159-8290.cd-14-0736

Koh KP et al (2011) Tet1 and Tet2 regulate 5-hydroxymethylcytosine production and cell lineage specification in mouse embryonic stem cells. Cell Stem Cell 8:200–213. https://doi.org/10.1016/j.stem.2011.01.008

Kondo M et al (2003) Biology of hematopoietic stem cells and progenitors: implications for clinical application. Annu Rev Immunol 21:759–806. https://doi.org/10.1146/annurev.immunol.21.120601.141007

Kunimoto H et al (2014) Tet2-mutated myeloid progenitors possess aberrant in vitro self-renewal capacity. Blood 123:2897–2899. https://doi.org/10.1182/blood-2014-01-552471

Ladle BH et al (2016) De novo DNA methylation by DNA methyltransferase 3a controls early effector CD8^{+} T-cell fate decisions following activation. Proc Natl Acad Sci U S A 113:10631–10636. https://doi.org/10.1073/pnas.1524490113

Laurie CC et al (2012) Detectable clonal mosaicism from birth to old age and its relationship to cancer. Nat Genet 44:642–650. https://doi.org/10.1038/ng.2271

Lee SC-W et al (2018) Synthetic lethal and convergent biological effects of cancer-associated spliceosomal gene mutations. Cancer Cell 34:225–241. https://doi.org/10.1016/j.ccell.2018.07.003

Lee-Six H et al (2018) Population dynamics of normal human blood inferred from somatic mutations. Nature 561:473–478. https://doi.org/10.1038/s41586-018-0497-0

Leoni C et al (2017) Dnmt3a restrains mast cell inflammatory responses. Proc Natl Acad Sci U S A 114:E1490–E1499. https://doi.org/10.1073/pnas.1616420114

Li Z et al (2011) Deletion of Tet2 in mice leads to dysregulated hematopoietic stem cells and subsequent development of myeloid malignancies. Blood 118: 4509–4518. https://doi.org/10.1182/blood-2010-12-325241

Liao M et al (2022) Aging-elevated inflammation promotes DNMT3A R878H-driven clonal hematopoiesis. Acta Pharm Sin B 12:678–691. https://doi.org/10.1016/j.apsb.2021.09.015

Liu X, Gong Y (2019) Isocitrate dehydrogenase inhibitors in acute myeloid leukemia. Biomarker Res 7:22. https://doi.org/10.1186/s40364-019-0173-z

Ljungström V et al (2022) Loss of Y and clonal hematopoiesis in blood—two sides of the same coin? Leukemia 36:889–891. https://doi.org/10.1038/s41375-021-01456-2

Loftfield E et al (2018) Predictors of mosaic chromosome Y loss and associations with mortality in the UK Biobank. Sci Rep 8:12316. https://doi.org/10.1038/s41598-018-30759-1

Loh P-R et al (2018) Insights into clonal haematopoiesis from 8,342 mosaic chromosomal alterations. Nature 559:350–355. https://doi.org/10.1038/s41586-018-0321-x

Loh P-R, Genovese G, McCarroll SA (2020) Monogenic and polygenic inheritance become instruments for clonal selection. Nature 584:136–141. https://doi.org/10.1038/s41586-020-2430-6

Lopez-Moyado IF et al (2019) Paradoxical association of TET loss of function with genome-wide DNA hypomethylation. Proc Natl Acad Sci U S A 116: 16933–16942. https://doi.org/10.1073/pnas.1903059116

Madzo J et al (2014) Hydroxymethylation at gene regulatory regions directs stem/early progenitor cell commitment during erythropoiesis. Cell Rep 6:231–244. https://doi.org/10.1016/j.celrep.2013.11.044

Malcovati L et al (2017) Clinical significance of somatic mutation in unexplained blood cytopenia. Blood 129: 3371–3378. https://doi.org/10.1182/blood-2017-01-763425

Marusyk A, Porter CC, Zaberezhnyy V, DeGregori J (2010) Irradiation selects for p53-deficient hematopoietic progenitors. PLoS Biol 8:e1000324. https://doi.org/10.1371/journal.pbio.1000324

Matatall KA et al (2016) Chronic infection depletes hematopoietic stem cells through stress-induced

terminal differentiation. Cell Rep 17:2584–2595. https://doi.org/10.1016/j.celrep.2016.11.031
McKerrell T et al (2015) Leukemia-associated somatic mutations drive distinct patterns of age-related clonal hemopoiesis. Cell Rep 10:1239–1245. https://doi.org/10.1016/j.celrep.2015.02.005
Meisel M et al (2018) Microbial signals drive pre-leukaemic myeloproliferation in a Tet2-deficient host. Nature 557:580–584. https://doi.org/10.1038/s41586-018-0125-z
Miller PG et al (2022) Association of clonal hematopoiesis with chronic obstructive pulmonary disease. Blood 139:357–368. https://doi.org/10.1182/blood.2021013531
Moran-Crusio K et al (2011) Tet2 loss leads to increased hematopoietic stem cell self-renewal and myeloid transformation. Cancer Cell 20:11–24. https://doi.org/10.1016/j.ccr.2011.06.001
Nagase R et al (2018) Expression of mutant Asxl1 perturbs hematopoiesis and promotes susceptibility to leukemic transformation. J Exp Med 215:1729–1747. https://doi.org/10.1084/jem.20171151
Niroula A et al (2021) Distinction of lymphoid and myeloid clonal hematopoiesis. Nat Med 27:1921–1927. https://doi.org/10.1038/s41591-021-01521-4
Okano M, Bell DW, Haber DA, Li E (1999) DNA methyltransferases Dnmt3a and Dnmt3b Are essential for de novo methylation and mammalian development. Cell 99:247–257. https://doi.org/10.1016/S0092-8674(00)81656-6
Olcaydu D et al (2009) A common JAK2 haplotype confers susceptibility to myeloproliferative neoplasms. Nat Genet 41:450–454. https://doi.org/10.1038/ng.341
Osorio FG et al (2018) Somatic mutations reveal lineage relationships and age-related mutagenesis in human hematopoiesis. Cell Rep 25:2308–2316. https://doi.org/10.1016/j.celrep.2018.11.014
Ostrander EL et al (2020) Divergent effects of Dnmt3a and Tet2 mutations on hematopoietic progenitor cell fitness. Stem Cell Rep 14:551–560. https://doi.org/10.1016/j.stemcr.2020.02.011
Pei W et al (2017) Polylox barcoding reveals haematopoietic stem cell fates realized in vivo. Nature 548:456–460. https://doi.org/10.1038/nature23653
Pierre RV, Hoagland HC (1972) Age-associated aneuploidy: loss of Y chromosome from human bone marrow cells with aging. Cancer 30:889–894. https://doi.org/10.1002/1097-0142(197210)30:4<889::aid-cncr2820300405>3.0.co;2-1
Pietras EM et al (2016) Chronic interleukin-1 exposure drives haematopoietic stem cells towards precocious myeloid differentiation at the expense of self-renewal. Nat Cell Biol 18:607–618. https://doi.org/10.1038/ncb3346
Pollyea DA et al (2019) Myelodysplastic syndrome-associated spliceosome gene mutations enhance innate immune signaling. Haematologica 104:e388–e392. https://doi.org/10.3324/haematol.2018.214155
Rasmussen KD et al (2015) Loss of TET2 in hematopoietic cells leads to DNA hypermethylation of active enhancers and induction of leukemogenesis. Genes Dev 29:910–922. https://doi.org/10.1101/gad.260174.115
Saiki R et al (2021) Combined landscape of single-nucleotide variants and copy number alterations in clonal hematopoiesis. Nat Med 27:1239–1249. https://doi.org/10.1038/s41591-021-01411-9
Sano S et al (2018a) CRISPR-mediated gene editing to assess the roles of Tet2 and Dnmt3a in clonal hematopoiesis and cardiovascular disease. Circ Res 123:335–341. https://doi.org/10.1161/circresaha.118.313225
Sano S et al (2018b) Tet2-Mediated clonal hematopoiesis accelerates heart failure through a mechanism involving the IL-1beta/NLRP3 inflammasome. J Am Coll Cardiol 71:875–886. https://doi.org/10.1016/j.jacc.2017.12.037
Sano S et al (2019) JAK2 (V617F)-mediated clonal hematopoiesis accelerates pathological remodeling in murine heart failure. JACC Basic Transl Sci 4:684–697. https://doi.org/10.1016/j.jacbts.2019.05.013
Seiler M et al (2018) H3B-8800, an orally available small-molecule splicing modulator, induces lethality in spliceosome-mutant cancers. Nat Med 24:497–504. https://doi.org/10.1038/nm.4493
Smith MA et al (2019) U2AF1 mutations induce oncogenic IRAK4 isoforms and activate innate immune pathways in myeloid malignancies. Nat Cell Biol 21:640–650. https://doi.org/10.1038/s41556-019-0314-5
Steensma DP et al (2015) Clonal hematopoiesis of indeterminate potential and its distinction from myelodysplastic syndromes. Blood 126:9–16. https://doi.org/10.1182/blood-2015-03-631747
Sun J et al (2014) Clonal dynamics of native haematopoiesis. Nature 514:322–327. https://doi.org/10.1038/nature13824
Terao C et al (2019) GWAS of mosaic loss of chromosome Y highlights genetic effects on blood cell differentiation. Nat Commun 10:4719. https://doi.org/10.1038/s41467-019-12705-5
Terao C et al (2020) Chromosomal alterations among age-related haematopoietic clones in Japan. Nature 584:130–135. https://doi.org/10.1038/s41586-020-2426-2
Thomas RM, Gamper CJ, Ladle BH, Powell JD, Wells AD (2012) De novo DNA methylation is required to restrict T helper lineage plasticity. J Biol Chem 287:22900–22909. https://doi.org/10.1074/jbc.m111.312785
Thompson DJ et al (2019) Genetic predisposition to mosaic Y chromosome loss in blood. Nature 575:652–657. https://doi.org/10.1038/s41586-019-1765-3
Uni M et al (2019) Modeling ASXL1 mutation revealed impaired hematopoiesis caused by derepression of p16Ink4a through aberrant PRC1-mediated histone modification. Leukemia 33:191–204. https://doi.org/10.1038/s41375-018-0198-6

Velten L et al (2021) Identification of leukemic and pre-leukemic stem cells by clonal tracking from single-cell transcriptomics. Nat Commun 12:1366. https://doi.org/10.1038/s41467-021-21650-1

Watson CJ, Blundell JR (2022) Mutation rates and fitness consequences of mosaic chromosomal alterations in blood. bioRxiv. https://doi.org/10.1101/2022.05.07.491016

Watson CJ et al (2020) The evolutionary dynamics and fitness landscape of clonal hematopoiesis. Science 367:1449–1454. https://doi.org/10.1126/science.aay9333

Welch JS et al (2012) The origin and evolution of mutations in acute myeloid leukemia. Cell 150:264–278. https://doi.org/10.1016/j.cell.2012.06.023

Williams N et al (2022) Life histories of myeloproliferative neoplasms inferred from phylogenies. Nature 602:162–168. https://doi.org/10.1038/s41586-021-04312-6

Wong TN et al (2018) Cellular stressors contribute to the expansion of hematopoietic clones of varying leukemic potential. Nat Commun 9:455. https://doi.org/10.1038/s41467-018-02858-0

Wright DJ et al (2017) Genetic variants associated with mosaic Y chromosome loss highlight cell cycle genes and overlap with cancer susceptibility. Nat Genet 49:674–679. https://doi.org/10.1038/ng.3821

Xie M et al (2014) Age-related mutations associated with clonal hematopoietic expansion and malignancies. Nat Med 20:1472–1478. https://doi.org/10.1038/nm.3733

Yamamoto K et al (2021) A histone modifier, ASXL1, interacts with NONO and is involved in paraspeckle formation in hematopoietic cells. Cell Rep 36:109576. https://doi.org/10.1016/j.celrep.2021.109576

Yamashita M, Passegue E (2019) TNF-alpha coordinates hematopoietic stem cell survival and myeloid regeneration. Cell Stem Cell 25:357–372. https://doi.org/10.1016/j.stem.2019.05.019

Young AL, Challen GA, Birmann BM, Druley TE (2016) Clonal haematopoiesis harbouring AML-associated mutations is ubiquitous in healthy adults. Nat Commun 7:12484. https://doi.org/10.1038/ncomms12484

Yura Y et al (2021) The cancer therapy-related clonal hematopoiesis driver gene Ppm1d promotes inflammation and non-ischemic heart failure in mice. Circ Res 129:684–698. https://doi.org/10.1161/circresaha.121.319314

Zekavat SM et al (2021) Hematopoietic mosaic chromosomal alterations increase the risk for diverse types of infection. Nat Med 27:1012–1024. https://doi.org/10.1038/s41591-021-01371-0

Zhang Q et al (2015) Tet2 is required to resolve inflammation by recruiting Hdac2 to specifically repress IL-6. Nature 525:389–393. https://doi.org/10.1038/nature15252

Zhang X et al (2016) DNMT3A and TET2 compete and cooperate to repress lineage-specific transcription factors in hematopoietic stem cells. Nat Genet 48:1014–1023. https://doi.org/10.1038/ng.3610

Zhang Q et al (2022a) Mosaic loss of chromosome Y promotes leukemogenesis and clonal hematopoiesis. JCI Insight 7:e153768. https://doi.org/10.1172/jci.insight.153768

Zhang S et al (2022b) Advanced strategies for therapeutic targeting of wild-type and mutant p53 in cancer. Biomolecules 12:548. https://doi.org/10.3390/biom12040548

Zhou W et al (2016) Mosaic loss of chromosome Y is associated with common variation near TCL1A. Nat Genet 48:563–568. https://doi.org/10.1038/ng.3545

Zhu F et al (2019) The GABA receptor GABRR1 is expressed on and functional in hematopoietic stem cells and megakaryocyte progenitors. Proc Natl Acad Sci U S A 116:18416–18422. https://doi.org/10.1073/pnas.1906251116

Zink F et al (2017) Clonal hematopoiesis, with and without candidate driver mutations, is common in the elderly. Blood 130:742–752. https://doi.org/10.1182/blood-2017-02-769869

Ex Vivo Expansion and Homing of Human Cord Blood Hematopoietic Stem Cells

6

Bin Guo, Xinxin Huang, Yandan Chen, and Hal E. Broxmeyer

Abstract

Cord blood (CB) has been proven to be an alternative source of haematopoietic stem cells (HSCs) for clinical transplantation and has multiple advantages, including but not limited to greater HLA compatibility, lower incidence of graft-versus-host disease (GvHD), higher survival rates and lower relapse rates among patients with minimal residual disease. However, the limited number of HSCs in a single CB unit limits the wider use of CB in clinical treatment. Many efforts have been made to enhance the efficacy of CB HSC transplantation, particularly by ex vivo expansion or enhancing the homing efficiency of HSCs. In this chapter, we will document the major advances regarding human HSC ex vivo expansion and homing and will also discuss the possibility of clinical translation of such laboratory work.

Keywords

Cord blood · Hematopoietic stem cell · Ex vivo expansion · Homing

Abbreviations

AGM	Aorta-gonad-mesonephros
ALDH	Aldehyde dehydrogenase
ANGPTL5	Angiopoietin-like 5
BM	Bone marrow
CAPE	Caffeic acid phenethyl ester
CaR	Calcium-sensing receptor
CB	Cord blood
CFU	Colony-forming unit
cGMP	Cyclic guanosine monophosphate
DMOG	Dimethyloxalylglycine
DPP4	Dipeptidyl peptidase 4
EPHOSS	Extra physiological oxygen stress/shock
FA	Fanconi anaemia
FGF-1	Fibroblast growth factor 1
Flt3L	Fms-like tyrosine kinase-3 ligand
G-CSF	Granulocyte colony-stimulating factor
GTP	Guanosine triphosphate
GvHD	Graft-versus-host disease

Bin Guo and Xinxin Huang contributed equally with all other contributors.

B. Guo (✉) · Y. Chen
Department of Pathophysiology, Key Laboratory of Cell Differentiation and Apoptosis of Chinese Ministry of Education, School of Medicine, Shanghai Jiao Tong University, Shanghai, China
e-mail: jybinguo@shsmu.edu.cn

X. Huang (✉)
Xuhui Hospital and Institutes of Biomedical Sciences, Fudan University, Shanghai, China
e-mail: xinxinhuang@fudan.edu.cn

H. E. Broxmeyer (✉)
Department of Microbiology and Immunology, School of Medicine, Indiana University, Indianapolis, IN, USA
e-mail: hbroxmey@iupui.edu

M. Zhao, P. Qian (eds.), *Hematopoietic Stem Cells*, Advances in Experimental Medicine and Biology 1442, https://doi.org/10.1007/978-981-99-7471-9_6

HAT	Histone acetyltransferase
HDAC	Histone deacetylase
HO-1	Heme oxygenase 1
HPCs	Haematopoietic progenitor cells
HSCs	Hematopoietic stem cells
IGF-2	Insulin-like growth factor 2
IGFBP2	Insulin-like growth factor binding protein 2
IL3	Interleukin-3
IL6	Interleukin-6
JNK	Jun N-terminal kinase
LSD1	Lysine-specific demethylase 1
m6A	N6-Methyladenosine
MAPK	Mitogen-activated protein kinase
MDS	Myelodysplastic syndromes
MNC	Mononuclear cell
mPB	Mobilized peripheral blood
MPP	Multipotent progenitors
MPTP	Mitochondria permeability transition pore
MSC	Mesenchymal stem/stromal cell
Msi2	Musashi 2
NO	Nitric oxide
PDE5	Phosphodiesterase 5
PGE2	Prostaglandin E2
PPAR	Peroxisome proliferator-activated receptor
RXR	Retinoid X receptor
SCF	Stem cell factor
sGC	Soluble guanylyl cyclase
SNP	Sodium nitroprusside
SR1	StemRegenin1
TPO	Thrombopoietin
VPA	Valproic acid

6.1 Introduction

Haematopoietic stem cells (HSCs), which reside at the apex of the haematopoietic hierarchy, have the capacity to regenerate the whole blood system of an organism (Morgan et al. 2017). HSC transplantation has been used to treat many diseases, including but not limited to haematological disorders (Copelan 2006). To date, there are three main sources of HSCs for clinical use: bone marrow (BM), mobilized peripheral blood (mPB) and cord blood (CB) (Chabannon et al. 2018). BM HSC transplantation is being performed less frequently due to excess pain suffered by donors and the requirement for medical care after surgery, sometimes for months afterwards (Bosi and Bartolozzi 2010). mPB remains the most widely used source of HSC. Engraftable HSCs can be efficiently mobilized by granulocyte colony-stimulating factor (G-CSF) and/or the CXCR4 antagonist AMD3100 (plerixafor) (Pelus and Broxmeyer 2018). However, G-CSF in some cases might result in malignant transformation in recipients of HSC donor cells (Connelly et al. 2012). In the 1980s, Broxmeyer et al. proposed that umbilical cord blood might be an alternative source of transplantable HSCs for clinical use (Ballen et al. 2013). This concept was proven by bench work in his laboratory at the Indiana University School of Medicine. In 1988, collaborating with Dr. Eliane Gluckman at the St. Louis Hospital in Paris, they successfully performed the first CB HSC transplantation to treat a 6-year-old Fanconi anaemia patient (Gluckman et al. 1989). This collaboration between a PhD scientist and an MD physician demonstrated for the first time that CB could definitely be used as a reliable source of functional HSC for clinical use. Eventually, people realized that cord blood-derived HSC transplantation showed multiple advantages, including but not limited to greater HLA compatibility, lower incidence of graft-versus-host disease (GvHD), higher survival rates and lower relapse rates among patients with minimal residual disease (Broxmeyer and Farag 2013; Broxmeyer 2016; Milano et al. 2016). However, the limited number of HSCs in single CB units limits the wider use of CB in clinical treatment. Efforts have been made to enhance the efficacy of CB HSC transplantation. This includes expansion or enhancement of the homing efficiency of HSCs (Huang et al. 2019). In this chapter, we document experience regarding human HSC ex vivo expansion and homing and discuss the possibility of clinical translation of such laboratory work.

6.2 Ex Vivo Expansion of CB HSC

Ex vivo expansion of CB HSCs aims to obtain more transplantable HSCs upon ex vivo culture. In vivo, one HSC can give rise to one daughter HSC and another daughter multipotent progenitors (MPP) through asymmetrical cell division. During ex vivo culture, theoretically, one initiating HSC will generate more HSCs over time through symmetrical division. Thus, ex vivo expansion of CB HSCs remains a promising method that might overcome the limited numbers of HSCs in single CB units (Takizawa et al. 2011).

6.2.1 Cytokine-Induced Expansion of CB HSCs

HSCs in vivo are nourished by complex cell-cell and cytokine-cell signalling networks in the bone marrow environmental niche (Boulais and Frenette 2015). In addition to mechanical contact with niche cells, cytokines generated by niche cells play significant roles in HSC survival, proliferation, self-renewal and differentiation (Zhang and Lodish 2008). Relatively well-characterized cytokines that regulate HSC maintenance include stem cell factor (SCF), thrombopoietin (TPO), Fms-like tyrosine kinase-3 ligand (Flt3L), granulocyte colony-stimulating factor (G-CSF) and interleukin-6 (IL6).

Without such added cytokines, HSCs and haematopoietic progenitor cells (HPCs) do not survive well and cannot be expanded ex vivo. SCF, which interacts with its cognate receptor c-Kit, is a major cytokine essential for HSC survival during ex vivo culture (Lennartsson and Ronnstrand 2012). SCF is mainly produced by bone marrow stromal cells and endothelial cells (Ding et al. 2012). It is interesting to observe that depletion of Scf from endothelial cells or leptin receptor (Lepr)-expressing perivascular stromal cells impaired HSC homeostasis, but the same was not seen for Scf depletion in haematopoietic cells (Ding et al. 2012). The SCF/c-Kit axis induces activation of the Akt pathway to promote cell survival and prevent apoptosis (Vajravelu et al. 2015). Adipocyte-derived SCF promotes the maintenance and regeneration of HSCs in the bone marrow (Zhou et al. 2017). TPO signalling is essential for the maintenance of HSCs (de Graaf and Metcalf 2011). Loss of function of TPO or its receptor c-Mpl causes severe impairment of the reconstituting capacity of HSCs and multipotent progenitors (MPPs), as revealed by competitive transplantation experiments (Kimura et al. 1998; Seita and Weissman 2010). Flt3L/Flt3 signalling is important for HSC survival and generation, while overactivation of the Flt3L/Flt3 pathway is frequently detected in leukaemia cells (Dolence et al. 2014; Grafone et al. 2012). Flt3L knockout (KO) mice showed significantly reduced myeloid and B-cell precursors, and BM from Flt3L KO mice failed to replenish the blood system in the recipient mice (Dolence et al. 2014). The combination of SCF, TPO and Flt3L has been widely used to expand human CB HSCs and HPCs ex vivo.

Other growth factors, including G-CSF, interleukin-3 (IL3), IL6, angiopoietin-like 5 (ANGPTL5), insulin-like growth factor binding protein 2 (IGFBP2), fibroblast growth factor 1 (FGF-1) and insulin-like growth factor 2 (IGF-2), have also been used to stimulate the expansion of HSCs in the presence of SCF, TPO and Flt3L (Schuettpelz et al. 2014; Testa et al. 1996; Tie et al. 2019; Zhang et al. 2006; Fan et al. 2014). Clinical trials have been performed to evaluate the efficacy of ex vivo expanded HSCs induced by growth factor cocktails. Unfortunately, clinical outcomes such as neutrophil recovery in most cases were not significantly improved when cytokine-expanded CB HSCs and HPCs were transplanted into recipient patients (Kiernan et al. 2017). Although growth factors promote ex vivo expansion of CB $CD34^+$ phenotypic HSCs and HPCs during ex vivo culturing, these phenotypic HSCs lose their reconstituting function in vivo. These studies demonstrate that many factors other than cytokines derived from the haematopoietic niche may be involved in the maintenance of HSC stemness and expansion.

6.2.2 Expansion of CB HSCs by Targeting Classical Cell Signal Transduction Pathways

The Notch signalling pathway is essential for HSC generation in the aorta-gonad-mesonephros (AGM) region during the embryonic stage, while it is dispensable for primitive haematopoiesis in the yolk sac and for adult haematopoietic stem cell self-renewal in the bone marrow (Lampreia et al. 2017; Wang et al. 2015; Duncan et al. 2005; Mancini et al. 2005). Notch ligand-mediated activation significantly promoted ex vivo expansion of CB HSCs and HPCs with rapid capacity of myeloid reconstitution in immunocompromised recipient mice (Delaney et al. 2010), but this has not yet been verified by others.

The role of Wingless (Wnt) signalling in HSC self-renewal and maintenance is dependent on species and developmental stage. The Wnt pathway is required for HSC emergence and expansion in zebrafish (Bigas et al. 2013). Wnt regulates HSC emergence from the mouse AGM but is dispensable for HSC homeostasis once HSCs express mature haematopoietic markers (Ruiz-Herguido et al. 2012). Loss of Wnt signalling severely impaired maintenance of the HSC pool in the mouse foetal liver and disrupted the repopulating capacity of the remaining HSCs (Luis et al. 2009). Wnt is dispensable for the self-renewal and differentiation of adult murine HSCs (Kabiri et al. 2015). Activation of the Wnt pathway in the mouse bone marrow niche limits HSC ectopic proliferation and promotes HSC maintenance (Fleming et al. 2008). Wnt activation in the presence of growth factors significantly enhances human CB $CD34^{+}CD38^{-}$ and $C133^{+}CD38^{-}$ HSC and HPC ex vivo expansion, probably by increasing the expression of pluripotency-related genes, including c-myc, nanog, oct3/4 and sox2 (Chotinantakul et al. 2013).

c-Jun N-terminal kinase (JNK) signalling regulates multiple biological processes, including cell proliferation, differentiation, cell death during development and tumorigenesis (Johnson and Nakamura 2007). JNK/c-Fos regulates the maintenance of murine HSC quiescence by negatively controlling cell cycle progression (Semba et al. 2020). Inhibiting JNK by JNK-IN-8 significantly promoted the expansion and self-renewal of human CB HSCs and HPCs by repressing c-Jun phosphorylation (Xiao et al. 2019).

p38 mitogen-activated protein kinase (p38) signalling plays an important role in the stress response (Han et al. 2020). p38 mitogen-activated protein kinase (MAPK) is ectopically activated in both myelodysplastic syndrome (MDS) and Fanconi anaemia (FA) haematopoietic cells (Navas et al. 2006; Svahn et al. 2015). Inhibition of p38 MAPK signalling promotes ex vivo expansion of human CB HSCs, but not HPCs, by preventing HSCs from undergoing senescence and apoptosis (Zou et al. 2012).

6.2.3 Stromal Cell-Based Expansion of CB HSC

HSCs are maintained in a microenvironmental niche in vivo (Gao et al. 2018). The HSC niche in the bone marrow is widely explored and is very complicated, while the HSC niche in the placenta or cord is less well understood.

Mesenchymal stem/stromal cells (MSCs) are a major component of bone marrow niches (Crippa and Bernardo 2018). MSCs also exist in placental and umbilical cord tissue and blood (Fajardo-Orduna et al. 2017). Coculture of CB $CD34^{+}$ cells with MSCs improves ex vivo expansion of CB HSCs and HPCs, which might help prevent the development of GvHD after transplantation (Li et al. 2007). Importantly, it seems that it is not necessary to isolate $CD34^{+}$ cells from CB since coculture of CB mononuclear cells (MNCs) with MSCs allows apparent expansion of $CD34^{+}$ HSCs and HPCs (Fajardo-Orduna et al. 2017).

Endothelial cells function as another important component of the HSC niche in vivo (Beerman et al. 2017). As feeder cells, endothelial cells are able to promote expansion of CB $CD34^{+}CD38^{-}$ HSCs and HPCs with long-term reconstituting capacity in recipient mice (Li et al. 2019). Engineered human foetal liver sinusoidal endothelial cells improve expansion of CB

$CD34^+CD38^-$ and $CD34^+CD38^-CD90^+$ HSCs and HPCs possibly by activating Notch signalling. During ex vivo culture, CB HSCs tend to interact with endothelial cells derived from CB endothelial precursors, and these endothelial cells probably function as a niche for CB HSCs during expansion by modulating both Notch and TGF-beta signalling (Li et al. 2019).

Considering the complexity of the haematopoietic niche in vivo, highly efficient stromal cell-based HSC expansion methods may require the establishment of a system that mimics the in vivo niche spatially and that comprises different cell components. Additionally, other factors, such as hypoxia and feedback inhibition effects, should also be taken into account for optimal expansion of HSCs ex vivo.

6.2.4 Small Molecule-Induced Expansion of CB HSCs

Small-molecule chemical compounds can be easily taken up by cells, and many functional proteins or signalling pathways can be monitored by treating cells with small molecules. Many small molecules, including diethylaminobenzaldehyde (DEAB), LG1506 (retinoic acid receptor antagonist), GW9662 (PPARG antagonist), SR1 (AhR antagonist), UM171, LSD inhibitor, BIO (GSK3β inhibitor), NR-101 (c-MPL agonist), HDAC inhibitor (trichostatin A and valproic acid), garcinol (GAR), nicotinamide riboside (NR) and CPI-203 (BET inhibitor), have been used to expand cord blood HSCs and HPCs ex vivo (de Lima et al. 2008; Horwitz et al. 2014; Safi et al. 2009; Guo et al. 2018a, b; Boitano et al. 2010; Chaurasia et al. 2014; Nishino et al. 2011; Fares et al. 2014; Subramaniam et al. 2020; Hua et al. 2020).

Aldehyde dehydrogenase (ALDH) is highly enriched in human HSCs and has been commonly used as a marker of stemness. Inhibition of ALDH by DEAB significantly promotes expansion of HSCs and HPCs by blocking retinoid signalling-induced cell differentiation (Chute et al. 2006). LG1506, a selective RXR (retinoid X receptor) modulator, largely enhances the expansion of human CB HSCs and HPCs during ex vivo culture by repressing RXR-RAR heterodimer activity (Safi et al. 2009). Interestingly, RXR dimerizes with many other nuclear receptors, including peroxisome proliferator-activated receptor (PPAR) family members. We found that antagonizing PPARgamma (PPARG) by GW9662 facilitates expansion of CB HSCs and HPCs by enhancing glucose metabolism. By RNA-seq analysis, FBP1 was identified as a downstream target of PPARG (Guo et al. 2018a). The PPARG-FBP1 axis thus functions as a negative regulator of human CB HSC and HPC glucose metabolism and stemness maintenance (Guo et al. 2018a, b).

StemRegenin1 (SR1) was the first powerful small molecule identified in a large-scale screen for agonists of human HSCs and HPCs using CD34 and CD133 as cell surface markers (Boitano et al. 2010). Coculture with 1 μM SR1 in serum-free media supplemented with TPO, SCF, FLT-3 L and interleukin-6 (IL6) demonstrated human phenotypic HSC expansion. SR1-treated CB $CD34^+$ cells showed significantly enhanced SCID repopulating cells (SRCs; a measure of the number of functional engrafting human HSCs) in sublethally irradiated immune-deficient NSG mice. Mechanistically, SR1 functions as an AhR antagonist and antagonizes AhR signalling in CB HSCs and HPCs. However, to date, the exact molecular mechanism by which SR1-AhR signalling regulates the ex vivo expansion and maintenance of human HSCs and HPCs remains unclear. SR1-expanded CB HSC expansion has been recently tested in a phase I/II clinical trial (Wagner Jr. et al. 2016). Simultaneous transplantation of SR1-expanded CB-derived $CD34^+$ cells and another unmanipulated CB unit significantly promoted faster recovery of neutrophils and platelets after infusion compared with control groups and led to successful engraftment in 17/17 patients, suggesting that SR1 might be a promising candidate drug for improving the clinical efficacy of HCT (Wagner Jr. et al. 2016). Please note that the authors did not transplant only SR1-expanded CB HSC/HPC into patients, so it's still hard to say if the SR1-expanded HSCs are long-term HSC based on this clinical study.

Another HSC chemical agonist is UM171. UM171 promotes ex vivo expansion of long-term repopulating HSCs in culture perhaps more significantly than SR1 (Fares et al. 2014). The clinical trial of UM171-expanded HSCs was recently published, and the preliminary data suggested that UM171-expanded CB HSCs are safe for transplantation and that a single UM171-expanded CB HSC showed prompt engraftment in patients (Cohen et al. 2020). Since the number of patients involved in this trial is just 22, how effective UM171 clinical transplantation is needs to be further studied. Lysine-specific demethylase 1 (LSD1) was reported as a regulator of murine bone marrow HSC differentiation. A recent study demonstrated that UM171 promotes HSC expansion by targeting LSD1 to block differentiation. UM171 treatment promoted LSD1-containing chromatin remodelling complex CoREST polyubiquitination and degradation (Subramaniam et al. 2020). In addition, eupalinilide E, a compound that promotes glycolysis in CB CD34^{+} cells, significantly enhances the effect of UM171 on CB HSC expansion (Zhang et al. 2020).

Short-term treatment of CB CD34^{+} cells with 6-bromoindirubin-3′-oxime (BIO), a GSK3β inhibitor, enhances engraftment of ex vivo expanded CB HSCs in immunodeficient recipient mice (Ko et al. 2011). GSK3β inhibitors activate Wnt signalling and promote the maintenance and self-renewal of both human and mouse embryonic stem cells (Ko et al. 2011). Inhibition of GSK3β downregulates genes involved in cell cycle progression, thus delaying cell division (Ko et al. 2011). The effect of GSK3β inhibitors on HSC expansion seems to be time- and dose dependent.

NR-101 is a c-MPL agonist that is the receptor of TPO. NR-101 promotes the expansion of CB HSCs and HPCs more significantly than TPO (Nishino et al. 2009). NR-101-expanded CB CD34^{+} cells show increased SRCs. Mechanistically, NR-101 treatment activated STAT5 but not STAT3 signalling. Meanwhile, NR-101 induces stabilization of HIF1alpha in CB CD34^{+} cells.

Histone deacetylase (HDAC) family members mainly function by removing acetyl groups from the lysine residues of histone proteins (de Ruijter et al. 2003). The HDAC family includes class I HDACs (HDACs 1/2/3 and 8), class II HDACs (HDAC 4/5/6/7/9/10) and class IV HDACs (HDAC 11) (de Ruijter et al. 2003). Trichostatin A, a potent and specific inhibitor of HDAC class I/II, promotes the expansion of CB CD34^{+}CD90^{+} cells and enhances the number of colony-forming units (CFUs; a measure of HPCs) (Obier et al. 2010). Trichostatin A-expanded CD34^{+}CD90^{+} cells show significantly enhanced SRCs. Another potent HDAC inhibitor that has been studied in CB HSC expansion is VPA. VPA treatment promotes the expansion of CB CD34^{+} cells and CD34^{+}CD90^{+} cells (Chaurasia et al. 2014). The expression of pluripotency genes, including SOX2, OCT4 and NANOG, is largely activated by VPA treatment (Chaurasia et al. 2014).

Garcinol (GAR), a histone acetyltransferase (HAT) inhibitor, was identified as an agonist of CB HSC and HPC expansion in an active natural compound screen (Nishino et al. 2011). GAR efficiently expands CB CD34^{+} cells and multipotent progenitors (colony-forming unit-granulocyte/erythrocyte/macrophage/megakaryocyte (CFU-GEMM)). GAR-expanded CD34^{+} cells showed significantly higher expression of AMICA1, BTG2 and HLF but had no obvious effect on genes involved in HSC self-renewal, such as HOXB4, BMI1, GATA2 and NOTCH1 (Nishino et al. 2011).

Nicotinamide (NAM), an NAD^{+}-boosting agent, inhibits differentiation and promotes maintenance of CB HSCs (Peled et al. 2012). Early clinical studies showed that NAM-expanded CB CD34^{+} cells have long-term multilineage engrafting capacity. Treatment with another similar compound, nicotinamide riboside (NR), also enhances the repopulating capacity of CB CD34^{+} cells (Vannini et al. 2019). NR facilitates mitochondrial clearance, downregulates mitochondrial metabolism of CB CD34^{+} cells and thus leads to increased asymmetric division of HSCs (Vannini et al. 2019).

CPI-201, a potent and specific bromodomain and extraterminal motif (BET) inhibitor, was recently reported as a novel CB HSC expansion activator (Hua et al. 2020). CPI-201 treatment expands both phenotypic CB HSCs and functional HSCs during ex vivo culture. Interestingly, unlike UM171, CPI-201 promotes CB megakaryocyte maturation (Hua et al. 2020). CPI-201 regulates the expression of HSC-related genes, including PROM1, HLF, CRHBP and MEIS1 (Hua et al. 2020). The exact mechanism of CPI-201 on CB HSC expansion remains unknown.

6.2.5 Expansion of CB HSCs by Manipulating Specific Genes Involved in Self-Renewal

Knowledge regarding the regulation of HSC self-renewal mainly comes from studies performed in mouse model systems, including bone marrow and foetal liver models. The mechanism behind human HSC self-renewal and maintenance remains relatively elusive, which is part of the reason why there is still no highly efficient ex vivo expansion system for CB HSCs that can be utilized in clinical therapy worldwide. It is of great significance to explore the molecular mechanisms underlying human CB HSC self-renewal and maintenance, especially under stress conditions, such as those induced during ex vivo expansion. Very limited numbers of genes involved in this process have been reported. These include HOXB4, MSI2, PPARG, YTHDF2 and MLLT3.

HOXB4 functions as a positive regulator of HSC self-renewal for both mouse and human HSCs (Antonchuk et al. 2002; Krosl et al. 2003). Overexpression of HOXB4 or incubation with HOXB4 protein both promotes ex vivo expansion of CB CD34$^+$ HSCs and HPCs (Krosl et al. 2003). Ectopic HOXB4 expression regulates Notch, Wnt/β-catenin, FGF, TGF-β and TNF-α signalling in HSCs and HPCs (Krosl et al. 2003). Activation of OCT4 expands CB HSCs and HPCs by enhancing HOXB4 expression (Huang et al. 2016).

The RNA-binding protein Musashi 2 (Msi2) is highly enriched in human HSCs, and its expression is decreased upon differentiation (Rentas et al. 2016). Overexpression of Msi2 promotes self-renewal and expansion of phenotypic and engraftable CB HSCs, probably by antagonizing AhR signalling (Rentas et al. 2016). USF2 and PLAG1 were identified as two key transcription factors that regulate Msi2 expression in CB HSCs (Belew et al. 2018). Co-overexpression of USF2 and PLAG1 mimicked the positive effects of MSI2 on CB HSC expansion and self-renewal (Belew et al. 2018).

PPARG expression significantly increases during ex vivo culture in RPMI1640 medium with 10% FBS and growth factors (SCF, TPO, Flt3L) (Guo et al. 2018b). Antagonizing PPARG promotes human CB HSC and HPC maintenance during ex vivo culture (Guo et al. 2018a). FBP1, a negative regulator of glycolysis, was identified as a downstream target of PPARG in CB HSCs/HPCs by RNA-seq analysis. Repression of FBP1 expands CB HSCs and HPCs compared with the vehicle control group. Inhibition of either PPARG or FBP1 apparently enhances glucose metabolism in CB HSCs and HPCs (Guo et al. 2018a). In StemSpan SFEM medium, PPARG and FBP1 expression in CB CD34$^+$ HSCs and HPCs was largely repressed compared with CB CD34$^+$ cells cultured in FBS-containing RPMI1640 medium. Consistently, blocking glycolysis with 2-deoxy-D-glucose (2-DG) somehow dampens the enhanced effect on CB HSC expansion by SFEM medium. Thus, repression of PPARG-FBP1 signalling might be a metabolic checkpoint of CB HSC maintenance during ex vivo culture.

N6-Methyladenosine (m6A) is a reversible mRNA modification that regulates multiple cellular processes by modulating RNA stability, splicing and translation (Guo et al. 2017a). Depletion of the m6A 'writer' METTL3 reduces global m6A modification levels and suppresses expansion of CB HSCs and HPCs by regulating cell differentiation (Li et al. 2018a; Huang and Broxmeyer 2018). Loss of function of the m6A reader Ythdf2 promotes ex vivo expansion of human CB HSCs and HPCs by increasing targeted decay of mRNA. Taken together, m6A modification functions as a positive regulator of human CB HSC self-renewal and maintenance.

The development of small molecules that target m6A modification might be more helpful and convenient for CB HSC expansion.

MLLT3 (AF9) is a proto-oncogene identified in leukaemia. An early study showed that MLLT3 promotes the commitment of erythroid and megakaryocytic progenitors (Pina et al. 2008). A recent study demonstrated that loss of MLLT3 impaired transplantable CB HSC maintenance and that overexpression of MLLT3 promoted functional CB HSC ex vivo expansion in culture (Calvanese et al. 2019). Enforced MLLT3 expression enhances H3K79me2 deposition in regulatory genes of HSC function and maintains HSC stemness during ex vivo expansion.

6.2.6 Opportunities and Challenges for CB HSC Expansion

Ex vivo expansion of CB HSCs is a potential means to overcome the rarity of HSCs in single CB units. Gene editing of HSCs is a very promising field, and it will broaden the clinical use of HSC transplantation-based gene therapy. To ensure effective editing of HSCs, but not HPCs, in CB, it is urgent to develop a reliable, stable and highly efficient ex vivo culture system of CB HSCs so that long-term eradication of blood diseases can be achieved. In addition, immune therapy of various cancer patients with engineered HSCs also requires sufficient numbers of stem cells. In theory, successful HSC expansion is beneficial to the clinical use of these treatments (Fig. 6.1).

People have tried many different ways to perform ex vivo expansion of CB HSCs, as discussed above (Table 6.1). However, there is still no FDA-approved ex vivo expansion method that can be utilized in clinical therapy. A major reason is that the mechanisms underlying CB HSC self-renewal and maintenance during ex vivo culture are poorly understood. Under stress conditions such as ex vivo culture or cytokine stimulation, CB HSCs quickly undergo differentiation and lose their long-term repopulating capacity. Another important issue is that the phenotype of HSCs can change during ex vivo expansion (Chen et al. 2019). In most cases, expansion of CD34$^+$ cells or even CD34$^+$CD90$^+$ cells actually means expansion of HPCs but not long-term engraftment and repopulation of functional HSCs. We need to know more about the molecular mechanisms of CB HSC self-renewal so that we can find a way to achieve enhanced expansion of functional HSCs.

6.3 Regulation of HSC Homing

HSC homing is generally defined as the process of HSCs migrating to and residing in a specialized microenvironmental niche, within which HSC proliferation, self-renewal and differentiation are elegantly balanced (Crane et al. 2017; Yu and Scadden 2016). HSC homing is a multistep physiologic process that includes leaving the peripheral circulation, interacting with the blood vessel lining endothelial cells and docking in specialized niches (Ratajczak and Suszynska 2016; Huang and Broxmeyer 2019). Two main axes have been identified during HSC homing (Fig. 6.2). The first axis involves chemokine CXCL12/SDF-1 and its receptor CXCR4 on the HSC surface (Nagasawa 2015; Zou et al. 1998). The CXCL12 gradient, produced by niche cells, serves as a chemoattractive signal to guide HSC migration towards their designated sites. The second axis is the vascular cell adhesion molecule 1 (VCAM1) and very late antigen 4 (VLA4, integrins $\alpha_4\beta_1$) axis (Rettig et al. 2012; Sanz-Rodriguez et al. 2001). VCAM1$^+$ macrophage-like cell populations direct HSC homing and retention in the niche (Li et al. 2018b). Genetic deletion of either the CXCL12/CXCR4 or VCAM1/VLA4 axis greatly impairs the HSC homing process. In addition, the bioactive sphingolipids sphingosine-1-phosphate (S1P) and ceramide-1-phosphate (C1P) are also chemotactic gradients involved in guiding HSCs to their BM microenvironment (Ratajczak et al. 2010; Juarez et al. 2012; Golan et al. 2012). HSC homing to the BM niche is the initial crucial step of clinic HSC transplantation, so the study of HSC homing has great potential in enhancing transplantation efficiency, especially when HSC

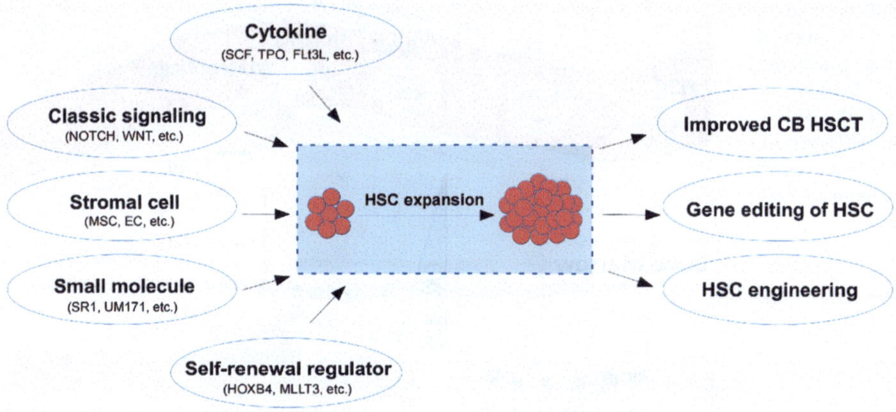

Fig. 6.1 Potential means to expand CB HSC. These include cytokine (SCF, TPO, Flt3L, etc.) treatment, modulating classic signalling transduction pathways (NOTCH, WNT, etc.), coculture with stromal cells (MSC, EC, etc.), stimulation with small-molecule compounds (SR1, UM171, etc.) and targeting self-renewal regulators (HOXB4, MLLT3, etc.). Establishing an efficient ex vivo expansion system of HSC can benefit clinical therapy with CB transplantation, gene editing of HSC and HSC engineering

Table 6.1 Potential means to expand CB HSCs

Different methods used to expand CB HSCs	Cytokines/pathways/small molecules/stromal cells/self-renewal regulators
Cytokines for CB HSC expansion	SCF, TPO, Flt3L, G-CSF, IL3, IL6, ANGPTL5, IFGBP2, FGF-1, IGF-2 (Lennartsson and Ronnstrand 2012; Ding et al. 2012; Vajravelu et al. 2015; Zhou et al. 2017; de Graaf and Metcalf 2011; Kimura et al. 1998; Seita and Weissman 2010; Dolence et al. 2014; Grafone et al. 2012; Schuettpelz et al. 2014; Testa et al. 1996; Tie et al. 2019; Zhang et al. 2006; Fan et al. 2014)
Classical signalling pathways as targets for CB HSC expansion	NOTCH, Wnt, JNK, MAPK (Delaney et al. 2010; Bigas et al. 2013; Ruiz-Herguido et al. 2012; Luis et al. 2009; Kabiri et al. 2015; Fleming et al. 2008; Chotinantakul et al. 2013; Johnson and Nakamura 2007; Semba et al. 2020; Xiao et al. 2019; Han et al. 2020; Navas et al. 2006; Svahn et al. 2015)
Stromal cells used for CB HSC expansion	MSCs, ECs (Li et al. 2007, 2019)
Small molecules for CB HSC expansion	DEAB, LG1506, GW9662, SR1, UM171, LSD1 inhibitor, BIO, NR-101, trichostatin A, VPA, garcinol, nicotinamide riboside, CPI-203 (de Lima et al. 2008; Horwitz et al. 2014; Safi et al. 2009; Guo et al. 2018a, b; Boitano et al. 2010; Chaurasia et al. 2014; Nishino et al. 2011, 2009; Fares et al. 2014; Subramaniam et al. 2020; Hua et al. 2020; Chute et al. 2006; Wagner Jr. et al. 2016; Cohen et al. 2020; Zhang et al. 2020; Ko et al. 2011; de Ruijter et al. 2003; Obier et al. 2010; Peled et al. 2012; Vannini et al. 2019)
Self-renewal regulators of CB HSC expansion	HOXB4, MSI2, PPARG, YTHDF2, MLLT3 (Krosl et al. 2003; Huang et al. 2016; Rentas et al. 2016; Belew et al. 2018; Guo et al. 2017a; Li et al. 2018a; Huang and Broxmeyer 2018; Pina et al. 2008; Calvanese et al. 2019)

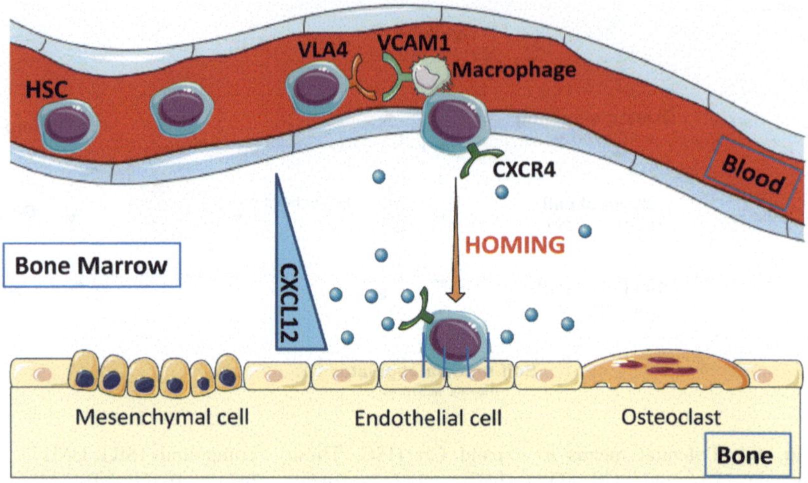

Fig. 6.2 HSC homing scheme. HSC homing is a multistep process including HSC leaving blood vessels and residing in the BM niche. VLA4-VCAM1 and CXCL12-CXCR4 axes are guiding signals that direct HSC homing

numbers are limited, as in the case of inadequate mPB or the case use of single CB units (Morgan et al. 2017; Broxmeyer et al. 1989).

6.3.1 Strategies to Enhance CB HSC Homing

In HSC transplantation, most infused HSCs fail to home to the BM microenvironment. To achieve higher engraftment, many efforts have been devoted to identifying means to enhance CB HSC homing efficacy to enable more HSCs to reach their BM destination. Researchers have reported multiple ways, forming several different layers to enhance CB HSC homing (from the cell membrane to the cytoplasm and inside the nucleus). Further efforts are needed to verify these strategies in clinical settings. Below are reported strategies to potentially enhance the CB HSC homing process.

6.3.1.1 Enhancing CB HSC Homing by Targeting Components on the Cell Surface

Dipeptidyl peptidase 4 (DPP4)/CD26 is a cell surface serine exopeptidase that cleaves the penultimate proline or alanine from the N-terminus polypeptides of targets, including numerous chemokines and cytokines (Ohnuma et al. 2008; Ropa and Broxmeyer 2020). DPP4 is widely expressed in many tissues, such as the liver, lung and kidney, in its membrane-bound form and in the serum in its soluble form. The enzymatic activity is critical for DPP4 to modulate many biological activities, such as immune, nervous and endocrine functions. Studies have demonstrated the roles of DPP4 in G-CSF-induced mobilization of HSCs, as well as HSC homing (Christopherson et al. 2003a, b). DPP4 cleaves the N-terminus of CXCL12, generating an inactive and truncated form of CXCL12. Inhibition or genetic deletion of DPP4 allows HSCs to home and engraft more efficiently in murine transplantation models (O'Leary et al. 2013; Campbell et al. 2007; Christopherson et al. 2004). Sitagliptin, an orally active DPP4 inhibitor, is widely used as an effective medication to lower high blood sugar in patients with diabetes. Sitagliptin administration has been shown to enhance single CB transplantation in patients with malignant blood disorders in a clinical trial (Farag et al. 2017). Therefore, manipulation of DPP4 activity shows potential in modulating CB HSC homing and needs further verification.

Prostaglandin E2 (PGE2) is an unsaturated fatty acid with various physiological activities (North et al. 2007). Studies have shown that pulse exposure of murine or human HSCs to PGE2 enhances their homing, survival and proliferation, ultimately leading to better engraftment (Hoggatt et al. 2009). PGE2 binds to specific G protein-coupled receptors (PTGER1–4/EP1–4), and murine and human HSCs consistently express these PGE2 receptors (Wang et al. 2017). Analysis has revealed that PGE2 treatment enhances CXCR4 mRNA levels and surface expression in HSCs, increases their migration to CXCL12 in vitro and elevates HSC homing to the BM microenvironment in vivo. Mechanistically, PGE2 treatment stably upregulates the protein levels of hypoxia-inducible factor 1 alpha (HIF1alpha), and HIF1alpha, as a DNA-binding transcription factor, binds to the hypoxia response elements on the CXCR4 transcription start site and promotes CXCR4 transcription. The CXCR4 selective antagonist AMD3100 blocks the effect of PGE2 on HSC homing, suggesting that CXCR4 mediates the effects of PGE2 on the homing and transplantation of HSCs. Serial transplantation experiments have shown that PGE2 pretreatment enhances HSC long-term reconstruction without lineage bias and does not alter HSC inherent competitiveness (Hoggatt et al. 2013). Since PGE2 pretreatment for HSC transplantation has been shown to be safe and effective in primates and in a phase I clinical trial (Cutler et al. 2013; Goessling et al. 2011), we await broad validation of PGE2 in CB HSC transplantation. The combination of PGE2 pretreatment and CD26 inhibition results in significantly enhanced competitive engraftment compared to either treatment alone, indicating that combination treatment could be an effective approach to further enhance clinical transplantation efficacy (Broxmeyer and Pelus 2014).

Calcium-sensing receptor (CaR), encoded by the CASR gene, is a G protein-coupled receptor that enables cells to sense small changes in extracellular calcium levels. CaR knockout HSCs show severe defects in homing and retention in the bone marrow microenvironment due to a cell autonomous defect in their capability to bind to bone marrow matrix molecules (Adams et al. 2006). Cinacalcet, a CaR agonist, is a medication used to treat high parathyroid hormone and high calcium levels in patients. Cinacalcet stimulation enhances the homing and engraftment of transplanted HSCs by increasing their binding to bone marrow matrix molecules and the responsiveness of HSCs to CXCL12 (Lam et al. 2011). In addition, cinacalcet treatment did not alter the expression of CXCR4 and VLA4, suggesting a unique strategy for CaR modulation in regulating HSC homing.

Lipid rafts are cholesterol-rich membrane microdomains that mediate many signal transductions (Ratajczak and Adamiak 2015). For optimal sensing of CXCL12, CXCR4 must be incorporated into cell membrane lipid rafts (Wysoczynski et al. 2005). Short-term hyperthermia (39 °C) treatment of HSCs results in increased membrane lipid raft formation, leading to more CXCR4 aggregation and colocalization with lipid rafts (Capitano et al. 2015). This study further confirms that hyperthermia treatment enhances HSC chemotaxis towards CXCL12 and promotes in vivo HSC homing and engraftment. Therefore, short-term hyperthermia treatment might be a simple and inexpensive method to enhance HSC homing and engraftment in patients. Additionally, Nlrp3 inflammasome signalling has been shown to regulate HSC homing and engraftment by promoting incorporation of CXCR4 into lipid rafts (Adamiak et al. 2020). Therefore, enhancing the integration of CXCR4 and lipid rafts could be an effective strategy to increase HSC homing.

6.3.1.2 Enhancing CB HSC Homing by Targeting Factors in the Cytoplasm

The free radical nitric oxide (NO) is a gaseous molecule that plays important roles in a variety of physiological regulatory processes (Sanders and Ward 2019). NO can freely diffuse across cellular membranes and activate its target enzyme in the cytoplasm, soluble guanylyl cyclase (sGC), to generate cyclic guanosine monophosphate (cGMP) from guanosine triphosphate (GTP) (Makrynitsa et al. 2019). cGMP acts as a second messenger to trigger important downstream signal transduction (Bork and Nikolaev 2018). It has recently been reported that NO signalling activation by sodium nitroprusside (SNP) results in significantly more HSC homing and engraftment in a murine transplantation model (Xu et al. 2020). Treating HSCs with riociguat, a sGC stimulator, also results in enhanced HSC homing and engraftment (Schwabl et al. 2018). Inside the cell, cGMP is degraded by phosphodiesterase 5 (PDE5), so a PDE5 inhibitor would suppress cGMP breakdown and activate cGMP signalling (Vasita et al. 2019). Treatment of HSCs with the PDE5 inhibitor sildenafil significantly increased homing and engraftment. By performing RNA sequencing analysis, an upregulation of genes linked with cell migration was revealed. The list of genes upregulated after NO signalling activation includes CXCR4 and PTGS2. PTGS2, also known as cyclooxygenase-2, is an enzyme involved in the generation of PGE2, so it would be interesting to see if PGE2 signalling is downstream of the NO signalling pathway in the regulation of HSC homing and engraftment. Taken together, this study suggests that HSC homing and engraftment can be enhanced by activating the NO/cGMP signalling pathway. The compounds tested in the study, such as SNP, riociguat and sildenafil, are FDA-approved medications broadly used for myocardial infarction, pulmonary hypertension and erectile dysfunction. Therefore, utilization of these drugs in HSC transplantation should be practical.

Heme oxygenase 1 (HO-1), encoded by the HMOX1 gene, is an essential stress-response enzyme in haem catabolism (Alcaraz et al. 2003). It cleaves haem to form biliverdin and plays an important role in anti-inflammatory, antiapoptotic and antioxidative activities during ischaemia/reperfusion injury. HO-1 is also important for CXCL12 signalling and the proper function of BM niche cells, including endothelial cells. HO-1 seems to be a negative regulator of HSC homing, as evidenced by the results that HO-1 knockout HSCs have enhanced migration towards CXCL12 gradients (Wysoczynski et al. 2015). Therefore, HO-1 inhibition could have a positive effect on enhancing the responsiveness of HSCs towards CXCL12, thus promoting HSC homing and engraftment. Indeed, transient treatment with an HO-1 inhibitor (tin protoporphyrin IX dichloride, SnPPIX) increases the chemotaxis and homing of HSCs and haematopoietic progenitor cells (Adamiak et al. 2016). This interesting approach awaits further tests in a clinical setting.

6.3.1.3 Enhancing CB HSC Homing by Targeting Components in the Nucleus

Histone acetylation is a major epigenetic modification of lysine residues on histone tails and plays key regulatory roles in gene expression (Zhou et al. 2018; Van and Santos 2018). Histone acetylation neutralizes the positive charge on the histone side chain, modulates chromatin structure and generates docking sites for transcription factors to facilitate transcription. Histone acetyltransferases (HATs) and deacetylases (HDACs) are two major types of enzymes responsible for the regulation of histone acetylation (Peserico and Simone 2011). HDAC inhibitors upregulate CXCR4 surface expression on HSCs and enhance CB HSC homing (Huang et al. 2018a). The HAT inhibitors C646 and EML425 can suppress the effects of HDAC inhibitors on CXCR4 expression, indicating that the balance between HATs and HDACs is critical for CXCR4 regulation. Moreover, HDAC5 has been identified as the HDAC specifically involved in CXCR4 regulation, HSC homing and engraftment (Huang et al. 2018a). HDAC5 inhibition increases histone acetylation levels on the CXCR4 promoter region and acetylated p65 levels in the nucleus, leading to elevated CXCR4 transcription. HDAC5 belongs to class IIa HDACs, which can react to various extracellular stimuli by assembling into multiple protein complexes and shuttling between the cytoplasm and nucleus (Delcuve et al. 2012). Thus, the involvement of HDAC5 provides a mechanism connecting histone acetylation, HSC homing regulation and extracellular signals.

In addition, histone acetylation and HSC homing are regulated by glucocorticoids and their receptors. Glucocorticoids are members of the steroid hormone family that are involved in carbohydrate metabolism and the immune response. Glucocorticoids have been recently identified as enhancers of HSC homing by upregulating CXCR4 surface expression (Guo et al. 2017b). Glucocorticoids bind to glucocorticoid receptors, which are nuclear receptors that function as ligand-activated transcription factors (Tan and

Wahli 2016). Activated glucocorticoid receptors can bind to the glucocorticoid response elements on the promoter region of downstream genes, which mediate various physiological effects. In HSCs, activated glucocorticoid receptors translocate into the nucleus and bind to the CXCR4 promoter region, followed by recruiting the SRC1/p300 HAT complex that promotes histone H4K5 and H4K16 acetylation, which then facilitates CXCR4 transcription and HSC homing. Both HDAC5 inhibition and glucocorticoid treatment converge on histone acetylation at the CXCR4 promoter region, indicating a key role for histone acetylation in the regulation of CXCR4 expression and HSC homing.

HIF1alpha is a DNA-binding transcription factor and a master regulator of the cellular response to hypoxia (Huang et al. 2018b; Galbraith et al. 2013). HIF1alpha plays an essential role in regulating cell survival, cell fate decisions, energy metabolism and HSC activity. The HIF1alpha stabilization compound dimethyloxalylglycine (DMOG) increases CXCR4 transcription and promotes HSC homing (Speth et al. 2014). Analysis shows that HIF1alpha binds to the hypoxia response elements located at approximately −1.3 kb from the transcription start site at the CXCR4 promoter and recruits p300/CBP to promote CXCR4 transcription. Caffeic acid phenethyl ester (CAPE) induces HIF1alpha expression by inhibiting HIF prolyl hydroxylase and preventing HIF1alpha degradation. CAPE administration promotes HSC homing and engraftment by inducing the expression of HIF1alpha and CXCL12 in the BM (Chen et al. 2017). The HIF1alpha inhibitor PX-478 suppresses CAPE-mediated enhanced HSC homing. HSCs reside in a hypoxic microenvironment in vivo. Collecting and processing HSCs in ambient air has an effect on HSC function, a phenomenon termed extra physiological oxygen stress/shock (EPHOSS) (Mantel et al. 2015; Broxmeyer et al. 2015). By collecting and processing mouse BM, human CB (Mantel et al. 2015) and mPB by G-CSF and by G-CSF plus AMD3100 (Aljoufi et al. 2020) under lowered O_2 tension of 3% O_2, more functional HSC numbers are collected (Mantel et al. 2015). EPHOSS is mediated by a p53-cyclophilin D-mitochondria permeability transition pore (MPTP) axis, involving HIF1alpha and hypoxia-regulated microRNA miR-210. These results further support the important role of HIF1alpha in HSC homing and engraftment.

6.3.2 Opportunities and Challenges for HSC Homing

The efficacy of HSC transplantation is compromised when donor cell numbers are low, such as the case of single CB transplantation. In addition to expansion (Fig. 6.1), enhancing HSC homing is an alternative means to increase the efficacy of CB transplantation and achieve positive clinical outcomes. Various approaches have been identified to promote HSC homing (Table 6.2), mostly by targeting the CXCL12/CXCR4 axis. In the near future, we hope to see some of these strategies being successfully translated and employed in clinical settings. It is also practical to combine some of the methods mentioned above for better CB HSC homing and engraftment or combine one HSC homing means with an HSC expansion approach to achieve even higher HSC engrafting efficacy (Fig. 6.2). We also expect some novel and even more powerful strategies to be developed to enhance HSC homing, especially by targeting the VCAM1/VLA4 axis, which has been much less explored. The concept of collecting more HSCs through hypoxia collection/processing of cells (Mantel et al. 2015; Broxmeyer et al. 2015; Aljoufi et al. 2020), or by other means to mimic this by using combinations of antioxidant and/or epigenetic enzyme inhibitors (Cai et al. 2018), or cold collection/processing of cells (Broxmeyer et al. 2020) could provide even greater numbers of HSCs for clinical use after procedures to ex vivo expand them and/or to increase their homing efficiency. Recently, functional HSCs have been generated from endothelial cells and pluripotent stem cells (Sandler et al. 2014; Sugimura et al. 2017). It would be interesting to see whether the engraftment of these HSCs could be enhanced by fine-tuning mechanisms

Table 6.2 Cellular components and factors involved in regulating CB HSC homing

Regulatory component	Description	Function	References
DPP4	Cell surface serine exopeptidase	Inhibits HSC homing by cleaving N-terminal of CXCL12 and generating an inactive/truncated form of CXCL12	Christopherson et al. (2004)
Prostaglandin E2	Unsaturated fatty acid	Binds to specific receptors (PTGER1-4/EP1-4) and promotes CXCR4 transcription	Hoggatt et al. (2013)
Calcium-sensing receptor	G protein-coupled receptor	Increases the binding to bone marrow matrix molecules and the responsiveness of HSCs to CXCL12	Lam et al. (2011)
Lipid rafts	Cholesterol-rich membrane microdomains	Enhances HSC chemotaxis towards CXCL12 and promotes in vivo HSC homing	Capitano et al. (2015)
Nitric oxide	Gaseous signalling molecule	Diffuses across cellular membranes and activates its soluble guanylyl cyclase (sGC) to generate cGMP and promotes CXCR4 transcription	Xu et al. (2020)
Heme oxygenase 1	Stress-response enzyme in heme catabolism	Suppresses chemotactic responsiveness of HSC towards CXCL12 and inhibits HSC homing	Wysoczynski et al. (2015)
HDAC5	Class IIa histone deacetylase	Inhibits histone acetylation levels on the CXCR4 promoter region and acetylated p65 levels in the nucleus	Huang et al. (2018a)
Glucocorticoid	Steroid hormone family member	Binds to the glucocorticoid receptor and facilitates CXCR4 transcription	Guo et al. (2017b)
HIF1alpha	DNA-binding transcription factor	Binds to the hypoxia response elements located at the CXCR4 promoter and recruits p300/CBP to promote CXCR4 transcription	Speth et al. (2014)

regulating the expansion or homing of HSCs. The pathways identified in regulating HSC homing might also participate in tumour metastasis or homing processes of other stem cell types, such as mesenchymal stem/stromal cells.

Acknowledgements The published studies by these authors were supported by the National Key R&D Program of China Stem Cell and Translation Research (2019YFA0111800), Major Research Plan of National Natural Science Foundation of China (91957107) and General Program of National Natural Science Foundation of China (81970095) to BG and Fudan University Start-up Research Grant to XH and US Public Health Service Grants to HEB from the NIH (R35HL139599, R01DK109188, U54 DK106846).

References

Adamiak M, Moore JBT, Zhao J, Abdelbaset-Ismail A, Grubczak K, Rzeszotek S, Wysoczynski M, Ratajczak MZ (2016) Downregulation of heme oxygenase 1 (HO-1) activity in hematopoietic cells enhances their engraftment after transplantation. Cell Transplant 25:1265–1276

Adamiak M, Abdel-Latif A, Bujko K, Thapa A, Anusz K, Tracz M, Brzezniakiewicz-Janus K, Ratajczak J, Kucia M, Ratajczak MZ (2020) Nlrp3 inflammasome signaling regulates the homing and engraftment of hematopoietic stem cells (HSPCs) by enhancing incorporation of CXCR4 receptor into membrane lipid rafts. Stem Cell Rev Rep 16:954–967

Adams GB, Chabner KT, Alley IR, Olson DP, Szczepiorkowski ZM, Poznansky MC, Kos CH, Pollak MR, Brown EM, Scadden DT (2006) Stem cell engraftment at the endosteal niche is specified by the calcium-sensing receptor. Nature 439:599–603

Alcaraz MJ, Fernandez P, Guillen MI (2003) Anti-inflammatory actions of the heme oxygenase-1 pathway. Curr Pharm Des 9:2541–2551

Aljoufi A, Cooper S, Broxmeyer HE (2020) Collection and processing of mobilized mouse peripheral blood at lowered oxygen tension yields enhanced numbers of hematopoietic stem cells. Stem Cell Rev Rep 16:946–953

Antonchuk J, Sauvageau G, Humphries RK (2002) HOXB4-induced expansion of adult hematopoietic stem cells ex vivo. Cell 109:39–45

Ballen KK, Gluckman E, Broxmeyer HE (2013) Umbilical cord blood transplantation: the first 25 years and beyond. Blood 122:491–498

Beerman I, Luis TC, Singbrant S, Lo Celso C, Mendez-Ferrer S (2017) The evolving view of the hematopoietic stem cell niche. Exp Hematol 50:22–26

Belew MS, Bhatia S, Keyvani Chahi A, Rentas S, Draper JS, Hope KJ (2018) PLAG1 and USF2 co-regulate expression of Musashi-2 in human hematopoietic stem and progenitor cells. Stem Cell Rep 10:1384–1397

Bigas A, Guiu J, Gama-Norton L (2013) Notch and Wnt signaling in the emergence of hematopoietic stem cells. Blood Cells Mol Dis 51:264–270

Boitano AE, Wang J, Romeo R, Bouchez LC, Parker AE, Sutton SE, Walker JR, Flaveny CA, Perdew GH, Denison MS, Schultz PG, Cooke MP (2010) Aryl hydrocarbon receptor antagonists promote the expansion of human hematopoietic stem cells. Science 329: 1345–1348

Bork NI, Nikolaev VO (2018) cGMP signaling in the cardiovascular system-the role of compartmentation and its live cell imaging. Int J Mol Sci 19:801

Bosi A, Bartolozzi B (2010) Safety of bone marrow stem cell donation: a review. Transplant Proc 42:2192–2194

Boulais PE, Frenette PS (2015) Making sense of hematopoietic stem cell niches. Blood 125:2621–2629

Broxmeyer HE (2016) Enhancing the efficacy of engraftment of cord blood for hematopoietic cell transplantation. Transfus Apher Sci 54:364–372

Broxmeyer HE, Farag S (2013) Background and future considerations for human cord blood hematopoietic cell transplantation, including economic concerns. Stem Cells Dev 22:103–110

Broxmeyer HE, Pelus LM (2014) Inhibition of DPP4/CD26 and dmPGE(2) treatment enhances engraftment of mouse bone marrow hematopoietic stem cells. Blood Cells Mol Dis 53:34–38

Broxmeyer HE, Douglas GW, Hangoc G, Cooper S, Bard J, English D, Arny M, Thomas L, Boyse EA (1989) Human umbilical cord blood as a potential source of transplantable hematopoietic stem/progenitor cells. Proc Natl Acad Sci U S A 86:3828–3832

Broxmeyer HE, O'Leary HA, Huang X, Mantel C (2015) The importance of hypoxia and extra physiologic oxygen shock/stress for collection and processing of stem and progenitor cells to understand true physiology/pathology of these cells ex vivo. Curr Opin Hematol 22:273–278

Broxmeyer HE, Cooper S, Capitano ML (2020) Enhanced collection of phenotypic and engrafting human cord blood hematopoietic stem cells at 4 degrees C. Stem Cells 38:1326–1331

Cai Q, Capitano M, Huang X, Guo B, Cooper S, Broxmeyer HE (2018) Combinations of antioxidants and/or of epigenetic enzyme inhibitors allow for enhanced collection of mouse bone marrow hematopoietic stem cells in ambient air. Blood Cells Mol Dis 71:23–28

Calvanese V, Nguyen AT, Bolan TJ, Vavilina A, Su T, Lee LK, Wang Y, Lay FD, Magnusson M, Crooks GM, Kurdistani SK, Mikkola HKA (2019) MLLT3 governs human haematopoietic stem-cell self-renewal and engraftment. Nature 576:281–286

Campbell TB, Hangoc G, Liu Y, Pollok K, Broxmeyer HE (2007) Inhibition of CD26 in human cord blood CD34+ cells enhances their engraftment of nonobese diabetic/severe combined immunodeficiency mice. Stem Cells Dev 16:347–354

Capitano ML, Hangoc G, Cooper S, Broxmeyer HE (2015) Mild heat treatment primes human CD34(+) cord blood cells for migration toward SDF-1alpha and enhances engraftment in an NSG mouse model. Stem Cells 33:1975–1984

Chabannon C, Kuball J, Bondanza A, Dazzi F, Pedrazzoli P, Toubert A, Ruggeri A, Fleischhauer K, Bonini C (2018) Hematopoietic stem cell transplantation in its 60s: a platform for cellular therapies. Sci Transl Med 10:9630

Chaurasia P, Gajzer DC, Schaniel C, D'Souza S, Hoffman R (2014) Epigenetic reprogramming induces the expansion of cord blood stem cells. J Clin Invest 124: 2378–2395

Chen X, Han Y, Zhang B, Liu Y, Wang S, Liao T, Deng Z, Fan Z, Zhang J, He L, Yue W, Li Y, Pei X (2017) Caffeic acid phenethyl ester promotes haematopoietic stem/progenitor cell homing and engraftment. Stem Cell Res Ther 8:255

Chen Y, Yao C, Teng Y, Jiang R, Huang X, Liu S, Wan J, Broxmeyer HE, Guo B (2019) Phorbol ester induced ex vivo expansion of rigorously-defined phenotypic but not functional human cord blood hematopoietic stem cells: a cautionary tale demonstrating that phenotype does not always recapitulate stem cell function. Leukemia 33:2962–2966

Chotinantakul K, Prasajak P, Leeanansaksiri W (2013) Wnt1 accelerates an ex vivo expansion of human cord blood CD34(+)CD38(-) cells. Stem Cells Int 2013:909812

Christopherson KW, Cooper S, Broxmeyer HE (2003a) Cell surface peptidase CD26/DPPIV mediates G-CSF mobilization of mouse progenitor cells. Blood 101: 4680–4686

Christopherson KW, Cooper S, Hangoc G, Broxmeyer HE (2003b) CD26 is essential for normal G-CSF-induced progenitor cell mobilization as determined by CD26-/- mice. Exp Hematol 31:1126–1134

Christopherson KW, Hangoc G, Mantel CR, Broxmeyer HE (2004) Modulation of hematopoietic stem cell homing and engraftment by CD26. Science 305: 1000–1003

Chute JP, Muramoto GG, Whitesides J, Colvin M, Safi R, Chao NJ, McDonnell DP (2006) Inhibition of aldehyde dehydrogenase and retinoid signaling induces the expansion of human hematopoietic stem cells. Proc Natl Acad Sci U S A 103:11707–11712

Cohen S, Roy J, Lachance S, Delisle JS, Marinier A, Busque L, Roy DC, Barabe F, Ahmad I, Bambace N, Bernard L, Kiss T, Bouchard P, Caudrelier P, Landais S, Larochelle F, Chagraoui J, Lehnertz B, Corneau S, Tomellini E, van Kampen JJA, Cornelissen JJ, Dumont-Lagace M, Tanguay M, Li Q, Lemieux S, Zandstra PW, Sauvageau G (2020) Hematopoietic stem cell transplantation using single UM171-expanded cord blood: a single-arm, phase 1-2 safety and feasibility study. Lancet Haematol 7:e134–e145

Connelly JA, Choi SW, Levine JE (2012) Hematopoietic stem cell transplantation for severe congenital neutropenia. Curr Opin Hematol 19:44–51

Copelan EA (2006) Hematopoietic stem-cell transplantation. N Engl J Med 354:1813–1826

Crane GM, Jeffery E, Morrison SJ (2017) Adult haematopoietic stem cell niches. Nat Rev Immunol 17:573–590

Crippa S, Bernardo ME (2018) Mesenchymal stromal cells: role in the BM niche and in the support of hematopoietic stem cell transplantation. HemaSphere 2:e151

Cutler C, Multani P, Robbins D, Kim HT, Le T, Hoggatt J, Pelus LM, Desponts C, Chen YB, Rezner B, Armand P, Koreth J, Glotzbecker B, Ho VT, Alyea E, Isom M, Kao G, Armant M, Silberstein L, Hu P, Soiffer RJ, Scadden DT, Ritz J, Goessling W, North TE, Mendlein J, Ballen K, Zon LI, Antin JH, Shoemaker DD (2013) Prostaglandin-modulated umbilical cord blood hematopoietic stem cell transplantation. Blood 122:3074–3081

de Graaf CA, Metcalf D (2011) Thrombopoietin and hematopoietic stem cells. Cell Cycle 10:1582–1589

de Lima M, McMannis J, Gee A, Komanduri K, Couriel D, Andersson BS, Hosing C, Khouri I, Jones R, Champlin R, Karandish S, Sadeghi T, Peled T, Grynspan F, Daniely Y, Nagler A, Shpall EJ (2008) Transplantation of ex vivo expanded cord blood cells using the copper chelator tetraethylenepentamine: a phase I/II clinical trial. Bone Marrow Transplant 41: 771–778

de Ruijter AJ, van Gennip AH, Caron HN, Kemp S, van Kuilenburg AB (2003) Histone deacetylases (HDACs): characterization of the classical HDAC family. Biochem J 370:737–749

Delaney C, Heimfeld S, Brashem-Stein C, Voorhies H, Manger RL, Bernstein ID (2010) Notch-mediated expansion of human cord blood progenitor cells capable of rapid myeloid reconstitution. Nat Med 16:232–236

Delcuve GP, Khan DH, Davie JR (2012) Roles of histone deacetylases in epigenetic regulation: emerging paradigms from studies with inhibitors. Clin Epigenetics 4:5

Ding L, Saunders TL, Enikolopov G, Morrison SJ (2012) Endothelial and perivascular cells maintain haematopoietic stem cells. Nature 481:457–462

Dolence JJ, Gwin KA, Shapiro MB, Medina KL (2014) Flt3 signaling regulates the proliferation, survival, and maintenance of multipotent hematopoietic progenitors that generate B cell precursors. Exp Hematol 42:380–393 e383

Duncan AW, Rattis FM, DiMascio LN, Congdon KL, Pazianos G, Zhao C, Yoon K, Cook JM, Willert K, Gaiano N, Reya T (2005) Integration of Notch and Wnt signaling in hematopoietic stem cell maintenance. Nat Immunol 6:314–322

Fajardo-Orduna GR, Mayani H, Flores-Guzman P, Flores-Figueroa E, Hernandez-Estevez E, Castro-Manrreza M, Pina-Sanchez P, Arriaga-Pizano L, Gomez-Delgado A, Alarcon-Santos G, Balvanera-Ortiz O, Montesinos JJ (2017) Human mesenchymal stem/stromal cells from umbilical cord blood and placenta exhibit similar capacities to promote expansion of hematopoietic progenitor cells in vitro. Stem Cells Int 2017:6061729

Fan X, Gay FP, Lim FW, Ang JM, Chu PP, Bari S, Hwang WY (2014) Low-dose insulin-like growth factor binding proteins 1 and 2 and angiopoietin-like protein 3 coordinately stimulate ex vivo expansion of human umbilical cord blood hematopoietic stem cells as assayed in NOD/SCID gamma null mice. Stem Cell Res Ther 5:71

Farag SS, Nelson R, Cairo MS, O'Leary HA, Zhang S, Huntley C, Delgado D, Schwartz J, Zaid MA, Abonour R, Robertson M, Broxmeyer H (2017) High-dose sitagliptin for systemic inhibition of dipeptidylpeptidase-4 to enhance engraftment of single cord umbilical cord blood transplantation. Oncotarget 8:110350–110357

Fares I, Chagraoui J, Gareau Y, Gingras S, Ruel R, Mayotte N, Csaszar E, Knapp DJ, Miller P, Ngom M, Imren S, Roy DC, Watts KL, Kiem HP, Herrington R, Iscove NN, Humphries RK, Eaves CJ, Cohen S, Marinier A, Zandstra PW, Sauvageau G (2014) Cord blood expansion. Pyrimidoindole derivatives are agonists of human hematopoietic stem cell self-renewal. Science 345:1509–1512

Fleming HE, Janzen V, Lo Celso C, Guo J, Leahy KM, Kronenberg HM, Scadden DT (2008) Wnt signaling in the niche enforces hematopoietic stem cell quiescence and is necessary to preserve self-renewal in vivo. Cell Stem Cell 2:274–283

Galbraith MD, Allen MA, Bensard CL, Wang X, Schwinn MK, Qin B, Long HW, Daniels DL, Hahn WC, Dowell RD, Espinosa JM (2013) HIF1A employs CDK8-mediator to stimulate RNAPII elongation in response to hypoxia. Cell 153:1327–1339

Gao X, Xu C, Asada N, Frenette PS (2018) The hematopoietic stem cell niche: from embryo to adult. Development 145:139691

Gluckman E, Broxmeyer HA, Auerbach AD, Friedman HS, Douglas GW, Devergie A, Esperou H, Thierry D, Socie G, Lehn P et al (1989) Hematopoietic reconstitution in a patient with Fanconi's anemia by means of umbilical-cord blood from an HLA-identical sibling. N Engl J Med 321:1174–1178

Goessling W, Allen RS, Guan X, Jin P, Uchida N, Dovey M, Harris JM, Metzger ME, Bonifacino AC, Stroncek D, Stegner J, Armant M, Schlaeger T, Tisdale JF, Zon LI, Donahue RE, North TE (2011) Prostaglandin E2 enhances human cord blood stem cell xenotransplants and shows long-term safety in preclinical nonhuman primate transplant models. Cell Stem Cell 8:445–458

Golan K, Vagima Y, Ludin A, Itkin T, Cohen-Gur S, Kalinkovich A, Kollet O, Kim C, Schajnovitz A, Ovadya Y, Lapid K, Shivtiel S, Morris AJ, Ratajczak MZ, Lapidot T (2012) S1P promotes murine progenitor cell egress and mobilization via S1P1-mediated ROS signaling and SDF-1 release. Blood 119:2478–2488

Grafone T, Palmisano M, Nicci C, Storti S (2012) An overview on the role of FLT3-tyrosine kinase receptor in acute myeloid leukemia: biology and treatment. Oncol Rev 6:e8

Guo M, Liu X, Zheng X, Huang Y, Chen X (2017a) m(6)A RNA modification determines cell fate by regulating mRNA degradation. Cell Reprogram 19:225–231

Guo B, Huang X, Cooper S, Broxmeyer HE (2017b) Glucocorticoid hormone-induced chromatin remodeling enhances human hematopoietic stem cell homing and engraftment. Nat Med 23:424–428

Guo B, Huang X, Lee MR, Lee SA, Broxmeyer HE (2018a) Antagonism of PPAR-gamma signaling expands human hematopoietic stem and progenitor cells by enhancing glycolysis. Nat Med 24:360–367

Guo B, Huang X, Broxmeyer HE (2018b) Enhancing human cord blood hematopoietic stem cell engraftment by targeting nuclear hormone receptors. Curr Opin Hematol 25:245–252

Han J, Wu J, Silke J (2020) An overview of mammalian p38 mitogen-activated protein kinases, central regulators of cell stress and receptor signaling. F1000Res 9:653

Hoggatt J, Singh P, Sampath J, Pelus LM (2009) Prostaglandin E2 enhances hematopoietic stem cell homing, survival, and proliferation. Blood 113:5444–5455

Hoggatt J, Mohammad KS, Singh P, Pelus LM (2013) Prostaglandin E2 enhances long-term repopulation but does not permanently alter inherent stem cell competitiveness. Blood 122:2997–3000

Horwitz ME, Chao NJ, Rizzieri DA, Long GD, Sullivan KM, Gasparetto C, Chute JP, Morris A, McDonald C, Waters-Pick B, Stiff P, Wease S, Peled A, Snyder D, Cohen EG, Shoham H, Landau E, Friend E, Peleg I, Aschengrau D, Yackoubov D, Kurtzberg J, Peled T (2014) Umbilical cord blood expansion with nicotinamide provides long-term multilineage engraftment. J Clin Invest 124:3121–3128

Hua P, Hester J, Adigbli G, Li R, Psaila B, Roy A, Bataille CJR, Wynne GM, Jackson T, Milne TA, Russell A, Davies J, Roberts I, Issa F, Watt SM (2020) The BET inhibitor CPI203 promotes ex vivo expansion of cord blood long-term repopulating HSCs and megakaryocytes. Blood 136:2410–2415

Huang X, Broxmeyer HE (2018) m(6)A reader suppression bolsters HSC expansion. Cell Res 28:875–876

Huang X, Broxmeyer HE (2019) Progress towards improving homing and engraftment of hematopoietic stem cells for clinical transplantation. Curr Opin Hematol 26:266–272

Huang X, Lee MR, Cooper S, Hangoc G, Hong KS, Chung HM, Broxmeyer HE (2016) Activation of OCT4 enhances ex vivo expansion of human cord blood hematopoietic stem and progenitor cells by regulating HOXB4 expression. Leukemia 30:144–153

Huang X, Guo B, Liu S, Wan J, Broxmeyer HE (2018a) Neutralizing negative epigenetic regulation by HDAC5 enhances human haematopoietic stem cell homing and engraftment. Nat Commun 9:2741

Huang X, Trinh T, Aljoufi A, Broxmeyer HE (2018b) Hypoxia signaling pathway in stem cell regulation: good and evil. Curr Stem Cell Rep 4:149–157

Huang X, Guo B, Capitano M, Broxmeyer HE (2019) Past, present, and future efforts to enhance the efficacy of cord blood hematopoietic cell transplantation. F1000Res 8:1833

Johnson GL, Nakamura K (2007) The c-jun kinase/stress-activated pathway: regulation, function and role in human disease. Biochim Biophys Acta 1773:1341–1348

Juarez JG, Harun N, Thien M, Welschinger R, Baraz R, Pena AD, Pitson SM, Rettig M, DiPersio JF, Bradstock KF, Bendall LJ (2012) Sphingosine-1-phosphate facilitates trafficking of hematopoietic stem cells and their mobilization by CXCR4 antagonists in mice. Blood 119:707–716

Kabiri Z, Numata A, Kawasaki A, Edison, Tenen DG, Virshup DM (2015) Wnts are dispensable for differentiation and self-renewal of adult murine hematopoietic stem cells. Blood 126:1086–1094

Kiernan J, Damien P, Monaghan M, Shorr R, McIntyre L, Fergusson D, Tinmouth A, Allan D (2017) Clinical studies of ex vivo expansion to accelerate engraftment after umbilical cord blood transplantation: a systematic review. Transfus Med Rev 31:173–182

Kimura S, Roberts AW, Metcalf D, Alexander WS (1998) Hematopoietic stem cell deficiencies in mice lacking c-Mpl, the receptor for thrombopoietin. Proc Natl Acad Sci U S A 95:1195–1200

Ko KH, Holmes T, Palladinetti P, Song E, Nordon R, O'Brien TA, Dolnikov A (2011) GSK-3beta inhibition promotes engraftment of ex vivo-expanded hematopoietic stem cells and modulates gene expression. Stem Cells 29:108–118

Krosl J, Austin P, Beslu N, Kroon E, Humphries RK, Sauvageau G (2003) In vitro expansion of hematopoietic stem cells by recombinant TAT-HOXB4 protein. Nat Med 9:1428–1432

Lam BS, Cunningham C, Adams GB (2011) Pharmacologic modulation of the calcium-sensing receptor enhances hematopoietic stem cell lodgment in the adult bone marrow. Blood 117:1167–1175

Lampreia FP, Carmelo JG, Anjos-Afonso F (2017) Notch signaling in the regulation of hematopoietic stem cell. Curr Stem Cell Rep 3:202–209

Lennartsson J, Ronnstrand L (2012) Stem cell factor receptor/c-Kit: from basic science to clinical implications. Physiol Rev 92:1619–1649

Li N, Feugier P, Serrurrier B, Latger-Cannard V, Lesesve JF, Stoltz JF, Eljaafari A (2007) Human mesenchymal stem cells improve ex vivo expansion of adult human

CD34+ peripheral blood progenitor cells and decrease their allostimulatory capacity. Exp Hematol 35:507–515

Li Z, Qian P, Shao W, Shi H, He XC, Gogol M, Yu Z, Wang Y, Qi M, Zhu Y, Perry JM, Zhang K, Tao F, Zhou K, Hu D, Han Y, Zhao C, Alexander R, Xu H, Chen S, Peak A, Hall K, Peterson M, Perera A, Haug JS, Parmely T, Li H, Shen B, Zeitlinger J, He C, Li L (2018a) Suppression of m(6)A reader Ythdf2 promotes hematopoietic stem cell expansion. Cell Res 28:904–917

Li D, Xue W, Li M, Dong M, Wang J, Wang X, Li X, Chen K, Zhang W, Wu S, Zhang Y, Gao L, Chen Y, Chen J, Zhou BO, Zhou Y, Yao X, Li L, Wu D, Pan W (2018b) VCAM-1(+) macrophages guide the homing of HSPCs to a vascular niche. Nature 564:119–124

Li H, Pei H, Xie X, Wang S, Jia Y, Zhang B, Fan Z, Liu Y, Bai Y, Han Y, He L, Nan X, Yue W, Pei X (2019) Liver sinusoidal endothelial cells promote the expansion of human cord blood hematopoietic stem and progenitor cells. Int J Mol Sci 20:1985

Luis TC, Weerkamp F, Naber BA, Baert MR, de Haas EF, Nikolic T, Heuvelmans S, De Krijger RR, van Dongen JJ, Staal FJ (2009) Wnt3a deficiency irreversibly impairs hematopoietic stem cell self-renewal and leads to defects in progenitor cell differentiation. Blood 113:546–554

Makrynitsa GI, Zompra AA, Argyriou AI, Spyroulias GA, Topouzis S (2019) Therapeutic targeting of the soluble guanylate cyclase. Curr Med Chem 26:2730–2747

Mancini SJ, Mantei N, Dumortier A, Suter U, MacDonald HR, Radtke F (2005) Jagged1-dependent notch signaling is dispensable for hematopoietic stem cell self-renewal and differentiation. Blood 105:2340–2342

Mantel CR, O'Leary HA, Chitteti BR, Huang X, Cooper S, Hangoc G, Brustovetsky N, Srour EF, Lee MR, Messina-Graham S, Haas DM, Falah N, Kapur R, Pelus LM, Bardeesy N, Fitamant J, Ivan M, Kim KS, Broxmeyer HE (2015) Enhancing hematopoietic stem cell transplantation efficacy by mitigating oxygen shock. Cell 161:1553–1565

Milano F, Appelbaum FR, Delaney C (2016) Cord-blood transplantation in patients with minimal residual disease. N Engl J Med 375:2204–2205

Morgan RA, Gray D, Lomova A, Kohn DB (2017) Hematopoietic stem cell gene therapy: progress and lessons learned. Cell Stem Cell 21:574–590

Nagasawa T (2015) CXCL12/SDF-1 and CXCR4. Front Immunol 6:301

Navas TA, Mohindru M, Estes M, Ma JY, Sokol L, Pahanish P, Parmar S, Haghnazari E, Zhou L, Collins R, Kerr I, Nguyen AN, Xu Y, Platanias LC, List AA, Higgins LS, Verma A (2006) Inhibition of overactivated p38 MAPK can restore hematopoiesis in myelodysplastic syndrome progenitors. Blood 108:4170–4177

Nishino T, Miyaji K, Ishiwata N, Arai K, Yui M, Asai Y, Nakauchi H, Iwama A (2009) Ex vivo expansion of human hematopoietic stem cells by a small-molecule agonist of c-MPL. Exp Hematol 37:1364–1377 e1364

Nishino T, Wang C, Mochizuki-Kashio M, Osawa M, Nakauchi H, Iwama A (2011) Ex vivo expansion of human hematopoietic stem cells by garcinol, a potent inhibitor of histone acetyltransferase. PLoS One 6:e24298

North TE, Goessling W, Walkley CR, Lengerke C, Kopani KR, Lord AM, Weber GJ, Bowman TV, Jang IH, Grosser T, Fitzgerald GA, Daley GQ, Orkin SH, Zon LI (2007) Prostaglandin E2 regulates vertebrate haematopoietic stem cell homeostasis. Nature 447:1007–1011

Obier N, Uhlemann CF, Muller AM (2010) Inhibition of histone deacetylases by Trichostatin A leads to a HoxB4-independent increase of hematopoietic progenitor/stem cell frequencies as a result of selective survival. Cytotherapy 12:899–908

Ohnuma K, Takahashi N, Yamochi T, Hosono O, Dang NH, Morimoto C (2008) Role of CD26/dipeptidyl peptidase IV in human T cell activation and function. Front Biosci 13:2299–2310

O'Leary H, Ou X, Broxmeyer HE (2013) The role of dipeptidyl peptidase 4 in hematopoiesis and transplantation. Curr Opin Hematol 20:314–319

Peled T, Shoham H, Aschengrau D, Yackoubov D, Frei G, Rosenheimer GN, Lerrer B, Cohen HY, Nagler A, Fibach E, Peled A (2012) Nicotinamide, a SIRT1 inhibitor, inhibits differentiation and facilitates expansion of hematopoietic progenitor cells with enhanced bone marrow homing and engraftment. Exp Hematol 40:342–355 e341

Pelus LM, Broxmeyer HE (2018) Peripheral blood stem cell mobilization; a look ahead. Curr Stem Cell Rep 4:273–281

Peserico A, Simone C (2011) Physical and functional HAT/HDAC interplay regulates protein acetylation balance. J Biomed Biotechnol 2011:371832

Pina C, May G, Soneji S, Hong D, Enver T (2008) MLLT3 regulates early human erythroid and megakaryocytic cell fate. Cell Stem Cell 2:264–273

Ratajczak MZ, Adamiak M (2015) Membrane lipid rafts, master regulators of hematopoietic stem cell retention in bone marrow and their trafficking. Leukemia 29:1452–1457

Ratajczak MZ, Suszynska M (2016) Emerging strategies to enhance homing and engraftment of hematopoietic stem cells. Stem Cell Rev 12:121–128

Ratajczak MZ, Lee H, Wysoczynski M, Wan W, Marlicz W, Laughlin MJ, Kucia M, Janowska-Wieczorek A, Ratajczak J (2010) Novel insight into stem cell mobilization-plasma sphingosine-1-phosphate is a major chemoattractant that directs the egress of hematopoietic stem progenitor cells from the bone marrow and its level in peripheral blood increases during mobilization due to activation of complement cascade/membrane attack complex. Leukemia 24:976–985

Rentas S, Holzapfel N, Belew MS, Pratt G, Voisin V, Wilhelm BT, Bader GD, Yeo GW, Hope KJ (2016) Musashi-2 attenuates AHR signalling to expand human haematopoietic stem cells. Nature 532:508–511

Rettig MP, Ansstas G, DiPersio JF (2012) Mobilization of hematopoietic stem and progenitor cells using inhibitors of CXCR4 and VLA-4. Leukemia 26:34–53

Ropa J, Broxmeyer HE (2020) An expanded role for dipeptidyl peptidase 4 in cell regulation. Curr Opin Hematol 27:215–224

Ruiz-Herguido C, Guiu J, D'Altri T, Ingles-Esteve J, Dzierzak E, Espinosa L, Bigas A (2012) Hematopoietic stem cell development requires transient Wnt/beta-catenin activity. J Exp Med 209:1457–1468

Safi R, Muramoto GG, Salter AB, Meadows S, Himburg H, Russell L, Daher P, Doan P, Leibowitz MD, Chao NJ, McDonnell DP, Chute JP (2009) Pharmacological manipulation of the RAR/RXR signaling pathway maintains the repopulating capacity of hematopoietic stem cells in culture. Mol Endocrinol 23:188–201

Sanders KM, Ward SM (2019) Nitric oxide and its role as a non-adrenergic, non-cholinergic inhibitory neurotransmitter in the gastrointestinal tract. Br J Pharmacol 176:212–227

Sandler VM, Lis R, Liu Y, Kedem A, James D, Elemento O, Butler JM, Scandura JM, Rafii S (2014) Reprogramming human endothelial cells to haematopoietic cells requires vascular induction. Nature 511:312–318

Sanz-Rodriguez F, Hidalgo A, Teixido J (2001) Chemokine stromal cell-derived factor-1alpha modulates VLA-4 integrin-mediated multiple myeloma cell adhesion to CS-1/fibronectin and VCAM-1. Blood 97:346–351

Schuettpelz LG, Borgerding JN, Christopher MJ, Gopalan PK, Romine MP, Herman AC, Woloszynek JR, Greenbaum AM, Link DC (2014) G-CSF regulates hematopoietic stem cell activity, in part, through activation of Toll-like receptor signaling. Leukemia 28:1851–1860

Schwabl P, Brusilovskaya K, Supper P, Bauer D, Konigshofer P, Riedl F, Hayden H, Fuchs CD, Stift J, Oberhuber G, Aschauer S, Bonderman D, Gnad T, Pfeifer A, Uschner FE, Trebicka J, Rohr-Udilova N, Podesser BK, Peck-Radosavljevic M, Trauner M, Reiberger T (2018) The soluble guanylate cyclase stimulator riociguat reduces fibrogenesis and portal pressure in cirrhotic rats. Sci Rep 8:9372

Seita J, Weissman IL (2010) Hematopoietic stem cell: self-renewal versus differentiation. Wiley Interdiscip Rev Syst Biol Med 2:640–653

Semba T, Sammons R, Wang X, Xie X, Dalby KN, Ueno NT (2020) JNK signaling in stem cell self-renewal and differentiation. Int J Mol Sci 21:2613

Speth JM, Hoggatt J, Singh P, Pelus LM (2014) Pharmacologic increase in HIF1alpha enhances hematopoietic stem and progenitor homing and engraftment. Blood 123:203–207

Subramaniam A, Zemaitis K, Safaee Talkhoncheh M, Yudovich D, Backstrom A, Debnath S, Chen J, Jain MV, Galeev R, Gaetani M, Zubarev RA, Larsson J (2020) Lysine-specific demethylase 1A (LSD1) restricts ex vivo propagation of human HSCs and is a target of UM171. Blood 136:2151–2161

Sugimura R, Jha DK, Han A, Soria-Valles C, da Rocha EL, Lu YF, Goettel JA, Serrao E, Rowe RG, Malleshaiah M, Wong I, Sousa P, Zhu TN, Ditadi A, Keller G, Engelman AN, Snapper SB, Doulatov S, Daley GQ (2017) Haematopoietic stem and progenitor cells from human pluripotent stem cells. Nature 545:432–438

Svahn J, Lanza T, Rathbun K, Bagby G, Ravera S, Corsolini F, Pistorio A, Longoni D, Farruggia P, Dufour C, Cappelli E (2015) p38 Mitogen-activated protein kinase inhibition enhances in vitro erythropoiesis of Fanconi anemia, complementation group A-deficient bone marrow cells. Exp Hematol 43:295–299

Takizawa H, Schanz U, Manz MG (2011) Ex vivo expansion of hematopoietic stem cells: mission accomplished? Swiss Med Wkly 141:w13316

Tan CK, Wahli W (2016) A trilogy of glucocorticoid receptor actions. Proc Natl Acad Sci U S A 113:1115–1117

Testa U, Fossati C, Samoggia P, Masciulli R, Mariani G, Hassan HJ, Sposi NM, Guerriero R, Rosato V, Gabbianelli M, Pelosi E, Valtieri M, Peschle C (1996) Expression of growth factor receptors in unilineage differentiation culture of purified hematopoietic progenitors. Blood 88:3391–3406

Tie R, Li H, Cai S, Liang Z, Shan W, Wang B, Tan Y, Zheng W, Huang H (2019) Interleukin-6 signaling regulates hematopoietic stem cell emergence. Exp Mol Med 51:1–12

Vajravelu BN, Hong KU, Al-Maqtari T, Cao P, Keith MC, Wysoczynski M, Zhao J, Moore JBT, Bolli R (2015) C-kit promotes growth and migration of human cardiac progenitor cells via the PI3K-AKT and MEK-ERK pathways. PLoS One 10:e0140798

Van HT, Santos MA (2018) Histone modifications and the DNA double-strand break response. Cell Cycle 17:2399–2410

Vannini N, Campos V, Girotra M, Trachsel V, Rojas-Sutterlin S, Tratwal J, Ragusa S, Stefanidis E, Ryu D, Rainer PY, Nikitin G, Giger S, Li TY, Semilietof A, Oggier A, Yersin Y, Tauzin L, Pirinen E, Cheng WC, Ratajczak J, Canto C, Ehrbar M, Sizzano F, Petrova TV, Vanhecke D, Zhang L, Romero P, Nahimana A, Cherix S, Duchosal MA, Ho PC, Deplancke B, Coukos G, Auwerx J, Lutolf MP, Naveiras O (2019) The NAD-booster nicotinamide riboside potently stimulates hematopoiesis through increased mitochondrial clearance. Cell Stem Cell 24:405–418

Vasita E, Yasmeen S, Andoh J, Bridges LR, Kruuse C, Pauls MMH, Pereira AC, Hainsworth AH (2019) The cGMP-degrading enzyme phosphodiesterase-5 (PDE5) in cerebral small arteries of older people. J Neuropathol Exp Neurol 78:191–194

Wagner JE Jr, Brunstein CG, Boitano AE, DeFor TE, McKenna D, Sumstad D, Blazar BR, Tolar J, Le C, Jones J, Cooke MP, Bleul CC (2016) Phase I/II trial of StemRegenin-1 expanded umbilical cord blood hematopoietic stem cells supports testing as a stand-alone graft. Cell Stem Cell 18:144–155

Wang W, Yu S, Zimmerman G, Wang Y, Myers J, Yu VW, Huang D, Huang X, Shim J, Huang Y, Xin W, Qiao P, Yan M, Xin W, Scadden DT, Stanley P, Lowe JB, Huang AY, Siebel CW, Zhou L (2015) Notch receptor-ligand engagement maintains hematopoietic stem cell quiescence and niche retention. Stem Cells 33:2280–2293

Wang Y, Lai S, Tang J, Feng C, Liu F, Su C, Zou W, Chen H, Xu D (2017) Prostaglandin E2 promotes human CD34+ cells homing through EP2 and EP4 in vitro. Mol Med Rep 16:639–646

Wysoczynski M, Reca R, Ratajczak J, Kucia M, Shirvaikar N, Honczarenko M, Mills M, Wanzeck J, Janowska-Wieczorek A, Ratajczak MZ (2005) Incorporation of CXCR4 into membrane lipid rafts primes homing-related responses of hematopoietic stem/progenitor cells to an SDF-1 gradient. Blood 105:40–48

Wysoczynski M, Ratajczak J, Pedziwiatr D, Rokosh G, Bolli R, Ratajczak MZ (2015) Identification of heme oxygenase 1 (HO-1) as a novel negative regulator of mobilization of hematopoietic stem/progenitor cells. Stem Cell Rev 11:110–118

Xiao X, Lai W, Xie H, Liu Y, Guo W, Liu Y, Li Y, Li Y, Zhang J, Chen W, Shi M, Shang L, Yin M, Wang C, Deng H (2019) Targeting JNK pathway promotes human hematopoietic stem cell expansion. Cell Discov 5:2

Xu D, Yang M, Capitano M, Guo B, Liu S, Wan J, Broxmeyer HE, Huang X (2020) Pharmacological activation of nitric oxide signaling promotes human hematopoietic stem cell homing and engraftment. Leukemia 35:229–234

Yu VW, Scadden DT (2016) Heterogeneity of the bone marrow niche. Curr Opin Hematol 23:331–338

Zhang CC, Lodish HF (2008) Cytokines regulating hematopoietic stem cell function. Curr Opin Hematol 15:307–311

Zhang CC, Kaba M, Ge G, Xie K, Tong W, Hug C, Lodish HF (2006) Angiopoietin-like proteins stimulate ex vivo expansion of hematopoietic stem cells. Nat Med 12: 240–245

Zhang J, Huang X, Guo B, Cooper S, Capitano ML, Johnson TC, Siegel DR, Broxmeyer HE (2020) Effects of Eupalinilide E and UM171, alone and in combination on cytokine stimulated ex-vivo expansion of human cord blood hematopoietic stem cells. Blood Cells Mol Dis 84:102457

Zhou BO, Yu H, Yue R, Zhao Z, Rios JJ, Naveiras O, Morrison SJ (2017) Bone marrow adipocytes promote the regeneration of stem cells and haematopoiesis by secreting SCF. Nat Cell Biol 19:891–903

Zhou H, Wang B, Sun H, Xu X, Wang Y (2018) Epigenetic regulations in neural stem cells and neurological diseases. Stem Cells Int 2018:6087143

Zou YR, Kottmann AH, Kuroda M, Taniuchi I, Littman DR (1998) Function of the chemokine receptor CXCR4 in haematopoiesis and in cerebellar development. Nature 393:595–599

Zou J, Zou P, Wang J, Li L, Wang Y, Zhou D, Liu L (2012) Inhibition of p38 MAPK activity promotes ex vivo expansion of human cord blood hematopoietic stem cells. Ann Hematol 91:813–823

N^6-Methyladenosine RNA Modification in Normal and Malignant Hematopoiesis

7

Hengyou Weng, Huilin Huang, and Jianjun Chen

Abstract

Over 170 nucleotide variants have been discovered in messenger RNAs (mRNAs) and non-coding RNAs so far. However, only a few of them, including N^6-methyladenosine (m^6A), 5-methylcytidine (m^5C), and N^1-methyladenosine (m^1A), could be mapped in the transcriptome. These RNA modifications appear to be dynamically regulated, with writer, eraser, and reader proteins being identified for each modification. As a result, there is a growing interest in studying their biological impacts on normal bioprocesses and tumorigenesis over the past few years. As the most abundant internal modification in eukaryotic mRNAs, m^6A plays a vital role in the post-transcriptional regulation of mRNA fate via regulating almost all aspects of mRNA metabolism, including RNA splicing, nuclear export, RNA stability, and translation. Studies on mRNA m^6A modification serve as a great example for exploring other modifications on mRNA. In this chapter, we will review recent advances in the study of biological functions and regulation of mRNA modifications, specifically m^6A, in both normal hematopoiesis and malignant hematopoiesis. We will also discuss the potential of targeting mRNA modifications as a treatment for hematopoietic disorders.

Keywords

m^6A RNA modifications · RNA epigenetics · Normal hematopoiesis · Malignant hematopoiesis · Targeted therapy

Abbreviations

2OG	2-Oxoglutarate
5-Aza	5-Azacytidine
ALKBH5	AlkB homolog 5
ALL	Acute lymphoblastic leukemia

Hengyou Weng and Huilin Huang contributed equally with all other contributors.

H. Weng (✉)
The First Affiliated Hospital, The Fifth Affiliated Hospital, Guangzhou Laboratory, Guangzhou Medical University, Guangzhou, China

Bioland Laboratory, Guangzhou, China
e-mail: weng_hengyou@gzlab.ac.cn

H. Huang (✉)
State Key Laboratory of Oncology in South China, Guangdong Provincial Clinical Research Center for Cancer, Sun Yat-sen University Cancer Center, Guangzhou, China
e-mail: huanghl1@sysucc.org.cn

J. Chen (✉)
Department of Systems Biology, Beckman Research Institute of City of Hope, Monrovia, CA, USA

Gehr Family Center for Leukemia Research and City of Hope Comprehensive Cancer Center, City of Hope, Duarte, CA, USA
e-mail: jianchen@coh.org

M. Zhao, P. Qian (eds.), *Hematopoietic Stem Cells*, Advances in Experimental Medicine and Biology 1442, https://doi.org/10.1007/978-981-99-7471-9_7

AML	Acute myeloid leukemia
Ara-C	Cytosine arabinoside
ATRA	All-trans retinoic acid
CLP	Common lymphoid progenitor
CML	Chronic myeloid leukemia
CMP	Common myeloid progenitor
CTCL	Cutaneous T-cell lymphoma
DAC	Decitabine
DLBCL	Diffuse large B-cell lymphoma
DNR	Daunorubicin
EHT	Endothelial-to-hematopoietic transition
ELAVL1	ELAV-like RNA-binding protein 1
FTO	Fat mass and obesity-associated
GMP	Granulocyte/macrophage progenitor
HNRNP	Heterogeneous nuclear ribonucleoprotein
HSCs	Hematopoietic stem cells
HSPCs	Hematopoietic stem and progenitor cells
IDH	Isocitrate dehydrogenase
IGF2BP	Insulin-like growth factor 2 mRNA-binding proteins
KH	K homology
lncRNA	Long non-coding RNA
LSCs/LICs	Leukemic stem cells/leukemia-initiating cells
m^1A	N^1-Methyladenosine
m^5C	5-Methylcytidine
m^6A	N^6-Methyladenosine
METTL14	Methyltransferase-like 14
METTL3	Methyltransferase-like 3
METTL5	Methyltransferase-like 5
MTase	Methyltransferase
MTC	Methyltransferase complex
NKTCL	Natural killer/T-cell lymphoma
R-2HG	R-2-Hydroxyglutarate
RBM15	RNA-binding motif protein 15
rRNA	Ribosomal RNA
SAM	S-Adenosyl-L-methionine
SNPs	Single-nucleotide polymorphisms
snRNA	Small nuclear RNA
TF	Transcription factor
TKI	Tyrosine kinase inhibitor
tRNA	Transfer RNA
VIRMA	Vir-like m^6A methyltransferase associated
WTAP	Wilms' tumor 1-associating protein
YTH	YT521-B homology
ZC3H13	Zinc finger CCCH domain-containing protein 13
ZCCHC4	Zinc finger CCHC-type-containing 4
α-KG	α-Ketoglutarate
ψ	Pseudouridine

7.1 Introduction

The first modified RNA nucleotide variant, pseudouridine (ψ), was discovered as the "fifth RNA nucleotide" in the 1950s (Davis and Allen 1957). Since then, over 170 types of RNA chemical modifications have been identified in both protein-coding and noncoding RNAs (Adams and Cory 1975; Alarcon et al. 2015b; Amort et al. 2013; Carlile et al. 2014; Charette and Gray 2000; Cohn and Volkin 1951; Deng et al. 2018c; Dunn 1961; El Yacoubi et al. 2012; Fu et al. 2014a; Hall 1963; Huang et al. 2020c; Huber et al. 2015; Krug et al. 1976; Roundtree et al. 2017a; Squires et al. 2012; Wei and Moss 1977). However, most of the previous research on RNA modification has focused on non-coding RNAs, such as transfer RNA (tRNA), ribosomal RNA (rRNA), and small nuclear RNA (snRNA). In 2011, Dr. He and colleagues reported that the m^6A modification on mRNA can be reversibly removed by FTO (Jia et al. 2011). This groundbreaking discovery has revived the field of RNA modification research. Newly developed methods for isolating RNAs containing specific types of modified nucleosides, coupled with high-throughput sequencing, now enable the mapping of landscapes of RNA modifications, such as m^6A, m^5C, m^1A, inosine, and pseudouridine (ψ), in mRNA across various cellular contexts (Carlile et al. 2014; Dominissini et al. 2012, 2016; Huang et al. 2020a; Li et al. 2015; Lovejoy et al. 2014; Meyer et al. 2012; Safra et al. 2017; Schwartz et al. 2014a; Squires et al. 2012; Suzuki et al. 2015). Meanwhile, significant progress has been made, and efforts are ongoing to identify regulators of RNA modifications (Huang et al. 2020c). Similar to modifications in histones and DNA methylation, reversible modifications in RNA involve three categories of regulatory proteins: "writer" proteins that deposit the modification, "eraser" proteins that remove the modification, and "reader" proteins that recognize the modification and mediate RNA fate decisions (Huang et al. 2020b).

The RNA m^{6}A modification has attracted the most attention in this field over the past decade. The discovery of m^{6}A modification can be traced back to the 1970s (Adams and Cory 1975; Krug et al. 1976; Wei and Moss 1977). It is now well acknowledged that m^{6}A is the most prevalent and abundant internal modification in eukaryotic mRNA. With the development of an antibody-based pull-down coupled with high-throughput sequencing method, it has become clear that m^{6}A modification in the transcriptome exhibits a unique pattern, showing an enrichment around the stop codons of mRNAs and a consensus sequence of RRACH (R = G or A; H = A, C, or U) (Dominissini et al. 2012; Meyer et al. 2012). Further investigation suggests that m^{6}A marks are installed co-transcriptionally into nascent RNAs and that histone H3K36me3 modification, along with other regulatory factors, plays a critical role in determining the site selection for m^{6}A deposition (Huang et al. 2019a). On the other hand, functional studies have demonstrated the involvement of m^{6}A modification in controlling normal biological and pathological processes, including stem cell biology, tissue development, circadian rhythm, sex determination, tumorigenesis, and drug response (Alarcon et al. 2015b; Barbieri et al. 2017; Chen et al. 2015; Deng et al. 2018a, b, c; Dong et al. 2021; Geula et al. 2015; Huang et al. 2018, 2020b; Li et al. 2017b; Su et al. 2018; Vu et al. 2017; Wang et al. 2014b; Weng et al. 2018; Xiang et al. 2017; Zhang et al. 2017a, b; Zhao et al. 2014, 2017a, b; Zheng et al. 2013; Zhou et al. 2015). Here, we summarize recent advances in understanding the biological functions and regulation of m^{6}A in both normal and malignant hematopoiesis. Additionally, we discuss the potential of targeting m^{6}A modifications as a treatment for hematopoietic disorders.

7.2 Regulators of m^{6}A Modification

The main "writer" of mRNA m^{6}A modification is a large multicomponent methyltransferase complex (MTC), in which methyltransferase-like 3 (METTL3) and methyltransferase-like 14 (METTL14) proteins form a core heterodimer, while other components, including Wilms' tumor 1-associating protein (WTAP), vir-like m^{6}A methyltransferase associated (VIRMA, also known as KIAA1429), RNA-binding motif protein 15 (RBM15), and zinc finger CCCH domain-containing protein 13 (ZC3H13), act as regulatory subunits to facilitate m^{6}A installation in cells (Bokar et al. 1997; Guo et al. 2018; Knuckles et al. 2018; Liu et al. 2014; Patil et al. 2016; Ping et al. 2014; Schwartz et al. 2014b; Wang et al. 2014b; Wen et al. 2018). Although both METTL3 and METTL14 belong to the MT-A70 family of S-adenosyl-L-methionine (SAM)-dependent methyltransferases (MTases), structural studies suggest that METTL3 is the only catalytic subunit in the MTC, while METTL14 provides an RNA-binding scaffold that allosterically activates and enhances the catalytic activity of METTL3 (Sledz and Jinek 2016; Wang et al. 2016a, b). METTL16 was initially identified as a methyltransferase for several structured RNAs (e.g., U6 snRNA) and pre-mRNA (Brown et al. 2016; Mendel et al. 2018; Pendleton et al. 2017). However, it has been recently shown that METTL16 could also methylate hundreds of mRNA transcripts in the nucleus, in addition to its methyltransferase-independent role in the cytosol as a facilitator of translation-initiation (Su et al. 2022). Two other m^{6}A writers, zinc finger CCHC-type-containing 4 (ZCCHC4) and methyltransferase-like 5 (METTL5), can independently catalyze m^{6}A modifications on 28S and 18S ribosomal RNAs (rRNAs), respectively (Ma et al. 2019; Pinto et al. 2020; van Tran et al. 2019).

Fat mass and obesity-associated (FTO) and AlkB homolog 5 (ALKBH5) are two "eraser" proteins that have been discovered so far to catalyze the removal of m^{6}A modification. They both belong to the AlkB subfamily of the Fe(II)/2-oxoglutarate (2OG) dioxygenase superfamily. This subfamily requires α-ketoglutarate (α-KG) and molecular oxygen as co-substrates, as well as ferrous iron Fe(II) as a cofactor, to catalyze the oxidation and demethylation of a substrate (Gerken et al. 2007; Kurowski et al. 2003). FTO was identified as the first m^{6}A demethylase that could demethylate m^{6}A in both DNA and RNA in vivo (Jia et al. 2011). Later on, it was reported that FTO also demethylates m^{6}A$_m$ (Mauer et al. 2017), a modification exclusively found at the first encoded nucleotide after the 5′ methylguanosine cap of mammalian mRNAs but with considerably lower overall abundance compared to m^{6}A (Su et al. 2018; Wei et al. 2018). Different from FTO, ALKBH5 catalyzes the direct removal of m^{6}A (Fu et al. 2014b).

The "reader" proteins act as effectors that mediate the biological consequences of m^{6}A modification by selectively binding to m^{6}A-modified RNAs (Deng et al. 2018c; Yang et al. 2018; Zhao et al. 2017a). Many m^{6}A readers have been identified so far, each with diverse mechanisms for recognizing of m^{6}A and resulting in various consequences on RNA metabolism. The YT521-B homology (YTH) domain family of proteins, including YTHDF1, YTHDF2, YTHDF3, YTHDC1, and YTHDC2, utilizes their conserved m^{6}A-binding pocket within the YTH domain to directly bind the m^{6}A base (Dominissini et al. 2012; Hsu et al. 2017; Luo and Tong 2014; Wang et al. 2014a, 2015; Xiao et al. 2016; Xu et al. 2014; Zhu et al. 2014). Among them, YTHDF2 was the first identified and the most studied m^{6}A reader protein that promotes the degradation of m^{6}A-modified target mRNAs (Du et al. 2016; Wang et al. 2014a). On the other hand, studies have shown that YTHDF1 promotes the translation of m^{6}A-modified mRNAs (Wang et al. 2015). In contrast, YTHDF3 and YTHDC2 can mediate mRNA decay while also enhancing translation (Bailey et al. 2017; Hsu et al. 2017; Jain et al. 2018; Li et al. 2017a; Shi et al. 2017; Wojtas et al. 2017). Unlike the reader proteins mentioned above that are located in the cytoplasm, YTHDC1 is primarily located in the nucleus. It plays a crucial role in regulating splicing, XIST-mediated X chromosome silencing, and nuclear export of m^{6}A-modified mRNAs (Patil et al. 2016; Roundtree et al. 2017b; Xiao et al. 2016). Additionally, it controls the integrity of heterochromatin by recognizing m^{6}A modifications on transposon-derived RNAs (Chen et al. 2021a; Liu et al. 2021). In contrast to the YTH family of proteins, insulin-like growth factor 2 mRNA-binding proteins (IGF2BPs), which include IGF2BP1/2/3, have been identified as a new family of m^{6}A readers. These proteins promote stability and translation of their target mRNAs (Huang et al. 2018). IGF2BP proteins use their K homology (KH) domains (especially the KH3-4 di-domain) and possibly the flanking sequence to recognize m^{6}A. They play a crucial role in determining the fate of mRNA by recruiting mRNA stabilizers like ELAV-like RNA-binding protein 1 (ELAVL1), also known as HuR (Huang et al. 2018). Several heterogeneous nuclear ribonucleoproteins (HNRNPs) have also been reported as m^{6}A readers. HNRNPC and HNRNPG have been shown to recognize m^{6}A-induced changes in mRNA secondary structures and facilitate alternative splicing of target mRNAs (Liu et al. 2015; Zhou et al. 2019). HNRNPA2B1 was previously shown to regulate alternative splicing and primary microRNA processing as an m^{6}A reader (Alarcon et al. 2015a). However, a later study suggested a mechanism called "m^{6}A switch" instead of direct binding to m^{6}A (Wu et al. 2018). The list of m^{6}A reader proteins is expanding, and other proteins are being proposed as m^{6}A interactors, including FMR1 and LRPPRC (Arguello et al. 2017; Edupuganti et al. 2017). However, additional mechanistic studies are necessary to more accurately categorize these proteins.

7.3 m^{6}A Modification in Normal Hematopoiesis

Hematopoiesis is defined as a tightly regulated process that produces mature blood cells from a small pool of multipotent hematopoietic stem cells (HSCs) (Doulatov et al. 2012; Rosenbauer and Tenen 2007). Decades of research have provided basic knowledge on the regulation of normal hematopoiesis, highlighting a critical role of hematopoietic transcription factors (TFs) in regulating the multistep normal hematopoiesis and in determining cell fate in the hematopoietic system (Goode et al. 2016; Koschmieder et al. 2005; Rosmarin et al. 2005). For example, PU.1 (also known as SPI1) and C/EBPα are essential in generating early myeloid progenitors (i.e., common myeloid progenitors, CMPs) and granulocyte/macrophage progenitors (GMPs), respectively (Dakic et al. 2005; Rosenbauer and Tenen 2007), while PAX5 is required for the early development of the B-cell lineage (Mikkola et al. 2002). In addition, epigenetic regulatory mechanisms, including DNA methylation, histone modifications, and non-coding RNAs, have been shown to contribute to HSC homeostasis and normal hematopoiesis (Butler and Dent 2013; Challen et al. 2014; Chen et al. 2010; Guillamot et al. 2016; Moran-Crusio et al. 2011; O'Connell et al. 2010; Ooi et al. 2010; Weng et al. 2019). Emerging as a new type of epigenetic regulation, m^6A RNA modification was demonstrated to be critical in governing HSC biology and hematopoiesis in recent years (Fig. 7.1).

7.3.1 METTL3

As the sole catalytic subunit in the m^6A MTC, METTL3 has been extensively studied in the hematopoietic system across a range of species, from zebrafish to mammals. Decreased levels of m^6A resulting from *mettl3* deficiency in zebrafish embryos cause blockage of HSPC emergence (Zhang et al. 2017a). Mechanistic studies have revealed that the reduction of m^6A on *notch1a* mRNA suppresses YTHDF2-mediated mRNA decay, leading to the continuous activation of the Notch signaling in arterial endothelial cells. This results in the blockage of endothelial-to-hematopoietic transition (EHT) and the repression of the earliest HSPC generation in *mettl3*-deficient zebrafish embryos (Zhang et al. 2017a). This mechanism appears to be conserved in mice, as knockdown of *Mettl3* in the aorta-gonad-mesonephros impairs colony formation, likely also through activation of the Notch1 signaling pathway (Zhang et al. 2017a). Yao et al. discovered that conditional ablation of Mettl3 in the mouse hematopoietic system significantly increased the frequency of HSCs in the bone marrow and suppressed self-renewal capability of HSCs in recipient mice undergoing bone marrow transplantation (BMT) (Yao et al. 2018). Vu and colleagues reported that the knockdown of *METTL3* expression in human HSPCs inhibited cell growth and increased myeloid differentiation. On the other hand, overexpression of wild-type METTL3, but not its catalytically dead mutant, promoted proliferation and colony formation and inhibited myeloid differentiation (Vu et al. 2017).

7.3.2 METTL14

In mouse bone marrow, *Mettl14* was found to be highly expressed in HSCs and $Lin^-Sca\text{-}1^+c\text{-}kit^+$ (LSK) cells and be responsible for the high m^6A level in these naïve cells (Weng et al. 2018). Notably, the expression of *Mettl14* was gradually downregulated during myelopoiesis, with the lowest expression observed in mature myeloid cells (Weng et al. 2018). Consistent with this expression pattern, the knockdown of *METTL14* in human HSPCs promoted myeloid differentiation in vitro. Moreover, conditional knockout of *Mettl14* in donor cells impaired the self-renewal ability of HSCs in the BMT recipient mice (Weng et al. 2018; Yao et al. 2018). Interestingly, SPI1, which is a transcriptional master regulator of myelopoiesis, was identified as a negative regulator that controls the transcription of *METTL14* in the hematopoietic system (Weng et al. 2018). Considering the role of SPI1 (Iwasaki et al. 2005), MYB (Mucenski et al. 1991; Sandberg et al. 2005), and MYC (Satoh et al. 2004; Wilson et al. 2004) transcription factors in regulating HSC self-renewal and differentiation, the SPI1-METTL14-m^6A-MYB/MYC regulation axis

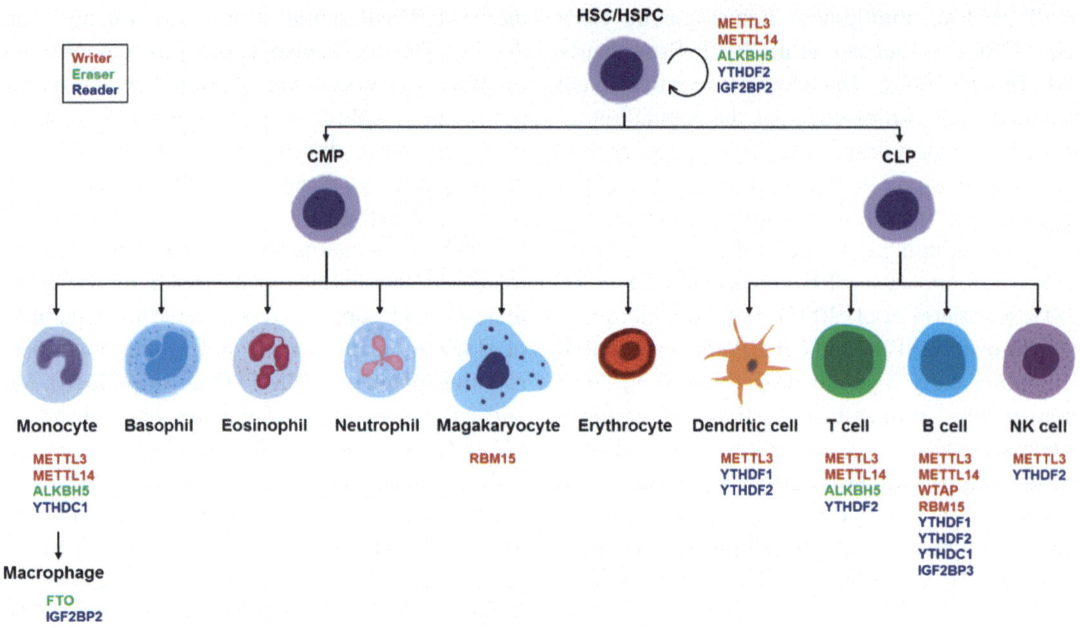

Fig. 7.1 The involvement of m^{6}A regulators in normal hematopoiesis. The m^{6}A modification and its associated m^{6}A regulators have been demonstrated to be crucial for the self-renewal of HSCs/HSPCs and for the proper development of blood cells. Writers are represented in red, erasers in green, and readers in blue

adds a new layer of complexity to the regulatory networks involved in normal hematopoiesis.

7.3.3 RBM15

RBM15, also known as *OTT1*, was initially identified as a fusion partner of the *MKL1* gene detected in infant acute megakaryocytic leukemia with t(1;22)(p13;q13). RBM15 was recently identified as a component of the m^{6}A MTC. Early studies have shown that conditional knockout of Rbm15 resulted in an increase in HSPCs and an expansion of myeloid and megakaryocytic cells in the spleen and bone marrow while blocking B-cell differentiation (Raffel et al. 2007). Ma et al. found that Rbm15 was expressed at the highest levels in HSCs and inhibited myeloid differentiation and megakaryocytic expansion by stimulating the Notch signaling (Ma et al. 2007). Consistently, Rbm15 is required for normal interactions between HSCs and their niche, as well as for normal megakaryocyte development, at least in part through the regulation of *MYC* expression (Niu et al. 2009).

7.3.4 ALKBH5

Two independent studies reported that ALKBH5 is dispensable for normal hematopoiesis (Shen et al. 2020; Wang et al. 2020). Using the Alkbh5 constitutive knockout mouse model, researchers demonstrated that Alkbh5 deletion did not result in significant changes in the total number of bone marrow cells or the percentages of different subpopulations of HSPCs or differentiated lineages in either bone marrow or peripheral blood. Moreover, the knockout of Alkbh5 did not affect the self-renewal capacity or the long-term function of HSCs (Shen et al. 2020; Wang et al. 2020).

7.3.5 YTHDF2

YTHDF2 is the first well-characterized m^{6}A reader that promotes the degradation of mRNA transcripts containing m^{6}A modification (Wang et al. 2014a). Li et al. reported a remarkable increase in functional HSCs in the bone marrow of *Ythdf2* conditional knockout mice and in human umbilical cord blood upon *YTHDF2* knockdown, which was at least partially attributed to the enhanced stability of mRNA transcripts that encode TFs critical for stem cell self-renewal (Li et al. 2018). It was also reported by another group that deficiency in YTHDF2 enhanced HSC activity, and YTHDF2 did not appear to be essential for normal HSC function (Paris et al. 2019). However, the same group recently investigated the long-term effects of YTHDF2 deletion on HSC maintenance and multi-lineage hematopoiesis, and the results suggest that YTHDF2 acts as a repressor of inflammatory pathways in HSCs and is a key factor for long-term HSC maintenance (Mapperley et al. 2021).

7.3.6 YTHDC1

Loss of Ythdc1 resulted in rapid hematopoietic failure in mice, leading to their death within 3 weeks (Sheng et al. 2021). Further studies have shown that induced deletion of Ythdc1 compromises hematopoiesis and HSC functions. The numbers of HSPCs, mature myeloid cells, and B cells all decrease dramatically in Ythdc1 KO recipient mice compared to Ythdc1 haploinsufficient or wild-type recipient mice (Sheng et al. 2021).

7.4 m^{6}A Modification in Malignant Hematopoiesis

The pioneering work by Ernest McCulloch and James Till in identifying HSC and characterizing their properties strongly suggests that maintaining HSC homeostasis requires a precise balance between self-renewal and multi-lineage differentiation (McCulloch and Till 2005; Siminovitch et al. 1963). Disrupting this balance, whether through the well-studied mutations or aberrant expression of TFs or through dysregulation of epigenetic modifications, places HSC at a higher risk of developing hematopoietic diseases, such as leukemia (Chen et al. 2010; Huang et al. 2013; Qing et al. 2021; Weng et al. 2019). Dysregulation of m^{6}A regulators has been observed in hematopoietic malignancies (Fig. 7.2).

7.4.1 Acute Myeloid Leukemia (AML)

AML is a common subtype of leukemia that is commonly diagnosed in both adults and children. Unfortunately, it has the lowest 5-year survival rate (<30%) among all types of leukemia. This makes AML a fatal subtype of leukemia. Accumulating evidence in the past few years has closely linked RNA m^{6}A modification with the initiation and development of AML.

FTO is the first gene known to play a role in malignant hematopoiesis, specifically AML, through an m^{6}A-dependent mechanism (Li et al. 2017b). Previously, *FTO* was recognized as a gene related to fat mass, adipogenesis, and body weight (Church et al. 2010; Fischer et al. 2009; Merkestein et al. 2015). However, recent large-scale epidemiology studies have revealed a connection between FTO and cancers such as leukemia and lymphoma, as evidenced by single-nucleotide polymorphisms (SNPs) in *FTO* among patients (Castillo et al. 2012; Hernandez-Caballero and Sierra-Ramirez 2015; Soderberg et al. 2009). We found that *FTO* is highly expressed in AML subtypes carrying t(11q23)/*MLL* rearrangements, t(15;17)/*PML-RARA*, *FLT3*-ITD, and/or *NPM1* mutations (Li et al. 2017b). Functioning as an m^{6}A eraser, FTO removes m^{6}A marks from the transcripts of *ASB2* and *RARA*, two genes with reported roles in myeloid cell differentiation and drug response of leukemia cells. This results in decreased mRNA stability of *ASB2* and *RARA*, therefore promoting AML cell survival and leukemogenesis (Li et al.

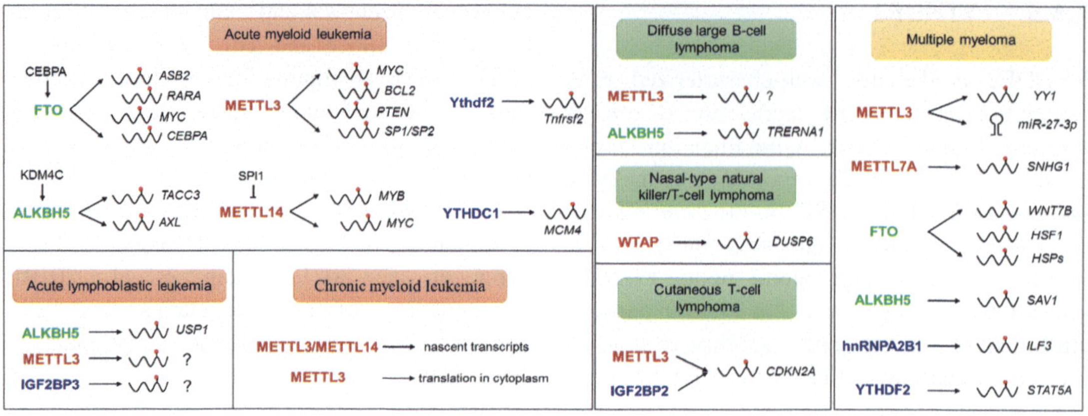

Fig. 7.2 Roles and mechanisms of m^{6}A regulators in malignant hematopoiesis. Deregulation of m^{6}A regulators has been identified in various hematopoietic disorders, including leukemia (acute myeloid leukemia, acute lymphoblastic leukemia, and chronic myeloid leukemia), lymphoma (diffuse large B-cell lymphoma, nasal-type natural killer/T-cell lymphoma, and cutaneous T-cell lymphoma), and multiple myeloma. Although some studies have yet to characterize the m^{6}A-modified target transcripts responsible for the function of m^{6}A regulators, critical target transcripts regulated by m^{6}A modification have been identified and illustrated in such hematopoietic malignancies

2017b). Interestingly, it was later discovered that FTO mediates the anti-leukemia activity of R-2-hydroxyglutarate (R-2HG), a metabolite produced in high levels by mutant isocitrate dehydrogenase 1/2 (IDH1/2) enzymes and that was previously reported to cause leukemia (Losman et al. 2013; Sasaki et al. 2012; Wang et al. 2013). R-2HG binds directly to FTO to inhibit its demethylase activity, leading to the hypomethylation and accelerated decay of *MYC* and *CEBPA* mRNAs (Su et al. 2018). Moreover, treatment with R-2HG or inhibition of FTO could sensitize AML cells to first-line chemotherapy drugs (Su et al. 2018). These studies collectively suggest that targeting FTO signaling could be a promising therapeutic approach for AML.

As another known m^{6}A eraser, *ALKBH5* has been found to be aberrantly overexpressed in AML, and its increased expression is associated with poor prognosis in AML patients (Shen et al. 2020; Wang et al. 2020). In line with this, *ALKBH5* was preferentially overexpressed in LSCs, and its m^{6}A demethylation activity was necessary for the survival of AML cells both in vitro and in vivo. Mechanistic study revealed that KDM4C depletion of the repressive histone mark H3K9me3 around the *ALKBH5* locus increases chromatin accessibility, resulting in the overexpression of *ALKBH5* in leukemia cells (Wang et al. 2020). *TACC3* and *AXL* have been identified as functional targets of ALKBH5. The stability of mRNA for both targets decreased when ALKBH5 was deleted. Importantly, loss of function of either *TACC3* or *AXL* could mimic the effects of ALKBH5 depletion, while the forced expression of *TACC3* or *AXL* could at least partially rescue the growth inhibition or clonogenic defect caused by ALKBH5 deficiency (Shen et al. 2020; Wang et al. 2020). Considering that ALKBH5 is dispensable for normal hematopoiesis, the essential role of ALKBH5 in AML cells and LSCs may represent a therapeutic vulnerability.

In AML, the oncogenic roles of METTL3 and METTL14 as m^{6}A writers have been reported by us and other groups. Both METTL3 and METTL14 were highly expressed in AML compared to the vast majority of other cancer types (Barbieri et al. 2017; Vu et al. 2017; Weng et al. 2018). METTL3 promoted AML cell proliferation and AML development in an m^{6}A-dependent manner by methylating its target transcripts, including *MYC*, *BCL2*, and *PTEN* (Vu et al. 2017). In addition, it has been reported that

METTL3 can be recruited to chromatin by CEBPZ, a transcription factor critical for hematopoietic differentiation (Barbieri et al. 2017). Promoter-bound METTL3 introduces m^{6}A modifications within the coding region of associated mRNA transcripts, such as *SP1* and *SP2*, and enhances their translation by relieving ribosome stalling (Barbieri et al. 2017). We found that METTL14 is aberrantly overexpressed in certain AML subtypes, including those carrying t(11q23)/*MLL* rearrangements, t(15;17)/*PML-RARA*, and t(8;21)/*AML1-ETO*, and that similar to *METTL3*, *METTL14* could be upregulated by corresponding oncofusion genes (Weng et al. 2018). In vitro and in vivo gain- and loss-of-function studies have demonstrated that METTL14 plays a critical oncogenic role in the initiation and development of AML (Weng et al. 2018). Importantly, *MYB* and *MYC*, two well-known TFs involved in leukemogenesis, were found to be direct targets of METTL14 and mediate the effect of *METTL14* knockdown/knockout in AML (Weng et al. 2018). Furthermore, we have identified SPI1 as a negative regulator of *METTL14* expression in both HSCs and AML cells (Weng et al. 2018). Taken together, these findings highlight the critical roles of the m^{6}A MTC core and m^{6}A modification in malignant hematopoiesis.

In addition to erasers and writers, the roles of m^{6}A readers have also been investigated in AML. *YTHDF2* has recently been proposed as another potential therapeutic target for AML. *YTHDF2* was found to be overexpressed in a wide range of human AML cases, and its expression was found to be correlated with LSC activity (Paris et al. 2019). Depletion of Ythdf2 selectively compromises AML initiation and propagation by extending the half-life of diverse m^{6}A-containing transcripts that contribute to the overall integrity of LSC function, including the tumor necrosis factor receptor Tnfrsf2 (Paris et al. 2019). In AML with t(8;21), expression of YTHDF2 was upregulated by the AML1/ETO-HIF1 alpha loop, and high expression of YTHDF2 was associated with a higher risk of relapse and inferior relapse-free survival (Chen et al. 2021b). YTHDC1 was found to be significantly upregulated in AML samples compared to healthy controls and is required for the growth of AML cells in vitro and AML development in vivo (Sheng et al. 2021). Mechanistic studies have revealed that YTHDC1 can stabilize mRNAs (e.g., MCM4) of cell cycle-associated genes to support the proliferation of malignant cells (Sheng et al. 2021). IGF2BP2 has recently been shown to be aberrantly overexpressed in AML cells, especially in LSCs, likely due to transcriptional activation by the AML-associated oncoproteins, especially MLL fusions. By recognizing m^{6}A modifications on key transcripts in the glutamate metabolism pathway, such as *MYC*, *SLC1A5*, and *GPT2*, IGF2BP2 stabilizes these mRNAs and enhances their translation. As a result, IGF2BP2 supports the high demand for glutamate in AML cells, thereby promoting the initiation and development of AML (Weng et al. 2022).

7.4.2 Acute Lymphoblastic (or Lymphocytic) Leukemia (ALL)

ALL is a type of cancer that originates in the bone marrow. It is characterized by the abnormal proliferation of early lymphoid precursors, which replace the normal hematopoietic cells in the marrow. ALL accounts for approximately 80% of childhood leukemia. Although the 5-year survival rate has increased to over 70% due to significant advances in treatment protocols in recent years, recurrence still occurs in 15–20% of cases.

High expression of USP1 is correlated with poor prognosis in T-ALL patients. It mediates T-ALL chemoresistance by interacting with and deubiquitinating Aurora B. It has recently been discovered that ALKBH5 can enhance the expression of USP1 by decreasing the m^{6}A level and mRNA stability of *USP1* mRNA. Therefore, inhibiting ALKBH5 reduced USP1 levels and improved glucocorticoid resistance in T-ALL cells by suppressing Aurora B expression (Gong et al. 2021). This suggests a new therapeutic approach for ALL.

Other regulators of m^{6}A modification have been shown to be dysregulated in ALL. For example, a study by Sun et al. found that ETV6/RUNX1-positive B-ALL had low expression of METTL3, which was associated with a high recurrence rate (Sun et al. 2019). However, it has yet to be investigated whether these regulators play a role in ALL.

7.4.3 Chronic Myeloid Leukemia (CML)

CML is a type of myeloproliferative neoplasm that is characterized by the presence of the Philadelphia chromosome and the resulting BCR-ABL1 oncofusion protein. Although tyrosine kinase inhibitors (TKIs) have been developed to successfully block the enzymatic activity of BCR-ABL1, these medications must be taken for life to keep the disease under control (Ben-Neriah et al. 1986; Druker et al. 2006, 1996; Goldman and Melo 2001; Rowley 1973). In addition, resistance to TKIs remains a challenge in curing CML patients. Therefore, further mechanistic studies are necessary to develop improved therapeutic options for CML.

Recently, Ianniello and colleagues found that the expression of METTL3/METTL14 was upregulated in CML patients. This upregulation was required for the proliferation of primary CML cells or CML cell lines, regardless of their sensitivity to the first-generation TKI imatinib (Ianniello et al. 2021). They further proposed a model in which METTL3/METTL14 modifies nascent transcripts in the nucleus, while METTL3 promotes translation of certain transcripts in the cytoplasm independent of its catalytic activity, highlighting a role of METTL3/METTL14 in CML and suggesting that inhibiting the METTL3/METTL14 complex could hold therapeutic potential for CML cells that have escaped from TKI treatment. Another group has reported that the downregulation of the lncRNA nuclear-enriched abundant transcript 1 (NEAT1) in CML is partly attributed to the METTL3-mediated m^{6}A modification, which in turn promotes the progression of CML (Yao et al. 2021).

7.4.4 Lymphoma

Expression of METTL3, as well as m^{6}A level, was found to be upregulated in diffuse large B-cell lymphoma (DLBCL), the most common subtype of lymphoma (Cheng et al. 2020). By regulating the m^{6}A level of pigment epithelium-derived factor (PEGF), METTL3 promotes the proliferation of DLBCL cells in vitro and the progression of DLBCL cells in nude mice (Cheng et al. 2020). ALKBH5 mediates the demethylation of lncRNA *TRERNA1*, promoting its expression in DLBCL. *TRERNA1* acts as a scaffold, recruiting EZH2 to epigenetically silence p21, thereby contributing to DLBCL cell proliferation both in vitro and in vivo (Song et al. 2022). In contrast to that in DLBCL, the expression of METTL3 was significantly downregulated in cutaneous T-cell lymphoma (CTCL) cells. The decline of METTL3 resulted in reduced methylation on *CDKN2A* mRNA, which in turn blocked the interaction between IGF2BP2 and *CDKN2A*, leading to the degradation of *CDKN2A* mRNA transcripts and the proliferation and migration of CTCL cells (Wang et al. 2022a). In nasal-type natural killer/T-cell lymphoma (NKTCL), WTAP increases the m^{6}A level of the dual-specificity phosphatase 6 (*DUSP6*) mRNA transcript, leading to enhanced mRNA stability of *DUSP6*. This contributes to the progression of NKTCL and chemotherapy resistance to cisplatin (Ma et al. 2021).

7.4.5 Multiple Myeloma (MM)

Studies on RNA m^{6}A modification in MM have emerged in the past 2 years. METTL3 has been reported to be upregulated in MM and to promote tumorigenesis by enhancing the stability of *Yin Yang 1* (*YY1*) mRNA and facilitating the maturation of primary miR-27a-3p in an m^{6}A-dependent manner (Che et al. 2022). METTL7A was identified as an RNA methyltransferase in MM cells. It mediates m^{6}A methylation of lncRNAs LOC606724 and SNHG1, promoting their packaging into adipocyte exosomes (Wang et al. 2022b). The increased levels of lncRNA in MM cells have been found to be positively correlated with poor prognosis in MM patients (Wang et al. 2022b). Isocitrate dehydrogenase 2 (IDH2) activates FTO in MM, which leads to a decrease in m^{6}A levels on *WNT7B* mRNA transcripts, resulting in the increase in WNT7B expression and the activation of the Wnt signaling pathway to promote tumorigenesis and progression of MM (Song et al. 2021). Upregulation of FTO has been reported by another group to be responsible for the decreased m^{6}A level in plasma cells from MM patients (Xu et al. 2022). FTO promotes MM cell proliferation, migration, and invasion by targeting HSF1/HSPs in a YTHDF2-dependent manner. Additionally, inhibiting FTO showed a synergistic effect with bortezomib treatment in inhibiting extramedullary myeloma formation. These findings highlight the FTO-HSF1/HSP axis as a potential therapeutic target in MM (Xu et al. 2022). The other m^{6}A eraser, ALKBH5, was also found to play a role in MM. Yu and colleagues found that ALKBH5 was highly expressed in CD138$^+$ plasma cells from MM patients. They also found that reducing ALKBH5 expression impeded MM cell survival and invasion, most likely through restoring the m^{6}A level of the *SAV1* transcript (Yu et al. 2022). HNRNPA2B1 has been found to be elevated in MM patients and is associated with poor prognosis. By recognizing m^{6}A modifications on *ILF3* mRNA, HNRNPA2B1 stabilizes *ILF3* mRNA and promotes *AKT3* expression in an ILF3-dependent manner (Jiang et al. 2021). The same group also reported an oncogenic role of YTHDF2 in MM, in which YTHDF2 destabilizes its target transcript *STAT5A* in an m^{6}A-dependent manner to activate the ERK signaling pathway (Hua et al. 2022).

7.5 Targeting of m^{6}A Modification in Malignant Hematopoiesis

Given the significant roles of m^{6}A regulators in hematopoietic malignancies, there has been a recent development of inhibitors that target these m^{6}A regulators. As the first m^{6}A regulator linked to malignant hematopoiesis, FTO has become the most attractive target for developing inhibitors to treat hematopoietic malignancies (Deng et al. 2018b). R-2HG, a major metabolic product of mutant IDH1/2, has been found to inhibit the m^{6}A demethylase activity of FTO and thereby exhibit anti-leukemia efficacy both in vitro and in vivo (Su et al. 2018). In addition, R-2HG could sensitize AML cells to a range of first-line therapeutic agents, including all-trans retinoic acid (ATRA), 5-azacytidine (5-Aza), decitabine (DAC), daunorubicin (DNR), and cytosine arabinoside (Ara-C) (Su et al. 2018). More potent FTO inhibitors, such as FB23-2, CS1 (also known as bisantrene), and CS2 (also known as brequinar), have been developed later. These inhibitors have shown remarkable suppression of AML progression in xeno-transplanted mice, including PDX AML models (Huang et al. 2019b; Su et al. 2020). Recently, inhibitors of METTL3 have also been reported. Among them, STM2457 has been shown to be a highly potent and selective first-in-class catalytic inhibitor of METTL3 and could specifically target stem cell subpopulations of AML (Yankova et al. 2021). In addition to targeting m^{6}A erasers and writers, inhibitors for m^{6}A readers have recently been developed. An inhibitor named CWI1-2, which targets IGF2BP2, has been developed and shown to suppress glutamate metabolism and inhibit the growth of AML cells both in vitro and in vivo (Weng et al. 2022). Such findings demonstrate that m^{6}A regulators hold promise as therapeutic targets for the treatment of hematopoietic malignancies.

7.6 Conclusions and Perspectives

In the past few years, it has been shown that RNA m^6A modification plays critical roles in almost every type of cancers, particularly in hematopoietic malignancies. However, whether the level of total cellular m^6A correlates with cancer states remains controversial in the field of RNA cancer epigenetics. In leukemia studies, evidence is accumulating that the abundance of m^6A on specific target transcripts, as well as on special target sites of certain target transcripts with important functions, is important for AML pathogenesis, rather than the total level of cellular m^6A (Deng et al. 2018c). This could explain the oncogenic roles of METTL3/METTL14 and FTO/ALKBH5, which represent m^6A writers and erasers, respectively, in AML. It suggests that tipping the balance of methylation and demethylation toward either direction could lead to deleterious effects for cancers. This information could be utilized for the development of novel targeted therapies for hematopoietic malignancies and other cancer types. Future investigations should focus on identifying critical m^6A sites on specific target transcripts. This will enable the manipulation of a single m^6A site for cancer therapy, rather than influencing the total cellular m^6A level. In addition, it is still unclear how various m^6A reader proteins recognize m^6A sites within the same cellular context and how this impacts the development of cancer.

Acknowledgment This work was supported in part by the National Key R&D Program of China 2020YFA0112403 (H.W.) from the Ministry of Science and Technology of the People's Republic of China, 82173058 (H.H.) and 82270168 (H.W.) from the National Natural Science Foundation of China, 2021A1515010425 (H.H.) from the Guangdong Basic and Applied Basic Research Foundation, the Pearl River Talent Recruitment Program 2021ZT09Y233 (H.W.), and the National Institutes of Health (NIH) grants R01 CA271497 (J.C.), R01 CA243386 (J.C.), R01 CA236399 (J.C.), and DK124166 (J.C.), as well as the Simms/Mann Family Foundation (J.C.). J.C. is a Leukemia & Lymphoma Society (LLS) scholar.

Competing Financial Interests All the authors have no competing interests.

References

Adams JM, Cory S (1975) Modified nucleosides and bizarre 5'-termini in mouse myeloma mRNA. Nature 255:28–33

Alarcon CR, Goodarzi H, Lee H, Liu X, Tavazoie S, Tavazoie SF (2015a) HNRNPA2B1 is a mediator of m(6)A-dependent nuclear RNA processing events. Cell 162:1299–1308. https://doi.org/10.1016/j.cell.2015.08.011

Alarcon CR, Lee H, Goodarzi H, Halberg N, Tavazoie SF (2015b) N6-methyladenosine marks primary microRNAs for processing. Nature 519:482–485. https://doi.org/10.1038/nature14281

Amort T, Souliere MF, Wille A, Jia XY, Fiegl H, Worle H, Micura R, Lusser A (2013) Long non-coding RNAs as targets for cytosine methylation. RNA Biol 10:1003–1008. https://doi.org/10.4161/rna.24454

Arguello AE, DeLiberto AN, Kleiner RE (2017) RNA Chemical proteomics reveals the N(6)-methyladenosine (m(6)A)-regulated protein-RNA interactome. J Am Chem Soc 139:17249–17252. https://doi.org/10.1021/jacs.7b09213

Bailey AS, Batista PJ, Gold RS, Chen YG, de Rooij DG, Chang HY, Fuller MT (2017) The conserved RNA helicase YTHDC2 regulates the transition from proliferation to differentiation in the germline. elife 6: e26116. https://doi.org/10.7554/eLife.26116

Barbieri I, Tzelepis K, Pandolfini L, Shi J, Millan-Zambrano G, Robson SC, Aspris D, Migliori V, Bannister AJ, Han N et al (2017) Promoter-bound METTL3 maintains myeloid leukaemia by m(6)A-dependent translation control. Nature 552:126–131. https://doi.org/10.1038/nature24678

Ben-Neriah Y, Daley GQ, Mes-Masson AM, Witte ON, Baltimore D (1986) The chronic myelogenous leukemia-specific P210 protein is the product of the bcr/abl hybrid gene. Science 233:212–214. https://doi.org/10.1126/science.3460176

Bokar JA, Shambaugh ME, Polayes D, Matera AG, Rottman FM (1997) Purification and cDNA cloning of the AdoMet-binding subunit of the human mRNA (N6-adenosine)-methyltransferase. RNA 3:1233–1247

Brown JA, Kinzig CG, DeGregorio SJ, Steitz JA (2016) Methyltransferase-like protein 16 binds the 3′-terminal triple helix of MALAT1 long noncoding RNA. Proc Natl Acad Sci U S A 113:14013–14018. https://doi.org/10.1073/pnas.1614759113

Butler JS, Dent SY (2013) The role of chromatin modifiers in normal and malignant hematopoiesis. Blood 121: 3076–3084. https://doi.org/10.1182/blood-2012-10-451237

Carlile TM, Rojas-Duran MF, Zinshteyn B, Shin H, Bartoli KM, Gilbert WV (2014) Pseudouridine profiling reveals regulated mRNA pseudouridylation in yeast and human cells. Nature 515:143–146. https://doi.org/10.1038/nature13802

Castillo JJ, Mull N, Reagan JL, Nemr S, Mitri J (2012) Increased incidence of non-Hodgkin lymphoma, leukemia, and myeloma in patients with diabetes mellitus type 2: a meta-analysis of observational studies. Blood 119:4845–4850. https://doi.org/10.1182/blood-2011-06-362830

Challen GA, Sun D, Mayle A, Jeong M, Luo M, Rodriguez B, Mallaney C, Celik H, Yang L, Xia Z et al (2014) Dnmt3a and Dnmt3b have overlapping and distinct functions in hematopoietic stem cells. Cell Stem Cell 15:350–364. https://doi.org/10.1016/j.stem.2014.06.018

Charette M, Gray MW (2000) Pseudouridine in RNA: what, where, how, and why. IUBMB Life 49:341–351. https://doi.org/10.1080/152165400410182

Che F, Ye X, Wang Y, Wang X, Ma S, Tan Y, Mao Y, Luo Z (2022) METTL3 facilitates multiple myeloma tumorigenesis by enhancing YY1 stability and pri-microRNA-27 maturation in m(6)A-dependent manner. Cell Biol Toxicol. https://doi.org/10.1007/s10565-021-09690-1

Chen J, Odenike O, Rowley JD (2010) Leukaemogenesis: more than mutant genes. Nat Rev Cancer 10:23–36. https://doi.org/10.1038/nrc2765

Chen T, Hao YJ, Zhang Y, Li MM, Wang M, Han W, Wu Y, Lv Y, Hao J, Wang L et al (2015) m(6)A RNA methylation is regulated by microRNAs and promotes reprogramming to pluripotency. Cell Stem Cell 16: 289–301. https://doi.org/10.1016/j.stem.2015.01.016

Chen C, Liu W, Guo J, Liu Y, Liu X, Liu J, Dou X, Le R, Huang Y, Li C et al (2021a) Nuclear m(6)A reader YTHDC1 regulates the scaffold function of LINE1 RNA in mouse ESCs and early embryos. Protein Cell 12:455–474. https://doi.org/10.1007/s13238-021-00837-8

Chen Z, Shao YL, Wang LL, Lin J, Zhang JB, Ding Y, Gao BB, Liu DH, Gao XN (2021b) YTHDF2 is a potential target of AML1/ETO-HIF1alpha loop-mediated cell proliferation in t(8;21) AML. Oncogene 40:3786–3798. https://doi.org/10.1038/s41388-021-01818-1

Cheng Y, Fu Y, Wang Y, Wang J (2020) The m6A methyltransferase METTL3 is functionally implicated in DLBCL development by regulating m6A modification in PEDF. Front Genet 11:955. https://doi.org/10.3389/fgene.2020.00955

Church C, Moir L, McMurray F, Girard C, Banks GT, Teboul L, Wells S, Bruning JC, Nolan PM, Ashcroft FM, Cox RD (2010) Overexpression of Fto leads to increased food intake and results in obesity. Nat Genet 42:1086–1092. https://doi.org/10.1038/ng.713

Cohn WE, Volkin E (1951) Nucleoside-5′-phosphates from ribonucleic acid. Nature 167:483–484

Dakic A, Metcalf D, Di Rago L, Mifsud S, Wu L, Nutt SL (2005) PU.1 regulates the commitment of adult hematopoietic progenitors and restricts granulopoiesis. J Exp Med 201:1487–1502. https://doi.org/10.1084/jem.20050075

Davis FF, Allen FW (1957) Ribonucleic acids from yeast which contain a fifth nucleotide. J Biol Chem 227:907–915

Deng X, Su R, Feng X, Wei M, Chen J (2018a) Role of N(6)-methyladenosine modification in cancer. Curr Opin Genet Dev 48:1–7. https://doi.org/10.1016/j.gde.2017.10.005

Deng X, Su R, Stanford S, Chen J (2018b) Critical enzymatic functions of FTO in obesity and cancer. Front Endocrinol 9:396. https://doi.org/10.3389/fendo.2018.00396

Deng X, Su R, Weng H, Huang H, Li Z, Chen J (2018c) RNA N(6)-methyladenosine modification in cancers: current status and perspectives. Cell Res 28:507–517. https://doi.org/10.1038/s41422-018-0034-6

Dominissini D, Moshitch-Moshkovitz S, Schwartz S, Salmon-Divon M, Ungar L, Osenberg S, Cesarkas K, Jacob-Hirsch J, Amariglio N, Kupiec M et al (2012) Topology of the human and mouse m6A RNA methylomes revealed by m6A-seq. Nature 485:201–206. https://doi.org/10.1038/nature11112

Dominissini D, Nachtergaele S, Moshitch-Moshkovitz S, Peer E, Kol N, Ben-Haim MS, Dai Q, Di Segni A, Salmon-Divon M, Clark WC et al (2016) The dynamic N(1)-methyladenosine methylome in eukaryotic messenger RNA. Nature 530:441–446. https://doi.org/10.1038/nature16998

Dong S, Wu Y, Liu Y, Weng H, Huang H (2021) N(6)-methyladenosine steers RNA metabolism and regulation in cancer. Cancer Commun 41:538–559. https://doi.org/10.1002/cac2.12161

Doulatov S, Notta F, Laurenti E, Dick JE (2012) Hematopoiesis: a human perspective. Cell Stem Cell 10:120–136. https://doi.org/10.1016/j.stem.2012.01.006

Druker BJ, Tamura S, Buchdunger E, Ohno S, Segal GM, Fanning S, Zimmermann J, Lydon NB (1996) Effects of a selective inhibitor of the Abl tyrosine kinase on the growth of Bcr-Abl positive cells. Nat Med 2:561–566. https://doi.org/10.1038/nm0596-561

Druker BJ, Guilhot F, O'Brien SG, Gathmann I, Kantarjian H, Gattermann N, Deininger MW, Silver RT, Goldman JM, Stone RM et al (2006) Five-year follow-up of patients receiving imatinib for chronic myeloid leukemia. N Engl J Med 355:2408–2417. https://doi.org/10.1056/NEJMoa062867

Du H, Zhao Y, He J, Zhang Y, Xi H, Liu M, Ma J, Wu L (2016) YTHDF2 destabilizes m(6)A-containing RNA through direct recruitment of the CCR4-NOT deadenylase complex. Nat Commun 7:12626. https://doi.org/10.1038/ncomms12626

Dunn DB (1961) The occurrence of 1-methyladenine in ribonucleic acid. Biochim Biophys Acta 46:198–200

Edupuganti RR, Geiger S, Lindeboom RGH, Shi H, Hsu PJ, Lu Z, Wang SY, Baltissen MPA, Jansen P, Rossa M et al (2017) N(6)-methyladenosine (m(6)A) recruits and repels proteins to regulate mRNA homeostasis. Nat Struct Mol Biol 24:870–878. https://doi.org/10.1038/nsmb.3462

El Yacoubi B, Bailly M, de Crecy-Lagard V (2012) Biosynthesis and function of posttranscriptional modifications of transfer RNAs. Annu Rev Genet 46:

69–95. https://doi.org/10.1146/annurev-genet-110711-155641

Fischer J, Koch L, Emmerling C, Vierkotten J, Peters T, Bruning JC, Ruther U (2009) Inactivation of the Fto gene protects from obesity. Nature 458:894–898. https://doi.org/10.1038/nature07848

Fu L, Guerrero CR, Zhong N, Amato NJ, Liu Y, Liu S, Cai Q, Ji D, Jin SG, Niedernhofer LJ et al (2014a) Tet-mediated formation of 5-hydroxymethylcytosine in RNA. J Am Chem Soc 136:11582–11585. https://doi.org/10.1021/ja505305z

Fu Y, Dominissini D, Rechavi G, He C (2014b) Gene expression regulation mediated through reversible m (6)A RNA methylation. Nat Rev Genet 15:293–306. https://doi.org/10.1038/nrg3724

Gerken T, Girard CA, Tung YC, Webby CJ, Saudek V, Hewitson KS, Yeo GS, McDonough MA, Cunliffe S, McNeill LA et al (2007) The obesity-associated FTO gene encodes a 2-oxoglutarate-dependent nucleic acid demethylase. Science 318:1469–1472. https://doi.org/10.1126/science.1151710

Geula S, Moshitch-Moshkovitz S, Dominissini D, Mansour AA, Kol N, Salmon-Divon M, Hershkovitz V, Peer E, Mor N, Manor YS et al (2015) Stem cells. m6A mRNA methylation facilitates resolution of naive pluripotency toward differentiation. Science 347:1002–1006. https://doi.org/10.1126/science.1261417

Goldman JM, Melo JV (2001) Targeting the BCR-ABL tyrosine kinase in chronic myeloid leukemia. N Engl J Med 344:1084–1086. https://doi.org/10.1056/NEJM200104053441409

Gong H, Liu L, Cui L, Ma H, Shen L (2021) ALKBH5-mediated m6A-demethylation of USP1 regulated T-cell acute lymphoblastic leukemia cell glucocorticoid resistance by Aurora B. Mol Carcinog 60:644–657. https://doi.org/10.1002/mc.23330

Goode DK, Obier N, Vijayabaskar MS, Lie ALM, Lilly AJ, Hannah R, Lichtinger M, Batta K, Florkowska M, Patel R et al (2016) Dynamic gene regulatory networks drive hematopoietic specification and differentiation. Dev Cell 36:572–587. https://doi.org/10.1016/j.devcel.2016.01.024

Guillamot M, Cimmino L, Aifantis I (2016) The impact of DNA methylation in hematopoietic malignancies. Trends Cancer 2:70–83. https://doi.org/10.1016/j.trecan.2015.12.006

Guo J, Tang HW, Li J, Perrimon N, Yan D (2018) Xio is a component of the Drosophila sex determination pathway and RNA N(6)-methyladenosine methyltransferase complex. Proc Natl Acad Sci U S A 115:3674–3679. https://doi.org/10.1073/pnas.1720945115

Hall RH (1963) Method for isolation of 2-O-methylribonucleosides and N1-methyladenosine from ribonucleic acid. Biochim Biophys Acta 68:278–283

Hernandez-Caballero ME, Sierra-Ramirez JA (2015) Single nucleotide polymorphisms of the FTO gene and cancer risk: an overview. Mol Biol Rep 42:699–704. https://doi.org/10.1007/s11033-014-3817-y

Hsu PJ, Zhu Y, Ma H, Guo Y, Shi X, Liu Y, Qi M, Lu Z, Shi H, Wang J et al (2017) Ythdc2 is an N(6)-methyladenosine binding protein that regulates mammalian spermatogenesis. Cell Res 27:1115–1127. https://doi.org/10.1038/cr.2017.99

Hua Z, Wei R, Guo M, Lin Z, Yu X, Li X, Gu C, Yang Y (2022) YTHDF2 promotes multiple myeloma cell proliferation via STAT5A/MAP2K2/p-ERK axis. Oncogene 41:1482–1491. https://doi.org/10.1038/s41388-022-02191-3

Huang H, Jiang X, Li Z, Li Y, Song CX, He C, Sun M, Chen P, Gurbuxani S, Wang J et al (2013) TET1 plays an essential oncogenic role in MLL-rearranged leukemia. Proc Natl Acad Sci U S A 110:11994–11999. https://doi.org/10.1073/pnas.1310656110

Huang H, Weng H, Sun W, Qin X, Shi H, Wu H, Zhao BS, Mesquita A, Liu C, Yuan CL et al (2018) Recognition of RNA N6-methyladenosine by IGF2BP proteins enhances mRNA stability and translation. Nat Cell Biol 20(3):285–295

Huang H, Weng H, Zhou K, Wu T, Zhao BS, Sun M, Chen Z, Deng X, Xiao G, Auer F et al (2019a) Histone H3 trimethylation at lysine 36 guides m(6)A RNA modification co-transcriptionally. Nature 567:414–419. https://doi.org/10.1038/s41586-019-1016-7

Huang Y, Su R, Sheng Y, Dong L, Dong Z, Xu H, Ni T, Zhang ZS, Zhang T, Li C et al (2019b) Small-molecule targeting of oncogenic FTO demethylase in acute myeloid leukemia. Cancer Cell 35:677–691. https://doi.org/10.1016/j.ccell.2019.03.006

Huang H, Weng H, Chen J (2020a) The biogenesis and precise control of RNA m(6)A methylation. Trends Genet 36:44–52. https://doi.org/10.1016/j.tig.2019.10.011

Huang H, Weng H, Chen J (2020b) m(6)A modification in coding and non-coding RNAs: roles and therapeutic implications in cancer. Cancer Cell 37:270–288. https://doi.org/10.1016/j.ccell.2020.02.004

Huang H, Weng H, Deng X, Chen J (2020c) RNA modifications in cancer: functions, mechanisms, and therapeutic implications. Annu Rev Cancer Biol 4: 221–240. https://doi.org/10.1146/annurev-cancerbio-030419-033357

Huber SM, van Delft P, Mendil L, Bachman M, Smollett K, Werner F, Miska EA, Balasubramanian S (2015) Formation and abundance of 5-hydroxymethylcytosine in RNA. Chembiochem 16: 752–755. https://doi.org/10.1002/cbic.201500013

Ianniello Z, Sorci M, Ceci Ginistrelli L, Iaiza A, Marchioni M, Tito C, Capuano E, Masciarelli S, Ottone T, Attrotto C et al (2021) New insight into the catalytic-dependent and -independent roles of METTL3 in sustaining aberrant translation in chronic myeloid leukemia. Cell Death Dis 12:870. https://doi.org/10.1038/s41419-021-04169-7

Iwasaki H, Somoza C, Shigematsu H, Duprez EA, Iwasaki-Arai J, Mizuno S, Arinobu Y, Geary K,

Zhang P, Dayaram T et al (2005) Distinctive and indispensable roles of PU.1 in maintenance of hematopoietic stem cells and their differentiation. Blood 106:1590–1600. https://doi.org/10.1182/blood-2005-03-0860

Jain D, Puno MR, Meydan C, Lailler N, Mason CE, Lima CD, Anderson KV, Keeney S (2018) Ketu mutant mice uncover an essential meiotic function for the ancient RNA helicase YTHDC2. elife 7:e30919. https://doi.org/10.7554/eLife.30919

Jia G, Fu Y, Zhao X, Dai Q, Zheng G, Yang Y, Yi C, Lindahl T, Pan T, Yang YG, He C (2011) N6-methyladenosine in nuclear RNA is a major substrate of the obesity-associated FTO. Nat Chem Biol 7: 885–887. https://doi.org/10.1038/nchembio.687

Jiang F, Tang X, Tang C, Hua Z, Ke M, Wang C, Zhao J, Gao S, Jurczyszyn A, Janz S et al (2021) HNRNPA2B1 promotes multiple myeloma progression by increasing AKT3 expression via m6A-dependent stabilization of ILF3 mRNA. J Hematol Oncol 14:54. https://doi.org/10.1186/s13045-021-01066-6

Knuckles P, Lence T, Haussmann IU, Jacob D, Kreim N, Carl SH, Masiello I, Hares T, Villasenor R, Hess D et al (2018) Zc3h13/Flacc is required for adenosine methylation by bridging the mRNA-binding factor Rbm15/Spenito to the m(6)A machinery component Wtap/Fl(2)d. Genes Dev 32:415–429. https://doi.org/10.1101/gad.309146.117

Koschmieder S, Rosenbauer F, Steidl U, Owens BM, Tenen DG (2005) Role of transcription factors C/EBPalpha and PU.1 in normal hematopoiesis and leukemia. Int J Hematol 81:368–377

Krug RM, Morgan MA, Shatkin AJ (1976) Influenza viral mRNA contains internal N6-methyladenosine and 5′--terminal 7-methylguanosine in cap structures. J Virol 20:45–53

Kurowski MA, Bhagwat AS, Papaj G, Bujnicki JM (2003) Phylogenomic identification of five new human homologs of the DNA repair enzyme AlkB. BMC Genomics 4:48. https://doi.org/10.1186/1471-2164-4-48

Li X, Zhu P, Ma S, Song J, Bai J, Sun F, Yi C (2015) Chemical pulldown reveals dynamic pseudouridylation of the mammalian transcriptome. Nat Chem Biol 11:592–597. https://doi.org/10.1038/nchembio.1836

Li A, Chen YS, Ping XL, Yang X, Xiao W, Yang Y, Sun HY, Zhu Q, Baidya P, Wang X et al (2017a) Cytoplasmic m(6)A reader YTHDF3 promotes mRNA translation. Cell Res 27:444–447. https://doi.org/10.1038/cr.2017.10

Li Z, Weng H, Su R, Weng X, Zuo Z, Li C, Huang H, Nachtergaele S, Dong L, Hu C et al (2017b) FTO plays an oncogenic role in acute myeloid leukemia as a N(6)-methyladenosine RNA demethylase. Cancer Cell 31: 127–141. https://doi.org/10.1016/j.ccell.2016.11.017

Li Z, Qian P, Shao W, Shi H, He XC, Gogol M, Yu Z, Wang Y, Qi M, Zhu Y et al (2018) Suppression of m(6) A reader Ythdf2 promotes hematopoietic stem cell expansion. Cell Res. https://doi.org/10.1038/s41422-018-0072-0

Liu J, Yue Y, Han D, Wang X, Fu Y, Zhang L, Jia G, Yu M, Lu Z, Deng X et al (2014) A METTL3-METTL14 complex mediates mammalian nuclear RNA N6-adenosine methylation. Nat Chem Biol 10: 93–95. https://doi.org/10.1038/nchembio.1432

Liu N, Dai Q, Zheng G, He C, Parisien M, Pan T (2015) N (6)-methyladenosine-dependent RNA structural switches regulate RNA-protein interactions. Nature 518:560–564. https://doi.org/10.1038/nature14234

Liu J, Gao M, He J, Wu K, Lin S, Jin L, Chen Y, Liu H, Shi J, Wang X et al (2021) The RNA m(6)A reader YTHDC1 silences retrotransposons and guards ES cell identity. Nature 591:322–326. https://doi.org/10.1038/s41586-021-03313-9

Losman JA, Looper RE, Koivunen P, Lee S, Schneider RK, McMahon C, Cowley GS, Root DE, Ebert BL, Kaelin WG Jr (2013) (R)-2-hydroxyglutarate is sufficient to promote leukemogenesis and its effects are reversible. Science 339:1621–1625. https://doi.org/10.1126/science.1231677

Lovejoy AF, Riordan DP, Brown PO (2014) Transcriptome-wide mapping of pseudouridines: pseudouridine synthases modify specific mRNAs in S. cerevisiae. PLoS One 9:e110799. https://doi.org/10.1371/journal.pone.0110799

Luo S, Tong L (2014) Molecular basis for the recognition of methylated adenines in RNA by the eukaryotic YTH domain. Proc Natl Acad Sci U S A 111:13834–13839. https://doi.org/10.1073/pnas.1412742111

Ma X, Renda MJ, Wang L, Cheng EC, Niu C, Morris SW, Chi AS, Krause DS (2007) Rbm15 modulates Notch-induced transcriptional activation and affects myeloid differentiation. Mol Cell Biol 27:3056–3064. https://doi.org/10.1128/MCB.01339-06

Ma H, Wang X, Cai J, Dai Q, Natchiar SK, Lv R, Chen K, Lu Z, Chen H, Shi YG et al (2019) N(6-) Methyladenosine methyltransferase ZCCHC4 mediates ribosomal RNA methylation. Nat Chem Biol 15:88–94. https://doi.org/10.1038/s41589-018-0184-3

Ma H, Shen L, Yang H, Gong H, Du X, Li J (2021) m6A methyltransferase Wilms' tumor 1-associated protein facilitates cell proliferation and cisplatin resistance in NK/T cell lymphoma by regulating dual-specificity phosphatases 6 expression via m6A RNA methylation. IUBMB Life 73:108–117. https://doi.org/10.1002/iub.2410

Mapperley C, van de Lagemaat LN, Lawson H, Tavosanis A, Paris J, Campos J, Wotherspoon D, Durko J, Sarapuu A, Choe J et al (2021) The mRNA m6A reader YTHDF2 suppresses proinflammatory pathways and sustains hematopoietic stem cell function. J Exp Med 218:e20200829. https://doi.org/10.1084/jem.20200829

Mauer J, Luo X, Blanjoie A, Jiao X, Grozhik AV, Patil DP, Linder B, Pickering BF, Vasseur JJ, Chen Q et al

(2017) Reversible methylation of m(6)Am in the 5′ cap controls mRNA stability. Nature 541:371–375. https://doi.org/10.1038/nature21022

McCulloch EA, Till JE (2005) Perspectives on the properties of stem cells. Nat Med 11:1026–1028. https://doi.org/10.1038/nm1005-1026

Mendel M, Chen KM, Homolka D, Gos P, Pandey RR, McCarthy AA, Pillai RS (2018) Methylation of structured RNA by the m(6)A writer METTL16 is essential for mouse embryonic development. Mol Cell 71:986–1000. https://doi.org/10.1016/j.molcel.2018.08.004

Merkestein M, Laber S, McMurray F, Andrew D, Sachse G, Sanderson J, Li M, Usher S, Sellayah D, Ashcroft FM, Cox RD (2015) FTO influences adipogenesis by regulating mitotic clonal expansion. Nat Commun 6:6792. https://doi.org/10.1038/ncomms7792

Meyer KD, Saletore Y, Zumbo P, Elemento O, Mason CE, Jaffrey SR (2012) Comprehensive analysis of mRNA methylation reveals enrichment in 3' UTRs and near stop codons. Cell 149:1635–1646. https://doi.org/10.1016/j.cell.2012.05.003

Mikkola I, Heavey B, Horcher M, Busslinger M (2002) Reversion of B cell commitment upon loss of Pax5 expression. Science 297:110–113. https://doi.org/10.1126/science.1067518

Moran-Crusio K, Reavie L, Shih A, Abdel-Wahab O, Ndiaye-Lobry D, Lobry C, Figueroa ME, Vasanthakumar A, Patel J, Zhao X et al (2011) Tet2 loss leads to increased hematopoietic stem cell self-renewal and myeloid transformation. Cancer Cell 20:11–24. https://doi.org/10.1016/j.ccr.2011.06.001

Mucenski ML, McLain K, Kier AB, Swerdlow SH, Schreiner CM, Miller TA, Pietryga DW, Scott WJ Jr, Potter SS (1991) A functional c-myb gene is required for normal murine fetal hepatic hematopoiesis. Cell 65:677–689

Niu C, Zhang J, Breslin P, Onciu M, Ma Z, Morris SW (2009) c-Myc is a target of RNA-binding motif protein 15 in the regulation of adult hematopoietic stem cell and megakaryocyte development. Blood 114:2087–2096. https://doi.org/10.1182/blood-2009-01-197921

O'Connell RM, Chaudhuri AA, Rao DS, Gibson WS, Balazs AB, Baltimore D (2010) MicroRNAs enriched in hematopoietic stem cells differentially regulate long-term hematopoietic output. Proc Natl Acad Sci U S A 107:14235–14240. https://doi.org/10.1073/pnas.1009798107

Ooi AG, Sahoo D, Adorno M, Wang Y, Weissman IL, Park CY (2010) MicroRNA-125b expands hematopoietic stem cells and enriches for the lymphoid-balanced and lymphoid-biased subsets. Proc Natl Acad Sci U S A 107:21505–21510. https://doi.org/10.1073/pnas.1016218107

Paris J, Morgan M, Campos J, Spencer GJ, Shmakova A, Ivanova I, Mapperley C, Lawson H, Wotherspoon DA, Sepulveda C et al (2019) Targeting the RNA m(6)A reader YTHDF2 selectively compromises cancer stem cells in acute myeloid leukemia. Cell Stem Cell 25:137–148. https://doi.org/10.1016/j.stem.2019.03.021

Patil DP, Chen CK, Pickering BF, Chow A, Jackson C, Guttman M, Jaffrey SR (2016) m(6)A RNA methylation promotes XIST-mediated transcriptional repression. Nature 537:369–373. https://doi.org/10.1038/nature19342

Pendleton KE, Chen B, Liu K, Hunter OV, Xie Y, Tu BP, Conrad NK (2017) The U6 snRNA m(6)A methyltransferase METTL16 regulates SAM synthetase intron retention. Cell 169:824–835. https://doi.org/10.1016/j.cell.2017.05.003

Ping XL, Sun BF, Wang L, Xiao W, Yang X, Wang WJ, Adhikari S, Shi Y, Lv Y, Chen YS et al (2014) Mammalian WTAP is a regulatory subunit of the RNA N6-methyladenosine methyltransferase. Cell Res 24:177–189. https://doi.org/10.1038/cr.2014.3

Pinto R, Vagbo CB, Jakobsson ME, Kim Y, Baltissen MP, O'Donohue MF, Guzman UH, Malecki JM, Wu J, Kirpekar F et al (2020) The human methyltransferase ZCCHC4 catalyses N6-methyladenosine modification of 28S ribosomal RNA. Nucleic Acids Res 48:830–846. https://doi.org/10.1093/nar/gkz1147

Qing Y, Su R, Chen J (2021) RNA modifications in hematopoietic malignancies: a new research frontier. Blood. https://doi.org/10.1182/blood.2019004263

Raffel GD, Mercher T, Shigematsu H, Williams IR, Cullen DE, Akashi K, Bernard OA, Gilliland DG (2007) Ott1 (Rbm15) has pleiotropic roles in hematopoietic development. Proc Natl Acad Sci U S A 104:6001–6006. https://doi.org/10.1073/pnas.0609041104

Rosenbauer F, Tenen DG (2007) Transcription factors in myeloid development: balancing differentiation with transformation. Nat Rev Immunol 7:105–117. https://doi.org/10.1038/nri2024

Rosmarin AG, Yang Z, Resendes KK (2005) Transcriptional regulation in myelopoiesis: hematopoietic fate choice, myeloid differentiation, and leukemogenesis. Exp Hematol 33:131–143. https://doi.org/10.1016/j.exphem.2004.08.015

Roundtree IA, Evans ME, Pan T, He C (2017a) Dynamic RNA modifications in gene expression regulation. Cell 169:1187–1200. https://doi.org/10.1016/j.cell.2017.05.045

Roundtree IA, Luo GZ, Zhang Z, Wang X, Zhou T, Cui Y, Sha J, Huang X, Guerrero L, Xie P et al (2017b) YTHDC1 mediates nuclear export of N(6)-methyladenosine methylated mRNAs. elife 6:e31311. https://doi.org/10.7554/eLife.31311

Rowley JD (1973) Letter: A new consistent chromosomal abnormality in chronic myelogenous leukaemia identified by quinacrine fluorescence and Giemsa staining. Nature 243:290–293. https://doi.org/10.1038/243290a0

Safra M, Sas-Chen A, Nir R, Winkler R, Nachshon A, Bar-Yaacov D, Erlacher M, Rossmanith W, Stern-Ginossar N, Schwartz S (2017) The m1A landscape on cytosolic and mitochondrial mRNA at single-base resolution. Nature 551:251–255. https://doi.org/10.1038/nature24456

Sandberg ML, Sutton SE, Pletcher MT, Wiltshire T, Tarantino LM, Hogenesch JB, Cooke MP (2005) c-Myb and p300 regulate hematopoietic stem cell proliferation and differentiation. Dev Cell 8:153–166

Sasaki M, Knobbe CB, Munger JC, Lind EF, Brenner D, Brustle A, Harris IS, Holmes R, Wakeham A, Haight J et al (2012) IDH1(R132H) mutation increases murine haematopoietic progenitors and alters epigenetics. Nature 488:656–659. https://doi.org/10.1038/nature11323

Satoh Y, Matsumura I, Tanaka H, Ezoe S, Sugahara H, Mizuki M, Shibayama H, Ishiko E, Ishiko J, Nakajima K, Kanakura Y (2004) Roles for c-Myc in self-renewal of hematopoietic stem cells. J Biol Chem 279:24986–24993. https://doi.org/10.1074/jbc.M400407200

Schwartz S, Bernstein DA, Mumbach MR, Jovanovic M, Herbst RH, Leon-Ricardo BX, Engreitz JM, Guttman M, Satija R, Lander ES et al (2014a) Transcriptome-wide mapping reveals widespread dynamic-regulated pseudouridylation of ncRNA and mRNA. Cell 159:148–162. https://doi.org/10.1016/j.cell.2014.08.028

Schwartz S, Mumbach MR, Jovanovic M, Wang T, Maciag K, Bushkin GG, Mertins P, Ter-Ovanesyan D, Habib N, Cacchiarelli D et al (2014b) Perturbation of m6A writers reveals two distinct classes of mRNA methylation at internal and 5′ sites. Cell Rep 8:284–296. https://doi.org/10.1016/j.celrep.2014.05.048

Shen C, Sheng Y, Zhu AC, Robinson S, Jiang X, Dong L, Chen H, Su R, Yin Z, Li W et al (2020) RNA demethylase ALKBH5 selectively promotes tumorigenesis and cancer stem cell self-renewal in acute myeloid leukemia. Cell Stem Cell 27:64–80. https://doi.org/10.1016/j.stem.2020.04.009

Sheng Y, Wei J, Yu F, Xu H, Yu C, Wu Q, Liu Y, Li L, Cui XL, Gu X et al (2021) A critical role of nuclear m6A reader YTHDC1 in leukemogenesis by regulating MCM complex-mediated DNA replication. Blood 138:2838–2852. https://doi.org/10.1182/blood.2021011707

Shi H, Wang X, Lu Z, Zhao BS, Ma H, Hsu PJ, Liu C, He C (2017) YTHDF3 facilitates translation and decay of N(6)-methyladenosine-modified RNA. Cell Res 27:315–328. https://doi.org/10.1038/cr.2017.15

Siminovitch L, McCulloch EA, Till JE (1963) The distribution of colony-forming cells among spleen colonies. J Cell Comp Physiol 62:327–336. https://doi.org/10.1002/jcp.1030620313

Sledz P, Jinek M (2016) Structural insights into the molecular mechanism of the m(6)A writer complex. elife 5:e18434. https://doi.org/10.7554/eLife.18434

Soderberg KC, Kaprio J, Verkasalo PK, Pukkala E, Koskenvuo M, Lundqvist E, Feychting M (2009) Overweight, obesity and risk of haematological malignancies: a cohort study of Swedish and Finnish twins. Eur J Cancer 45:1232–1238. https://doi.org/10.1016/j.ejca.2008.11.004

Song S, Fan G, Li Q, Su Q, Zhang X, Xue X, Wang Z, Qian C, Jin Z, Li B, Zhuang W (2021) IDH2 contributes to tumorigenesis and poor prognosis by regulating m6A RNA methylation in multiple myeloma. Oncogene 40:5393–5402. https://doi.org/10.1038/s41388-021-01939-7

Song W, Fei F, Qiao F, Weng Z, Yang Y, Cao B, Yue J, Xu J, Zheng M, Li J (2022) ALKBH5-mediated N(6)-methyladenosine modification of TRERNA1 promotes DLBCL proliferation via p21 downregulation. Cell Death Dis 8:25. https://doi.org/10.1038/s41420-022-00819-7

Squires JE, Patel HR, Nousch M, Sibbritt T, Humphreys DT, Parker BJ, Suter CM, Preiss T (2012) Widespread occurrence of 5-methylcytosine in human coding and non-coding RNA. Nucleic Acids Res 40:5023–5033. https://doi.org/10.1093/nar/gks144

Su R, Dong L, Li C, Nachtergaele S, Wunderlich M, Qing Y, Deng X, Wang Y, Weng X, Hu C et al (2018) R-2HG exhibits anti-tumor activity by targeting FTO/m(6)A/MYC/CEBPA signaling. Cell 172:90–105. https://doi.org/10.1016/j.cell.2017.11.031

Su R, Dong L, Li Y, Gao M, Han L, Wunderlich M, Deng X, Li H, Huang Y, Gao L et al (2020) Targeting FTO suppresses cancer stem cell maintenance and immune evasion. Cancer Cell 38:79–96. https://doi.org/10.1016/j.ccell.2020.04.017

Su R, Dong L, Li Y, Gao M, He PC, Liu W, Wei J, Zhao Z, Gao L, Han L et al (2022) METTL16 exerts an m(6)A-independent function to facilitate translation and tumorigenesis. Nat Cell Biol 24:205–216. https://doi.org/10.1038/s41556-021-00835-2

Sun C, Chang L, Liu C, Chen X, Zhu X (2019) The study of METTL3 and METTL14 expressions in childhood ETV6/RUNX1-positive acute lymphoblastic leukemia. Mol Genet Genom Med 7:e00933. https://doi.org/10.1002/mgg3.933

Suzuki T, Ueda H, Okada S, Sakurai M (2015) Transcriptome-wide identification of adenosine-to-inosine editing using the ICE-seq method. Nat Protoc 10:715–732. https://doi.org/10.1038/nprot.2015.037

van Tran N, Ernst FGM, Hawley BR, Zorbas C, Ulryck N, Hackert P, Bohnsack KE, Bohnsack MT, Jaffrey SR, Graille M, Lafontaine DLJ (2019) The human 18S rRNA m6A methyltransferase METTL5 is stabilized by TRMT112. Nucleic Acids Res 47:7719–7733. https://doi.org/10.1093/nar/gkz619

Vu LP, Pickering BF, Cheng Y, Zaccara S, Nguyen D, Minuesa G, Chou T, Chow A, Saletore Y, MacKay M et al (2017) The N(6)-methyladenosine (m(6)A)-forming enzyme METTL3 controls myeloid differentiation of normal hematopoietic and leukemia cells. Nat Med 23:1369–1376. https://doi.org/10.1038/nm.4416

Wang F, Travins J, DeLaBarre B, Penard-Lacronique V, Schalm S, Hansen E, Straley K, Kernytsky A, Liu W, Gliser C et al (2013) Targeted inhibition of mutant IDH2 in leukemia cells induces cellular differentiation.

Science 340:622–626. https://doi.org/10.1126/science.1234769
Wang X, Lu Z, Gomez A, Hon GC, Yue Y, Han D, Fu Y, Parisien M, Dai Q, Jia G et al (2014a) N6-methyladenosine-dependent regulation of messenger RNA stability. Nature 505:117–120. https://doi.org/10.1038/nature12730
Wang Y, Li Y, Toth JI, Petroski MD, Zhang Z, Zhao JC (2014b) N6-methyladenosine modification destabilizes developmental regulators in embryonic stem cells. Nat Cell Biol 16:191–198. https://doi.org/10.1038/ncb2902
Wang X, Zhao BS, Roundtree IA, Lu Z, Han D, Ma H, Weng X, Chen K, Shi H, He C (2015) N(6)-methyladenosine modulates messenger RNA translation efficiency. Cell 161:1388–1399. https://doi.org/10.1016/j.cell.2015.05.014
Wang P, Doxtader KA, Nam Y (2016a) Structural basis for cooperative function of Mettl3 and Mettl14 methyltransferases. Mol Cell 63:306–317. https://doi.org/10.1016/j.molcel.2016.05.041
Wang X, Feng J, Xue Y, Guan Z, Zhang D, Liu Z, Gong Z, Wang Q, Huang J, Tang C et al (2016b) Structural basis of N(6)-adenosine methylation by the METTL3-METTL14 complex. Nature 534:575–578. https://doi.org/10.1038/nature18298
Wang J, Li Y, Wang P, Han G, Zhang T, Chang J, Yin R, Shan Y, Wen J, Xie X et al (2020) Leukemogenic chromatin alterations promote AML leukemia stem cells via a KDM4C-ALKBH5-AXL signaling axis. Cell Stem Cell 27:81–97. https://doi.org/10.1016/j.stem.2020.04.001
Wang X, Hu M, Yu L, Wang X, Jiang X, Zhang G, Ding K (2022a) The "m6A writer" METTL3 and the "m6A reader" IGF2BP2 regulate cutaneous T-cell lymphomas progression via CDKN2A. Hematol Oncol. https://doi.org/10.1002/hon.3005
Wang Z, He J, Bach DH, Huang YH, Li Z, Liu H, Lin P, Yang J (2022b) Induction of m(6)A methylation in adipocyte exosomal LncRNAs mediates myeloma drug resistance. J Exp Clin Cancer Res 41:4. https://doi.org/10.1186/s13046-021-02209-w
Wei CM, Moss B (1977) Nucleotide sequences at the N6-methyladenosine sites of HeLa cell messenger ribonucleic acid. Biochemistry 16:1672–1676
Wei J, Liu F, Lu Z, Fei Q, Ai Y, He PC, Shi H, Cui X, Su R, Klungland A et al (2018) Differential m(6)A, m(6)Am, and m(1)A demethylation mediated by FTO in the cell nucleus and cytoplasm. Mol Cell 71(973-985): e975. https://doi.org/10.1016/j.molcel.2018.08.011
Wen J, Lv R, Ma H, Shen H, He C, Wang J, Jiao F, Liu H, Yang P, Tan L et al (2018) Zc3h13 regulates nuclear RNA m(6)A methylation and mouse embryonic stem cell self-renewal. Mol Cell 69:1028–1038. https://doi.org/10.1016/j.molcel.2018.02.015
Weng H, Huang H, Wu H, Qin X, Zhao BS, Dong L, Shi H, Skibbe J, Shen C, Hu C et al (2018) METTL14 inhibits hematopoietic stem/progenitor differentiation and promotes leukemogenesis via mRNA m(6)A modification. Cell Stem Cell 22:191–205. https://doi.org/10.1016/j.stem.2017.11.016
Weng H, Huang H, Chen J (2019) RNA N (6)-methyladenosine modification in normal and malignant hematopoiesis. Adv Exp Med Biol 1143:75–93. https://doi.org/10.1007/978-981-13-7342-8_4
Weng H, Huang F, Yu Z, Chen Z, Prince E, Kang Y, Zhou K, Li W, Hu J, Fu C et al (2022) The m(6)A reader IGF2BP2 regulates glutamine metabolism and represents a therapeutic target in acute myeloid leukemia. Cancer Cell 40:1566–1582. https://doi.org/10.1016/j.ccell.2022.10.004
Wilson A, Murphy MJ, Oskarsson T, Kaloulis K, Bettess MD, Oser GM, Pasche AC, Knabenhans C, MacDonald HR, Trumpp A (2004) c-Myc controls the balance between hematopoietic stem cell self-renewal and differentiation. Genes Dev 18:2747–2763. https://doi.org/10.1101/gad.313104
Wojtas MN, Pandey RR, Mendel M, Homolka D, Sachidanandam R, Pillai RS (2017) Regulation of m(6)A transcripts by the 3′→5′ RNA helicase YTHDC2 is essential for a successful meiotic program in the mammalian germline. Mol Cell 68:374–387. https://doi.org/10.1016/j.molcel.2017.09.021
Wu B, Su S, Patil DP, Liu H, Gan J, Jaffrey SR, Ma J (2018) Molecular basis for the specific and multivariant recognitions of RNA substrates by human hnRNP A2/B1. Nat Commun 9:420. https://doi.org/10.1038/s41467-017-02770-z
Xiang Y, Laurent B, Hsu CH, Nachtergaele S, Lu Z, Sheng W, Xu C, Chen H, Ouyang J, Wang S et al (2017) RNA m(6)A methylation regulates the ultraviolet-induced DNA damage response. Nature 543:573–576. https://doi.org/10.1038/nature21671
Xiao W, Adhikari S, Dahal U, Chen YS, Hao YJ, Sun BF, Sun HY, Li A, Ping XL, Lai WY et al (2016) Nuclear m(6)A reader YTHDC1 regulates mRNA splicing. Mol Cell 61:507–519. https://doi.org/10.1016/j.molcel.2016.01.012
Xu C, Wang X, Liu K, Roundtree IA, Tempel W, Li Y, Lu Z, He C, Min J (2014) Structural basis for selective binding of m6A RNA by the YTHDC1 YTH domain. Nat Chem Biol 10:927–929. https://doi.org/10.1038/nchembio.1654
Xu A, Zhang J, Zuo L, Yan H, Chen L, Zhao F, Fan F, Xu J, Zhang B, Zhang Y et al (2022) FTO promotes multiple myeloma progression by posttranscriptional activation of HSF1 in an m(6)A-YTHDF2-dependent manner. Mol Ther 30:1104–1118. https://doi.org/10.1016/j.ymthe.2021.12.012
Yang Y, Hsu PJ, Chen YS, Yang YG (2018) Dynamic transcriptomic m(6)A decoration: writers, erasers, readers and functions in RNA metabolism. Cell Res 28:616–624. https://doi.org/10.1038/s41422-018-0040-8
Yankova E, Blackaby W, Albertella M, Rak J, De Braekeleer E, Tsagkogeorga G, Pilka ES, Aspris D, Leggate D, Hendrick AG et al (2021) Small-molecule inhibition of METTL3 as a strategy against myeloid

leukaemia. Nature 593:597–601. https://doi.org/10.1038/s41586-021-03536-w

Yao QJ, Sang L, Lin M, Yin X, Dong W, Gong Y, Zhou BO (2018) Mettl3-Mettl14 methyltransferase complex regulates the quiescence of adult hematopoietic stem cells. Cell Res. https://doi.org/10.1038/s41422-018-0062-2

Yao FY, Zhao C, Zhong FM, Qin TY, Wen F, Li MY, Liu J, Huang B, Wang XZ (2021) m(6)A modification of lncRNA NEAT1 regulates chronic myelocytic leukemia progression via miR-766-5p/CDKN1A axis. Front Oncol 11:679634. https://doi.org/10.3389/fonc.2021.679634

Yu T, Yao L, Yin H, Teng Y, Hong M, Wu Q (2022) ALKBH5 promotes multiple myeloma tumorigenicity through inducing m(6)A-demethylation of SAV1 mRNA and myeloma stem cell phenotype. Int J Biol Sci 18:2235–2248. https://doi.org/10.7150/ijbs.64943

Zhang C, Chen Y, Sun B, Wang L, Yang Y, Ma D, Lv J, Heng J, Ding Y, Xue Y et al (2017a) m(6)A modulates haematopoietic stem and progenitor cell specification. Nature 549:273–276. https://doi.org/10.1038/nature23883

Zhang S, Zhao BS, Zhou A, Lin K, Zheng S, Lu Z, Chen Y, Sulman EP, Xie K, Bogler O et al (2017b) m(6)A demethylase ALKBH5 maintains tumorigenicity of glioblastoma stem-like cells by sustaining FOXM1 expression and cell proliferation program. Cancer Cell 31:591–606. https://doi.org/10.1016/j.ccell.2017.02.013

Zhao X, Yang Y, Sun BF, Shi Y, Yang X, Xiao W, Hao YJ, Ping XL, Chen YS, Wang WJ et al (2014) FTO-dependent demethylation of N6-methyladenosine regulates mRNA splicing and is required for adipogenesis. Cell Res 24:1403–1419. https://doi.org/10.1038/cr.2014.151

Zhao BS, Roundtree IA, He C (2017a) Post-transcriptional gene regulation by mRNA modifications. Nat Rev Mol Cell Biol 18:31–42. https://doi.org/10.1038/nrm.2016.132

Zhao BS, Wang X, Beadell AV, Lu Z, Shi H, Kuuspalu A, Ho RK, He C (2017b) m(6)A-dependent maternal mRNA clearance facilitates zebrafish maternal-to-zygotic transition. Nature 542:475–478. https://doi.org/10.1038/nature21355

Zheng G, Dahl JA, Niu Y, Fedorcsak P, Huang CM, Li CJ, Vagbo CB, Shi Y, Wang WL, Song SH et al (2013) ALKBH5 is a mammalian RNA demethylase that impacts RNA metabolism and mouse fertility. Mol Cell 49:18–29. https://doi.org/10.1016/j.molcel.2012.10.015

Zhou J, Wan J, Gao X, Zhang X, Jaffrey SR, Qian SB (2015) Dynamic m(6)A mRNA methylation directs translational control of heat shock response. Nature 526:591–594. https://doi.org/10.1038/nature15377

Zhou KI, Shi H, Lyu R, Wylder AC, Matuszek Z, Pan JN, He C, Parisien M, Pan T (2019) Regulation of co-transcriptional pre-mRNA splicing by m(6)A through the low-complexity protein hnRNPG. Mol Cell 76:70–81. https://doi.org/10.1016/j.molcel.2019.07.005

Zhu T, Roundtree IA, Wang P, Wang X, Wang L, Sun C, Tian Y, Li J, He C, Xu Y (2014) Crystal structure of the YTH domain of YTHDF2 reveals mechanism for recognition of N6-methyladenosine. Cell Res 24:1493–1496. https://doi.org/10.1038/cr.2014.152

Immune System Influence on Hematopoietic Stem Cells and Leukemia Development

8

John M. Perry

Abstract

Hematopoietic stem cells (HSCs) are the source for all blood cells, including immune cells, and they interact dynamically with the immune system. This chapter will explore the nature of stem cells, particularly HSCs, in the context of their immune microenvironment. The dynamic interactions between stem cells and the immune system can have profound implications for current and future therapies, particularly regarding a potential "immune-privileged" HSC microenvironment. Immune/stem cell interactions change during times of stress and injury. Recent advances in cancer immunotherapy have overturned the long-standing belief that, being derived from the self, cancer cells should be immunotolerant. Instead, an immunosurveillance system recognizes and eliminates emergent pre-cancerous cells. Only in the context of a failing immunosurveillance system does cancer fully develop. Combined with the knowledge that stem cells or their unique properties can be critically important for cancer initiation, persistence, and resistance to therapy, understanding the unique immune properties of stem cells will be critical for the development of future cancer therapies. Accordingly, the therapeutic implications for leukemic stem cells (LSCs) inheriting an immune-privileged state from HSCs will be discussed. Through their dynamic interactions with a diverse immune system, stem cells serve as the light and dark root of cancer prevention vs. development.

Keywords

Hematopoietic stem cell · Immune microenvironment · Leukemic stem cell · Leukemia

8.1 The Stem Cell Milieu

Hematopoiesis, literally the genesis of blood, is governed by HSCs, which produce all blood cells and cellular products. These include red blood cells, the most abundant cell type in the body, and platelets, which are essentially cellular fragments produced by megakaryocytes. HSCs are also responsible for production of the immune system (Abkowitz et al. 1990; Baron 2003; Lensch and Daley 2004; Orkin and Zon 2002, 2008). Less appreciated is the fact that HSCs and their immune progeny regulate each other, and this has profound consequences for homeostasis, response to stress and injury, and

J. M. Perry (✉)
Children's Mercy Kansas City, Kansas City, MO, USA

University of Kansas Medical Center, Kansas City, KS, USA

University of Missouri Kansas City School of Medicine, Kansas City, MO, USA
e-mail: jmperry@cmh.edu

M. Zhao, P. Qian (eds.), *Hematopoietic Stem Cells*, Advances in Experimental Medicine and Biology 1442, https://doi.org/10.1007/978-981-99-7471-9_8

tumorigenesis (Perry and Li 2007; Zhao et al. 2014; Wang et al. 2022).

8.1.1 Stem Cells as Architects of Their Own Niche

Stem cells are ultimately responsible for the maintenance of tissues during homeostasis and in conditions of stress such as injury, infection, and disease states such as cancer. Stem cells follow specific programs in communicating and interacting with their microenvironment or niche, and these programs dynamically change during transitions between homeostatic and regenerative states. How these interactive, communication networks change and potentially break down during response to stress and injury or cancer development has been essential to major advances in regenerative medicine and cancer therapy, although these issues remain largely mysterious.

At root, HSCs are defined by the ability to sustain life-long blood production, including all lineages, and to reconstitute these after radiation- or chemotherapy-induced cytotoxicity. However, HSCs are not a homogenous population. Instead, diverse subpopulations of HSCs coexist both during homeostasis and in response to stress or injury, although the subpopulations responding to and predominating in these different conditions are distinct. Subpopulations that generally do not participate in homeostatic hematopoiesis may be called up to regenerate the hematopoietic system after stress or injury, including chemo- and radiotherapy administered in cancer treatment or conditioning for bone marrow/stem cell transplant therapy. Decades of research have identified diverse cell surface marker panels for isolating phenotypically distinct subsets of HSCs with different properties. This diversity includes unique epigenetic and metabolic states, long-term and short-term self-renewal and regenerative capacities, cell cycle rates, as well as lineage potential and bias. Furthermore, this diversity is critically important during steady-state, emergency, and aging hematopoietic states (Uchida et al. 1993; Beerman et al. 2010; Benveniste et al. 2010; Benz et al. 2012; Copley et al. 2012; Li and Clevers 2010).

The basis of HSC heterogeneity has both regulative and stochastic aspects, which are determined intrinsically and extrinsically (Crisan and Dzierzak 2016). How and to what extent these mechanisms shape the functional diversification of HSCs continues to be enigmatic. However, known mechanisms include differential genetic and epigenetic programming during different stages of development along with differential localization to distinct microenvironments and reprogramming in response to diverse molecular and cellular stimuli (Guezguez et al. 2013).

8.1.2 Intrinsic Stem Cell Mechanisms of Immunomodulation

While the role of the local stem cell microenvironment has justifiably garnered significant attention over the past two decades, it is important to understand that, whether dormant or active, HSCs are not passive responders to their microenvironments but play a vital role in not only the creation but continuing regulation of the cells surrounding them. With respect to the immune system, antigen presentation has been shown to specifically protect the integrity of the HSC pool (Hernandez-Malmierca et al. 2022).

In particular, HSCs act as antigen-presenting cells, with immunogenic antigens activating a bidirectional interface with helper ($CD4^+$) T cells, which acts as a quality control mechanism removing defective or damaged HSCs. Over time, HSCs acquire oncogenic mutations, which can result in hematological malignancies. This antigen presentation by HSCs is constitutively accomplished via major histocompatibility complex class (MHC)-II. This neoantigen presentation can activate $CD4^+$ T cells, ultimately resulting in the differentiation and depletion of immunogenic HSCs. Importantly, this mechanism acts to prohibit the transformation of HSCs into leukemia-initiating LSCs, thus serving as a stem cell-specific immunosurveillance system (Hernandez-Malmierca et al. 2022; Fujisaki et al. 2011; Hirata et al. 2018).

While certain molecular signals maintaining stem cell integrity are known, mechanisms that preserve the safety of the stem cell pool by eliminating pre-cancerous HSCs are largely unknown. However, recently, it was shown that MHC-II expression on HSCs during malignant transformation activates communication between pre-leukemic HSCs and antigen-specific $CD4^+$ T cells. This results in propagation of an immune response to target unhealthy HSCs, thus attenuating the onset of leukemia (Hernandez-Malmierca et al. 2022). These results demonstrate that HSCs are not passive with response to immune signals. Instead, HSCs actively respond to adaptive immune responses as a protective mechanism to safeguard and preserve the healthy HSC pool and ultimately prevent leukemic transformation.

Given that HSCs persist for many decades, genomic aberrations accumulate and can represent a major threat to the entire hematopoietic system as all blood cells are ultimately derived from HSCs. This leads to increased frequency and high risk of leukemia development during aging and other causes of mutational accumulation (Genovese et al. 2014; Jaiswal et al. 2014). Sheltering HSCs from damage accumulated by reactive oxygen species (ROS) and replicative stress can be accomplished, in part, by enforced quiescence and a state of low metabolism (Ho et al. 2017; van Galen et al. 2014; Walter et al. 2015). This state might also be termed dormancy (Walter et al. 2015; Essers et al. 2009; Trumpp et al. 2010; Perry et al. 2013; Wilson et al. 2008). Of course, stress due to infections or injury will induce at least transient activation of stem and especially progenitor cells. This can be accomplished through the release of inflammatory signals, which boosts blood production; however, excessive inflammation results in replicative stress, increased DNA damage, and ultimately exhaustion of the stem cell pool (Walter et al. 2015; Essers et al. 2009; Sato et al. 2009; Takizawa et al. 2017; Zhang et al. 2016) To protect HSCs from intense inflammatory assault, $CD4^+$ regulatory T cells (Tregs) have been shown to help create an immune-privileged microenvironment within the bone marrow to maintain stem cell quiescence (Fujisaki et al. 2011; Hirata et al. 2018).

8.2 Distinctive Immunogenicity of Stem Cells

In addition to an adaptive immune role in stem cell regulation, vascular cell adhesion molecule-1 (VCAM1) has been shown to confer innate immunotolerance on HSCs and LSCs. Homing of HSCs to the bone marrow occurs, in part, through interactions with VCAM1 (Miyake et al. 1991; Simmons et al. 1992; Ulyanova et al. 2005). Upon entering the bone marrow microenvironment, HSCs are essentially vetted by phagocytes within the perivascular space. This vetting has been proposed to "license" HSCs for continued survival and proliferation. Related to this, healthy HSCs express VCAM1, which serves as a "don't-eat-me" signal when presented to MHC-I on perivascular phagocytic cells. Interestingly, VCAM1 deletion leads to reduced hematopoietic reconstitution in a haplotype-mismatched transplant setting. This is due to clearance of HSCs by perivascular phagocytic sentinels when VCAM1 expression is deficient (Pinho et al. 2022).

Notably, VCAM1 expression is upregulated on LSCs. VCAM1 overexpression is known to be utilized by certain cancer cells as an immune escape mechanism, including in hematological cancers such as acute myeloid leukemia (AML). Indeed, high expression of VCAM1 is indicative of poor outcomes due to VCAM1's promotion of leukemia progression. Furthermore, VCAM1 deletion or inhibition results in decreased tumor burden and prolonged survival (Pinho et al. 2022). These results indicate that VCAM1 serves as an important immune checkpoint regulating entry into the bone marrow microenvironment, which not coincidentally is the site of HSC and LSC niches. Since VCAM1 is essential for HSC engraftment, at least in mismatched transplant settings, and loss of VCAM1 reduces establishment and progression of AML, this suggests an immunotherapeutic strategy for targeting LSCs by inhibiting innate immunotolerance via

VCAM1 inhibition. Indeed, blockade of VCAM1 was shown to inhibit LSCs and synergize with single-agent chemotherapy in mouse models. It may also be interesting to test whether overexpression of VCAM1 might allow haplo-mismatched HSCs to be transplanted. That is, constitutive VCAM1 overexpression may force acceptance of mismatched donor HSCs for therapeutic purposes when matched donors are unavailable.

Considering the poor therapeutic outcomes due to relapse especially for AML, the use of immunotherapy could be the key to preventing tumorigenesis and continued growth and regrowth (Barkal et al. 2018; Havel et al. 2019; Jaiswal et al. 2009; Majeti et al. 2009; Woo et al. 2015). Interestingly, discovery of VCAM1 as a "don't-eat-me" signal used by stem cells, including healthy HSCs and cancer-driving LSCs, depended on the use of a haplotype-mismatched transplant model, which was only used inadvertently. Healthy HSCs typically do not require VCAM1 to avoid phagocytosis in an autologous or donor-matched setting. However, LSCs do require VCAM1, even when derived autologously. This indicates that LSCs express neoantigens perceived by perivascular bone marrow phagocytes as being non-self or otherwise recognized as in a disease state. Identification of such antigens, particularly if a discrete number unique to LSCs exists, would allow more durable cures to be achieved by targeting these rare cells responsible for relapse. Considering the synergy of anti-VCAM1 with the mutagenic chemotherapy drug cytarabine, it is interesting to speculate that cytarabine may aid in the induction or propagation of neoantigens. Interestingly, cytarabine preferentially induces mutations in a predictable manner, which can be identified in relapsed AML (Fordham et al. 2015). In addition, the fact that cells of the innate immune system are responsible for the clearance of LSCs suggests that more generalized, "off-the-shelf" therapies may eventually be developed for targeting LSCs.

8.2.1 Stem Cell Immune Privilege

The prevalence of autoimmune disorders demonstrates that the line between self and non-self is not absolute. For certain particularly vital cells and tissues, particularly those of extreme rarity (i.e., stem cells) or essential for the continuity of the species (i.e., germ line cells), immune privilege provides additional safe harbor from immune recognition, especially during times of inflammation due to infection or wound healing. Chronic inflammation may result in the acquisition of immune privilege to protect healthy tissues from pervasive immune onslaught. Indeed, the association of chronic inflammation and cancer development could be due to the intermediate development of immune privilege. Similarly, tumor development can employ immune privilege for cancer progression to occur. Indeed, immune privilege may be a necessary prerequisite for tumorigenic progression.

Stem cell immune privilege begins at conception. Being genetic hybrids also containing non-self (i.e., paternal) DNA, reimplantation embryos with totipotent stem cells must be protected from maternal allorejection. Indeed, embryonic stem cell lines retain immunosuppressive properties, which makes them especially attractive for therapy as matching requirements may be reduced, although their differentiated derivatives certainly cannot always completely subvert immune responses. Thus, immune privilege is a relative property that can be both acquired and lost.

Immune privilege was originally discovered by the ability to transplant allogeneic cells or tissues at certain sites without immune rejection despite the fact that the same tissues would be rejected when transplanted into other donor sites. Skin grafts are a classic example of this phenomenon (Simpson 2006; Niederkorn 2006). Other immune-privileged sites include the brain and anterior eye (Mrass and Weninger 2006; Streilein 2003). Most obviously, maternal-fetal tolerance of a semi-allogeneic "graft" in the uterine wall represents a form of immune privilege (Ichiryu and Fairchild 2013).

Beyond pregnancy and transplantation, it has become clear that many cells and tissues can be placed on a relative spectrum of immune privilege, and this relative state of tolerogenicity may be dynamic and context dependent. Of course, immune privilege can be compromised or fail entirely (Paus et al. 2005). More unexpectedly, immune privilege can be acquired and is mediated by immunosuppressive effector cells, particularly Treg ($CD4^+$ $FoxP3^+$) cells, which induce tolerance locally (Cobbold et al. 2006). Thus, immune privilege is not restricted to specific organs or anatomical locations and can theoretically be induced anywhere and at very specific, discrete microenvironments. Notably, this includes allografts, infected tissues, tumors, and stem cell niches (Ichiryu and Fairchild 2013; Cobbold et al. 2006).

Accordingly, researchers have sought opportunities for therapeutic intervention by immune manipulation. Immune privilege can be beneficial in the context of tissue or organ transplantation. Conferring immune privilege on cell, tissue, or organ allografts can allow successful long-term grafts without the ongoing necessity of systemic immunosuppression. Such treatment may also be considered for autoimmune conditions such as type 1 diabetes. To overcome immunological barriers and avoid rejection, the field of regenerative medicine is particularly interested in immune privilege. Unless patient-derived induced pluripotent stem cells (iPSCs) are used, cells and cell products used in regenerative medicine are at risk for rejection.

Adult stem cells populate many tissues in the body to maintain homeostasis and regeneration in response to injury. Being rare but extremely vital and long-lived, postnatal stem cells may retain special immune privilege to avoid autoimmune attack or injury from acute or chronic inflammation. Yet this immune-privileged state presents a danger. Germ cells are a classic example of adult stem cells having a particularly potent degree of immune privilege (Kalavska et al. 2020). Testicular germ cell tumors are the most common type of cancer among young men, and the link with germ cells being immune privileged is not likely to be coincidental. Indeed, as discussed in more detail below, immune privilege either innate to a particular tissue or acquired during tumorigenesis presents a major barrier to successful anticancer immunotherapy. As such, targeting the property of immune privilege in cancers, and particularly the cancer stem cells that fuel them, is expected to be critical in improving the efficacy of immunotherapy.

8.2.2 Extrinsic Stem Cell Mechanisms of Immunomodulation

Stem cells are not only architects of their own microenvironment; this microenvironment is also critical to regulating and maintaining stem cells. Classically, stem cells have been characterized as quiescent. This is, at best, an oversimplification. Regardless, quiescence is itself not an inert or passive state. Just as terminally differentiated cells require continuous, active inputs for their regulation and maintenance, so too do stem cells, even in the quiescent or dormant state.

Antigen-presenting cells (APCs) such as dendritic cells and B cells stimulate adaptive immune responses by presenting antigens through MHC-II to helper T cells (Neefjes et al. 2011; Roche and Furuta 2015). Extrinsic factors in the microenvironment, the state of APCs, and dialog between them coordinate to balance immunogenic vs. tolerogenic responses (Goodnow et al. 2005; Wakkach et al. 2003; Jurewicz and Stern 2019). High levels of MHC-II are constitutively expressed by professional APCs, but nonprofessional APCs (essentially all other cells in the body) develop antigen-presenting activity only when exposed to pro-inflammatory cytokines (Roche and Furuta 2015; Jakubzick et al. 2017; Kambayashi and Laufer 2014; Merad et al. 2013; Steinman 2007). For decades, reports have suggested that HSCs and other immature hematopoietic cells may express MHC-II, at least under certain conditions (Fitchen et al. 1981; Russell and van den Engh 1979; Sieff et al. 1982; Szer et al. 1985). Interestingly, HSCs and early progenitors were recently shown to constitutively present antigens through MHC-II. AML is an aggressive cancer with a hierarchical developmental structure driven by LSCs, which originate from pre-leukemia HSCs or myeloid progenitors. MHC-II is expressed heterogeneously in AML, and downregulation of MHC-II is associated with relapse (Christopher et al. 2018; Miale et al. 1982; Newman et al. 1983; Toffalori et al. 2019). Interestingly, relapse in AML after HSC transplantation was not found to be associated with immune-related neoantigen acquisition in at least one study. Instead, pathways that regulate immune function, such as those involved in antigen presentation, were dysregulated, and these were epigenetically rather than genetically altered. As such, they could be reversible with appropriate therapy (Christopher et al. 2018). Another study also found that deregulation of T cell pathways governing allorecognition was a specific driver of AML relapse following allogenic HSC transplantation (Toffalori et al. 2019). Indeed, HSCs directly interact with antigen-specific $CD4^+$ T cells, leading to HSC exhaustion by excessive differentiation. Activated $CD4^+$ T cells then regulate the bone marrow microenvironment to inhibit pro-inflammatory responses. Overall, this immunosurveillance was shown to inhibit leukemia development when HSCs are transformed into LSCs (Hernandez-Malmierca et al. 2022).

8.3 Immunosurveillance and Leukemia Initiation

Rare LSCs selectively persist following therapy-induced remission. Long-standing efforts have sought to understand the unique resistance mechanisms inherent in LSCs to conventional therapeutic strategies (Pardal et al. 2003; Lapidot et al. 1996; Lapidot et al. 1994; Reya et al. 2001; Shackleton et al. 2009). As LSCs represent the cellular origin of relapse, understanding the unique properties responsible for this persistence will be critical for future progress in treating leukemia. In particular, stem cell states ("stemness") improperly retained during differentiation or regenerated in progenitor or more mature cells may be expected to be the critical difference between LSCs and their bulk progeny (Becker and Jordan 2011; Ivanova et al. 2003; Malta et al. 2018; Plaks et al. 2015; Ramalho-Santos et al. 2002). Given the unique properties of stem cells with respect to the immune system, LSCs may be particularly potent at immune escape and immunotherapy resistance (Perry et al. 2011, 2020; Perry and Li 2020).

Immunosurveillance is a potent system for recognizing and eliminating emergent, pre-cancerous cells. Indeed, recent evidence indicates that failure of anticancer immunosurveillance is a prerequisite for tumorigenic progression. In this view, anticancer research should focus on restoring or preventing the failure of this system (Dunn et al. 2002; Finn 2018; Galluzzi et al. 2019; Lesokhin et al. 2015; Zitvogel et al. 2016). Currently, whether and how LSCs and other tumor-initiating cells resist anticancer immunity and how they escape immunosurveillance remains primarily unknown. However, recent evidence indicates that LSCs and other tumor-initiating cells possess unique properties allowing them to overcome immune barriers more potently than other non-stem or stem-like cancer cells.

A notable example of tumor-initiating cells exhibiting properties of immune escape was demonstrated in an epithelial squamous cell carcinoma mouse model. Carcinoma cells in this model could be effectively targeted by adoptive T cell transfer-based immunotherapy using chimeric antigen receptor (CAR) $CD8^+$ cells targeting tumor-specific antigens. Interestingly, a discrete subpopulation of tumor-initiating stem cells were particularly refractory to CAR T cell therapy, resulting in relapse. Considering that relapse is the main obstacle to successful CAR T cell therapy, understanding how tumor-initiating stem cells in this model escape immunosurveillance could lead to new treatments and increase the efficacy of CAR T cell therapy (Miao et al. 2019).

Mechanistically, at least three cellular and molecular events orchestrating immune escape of cancer-initiating cells can be envisioned.

First, CSCs must endure T cell attack, perhaps via immune checkpoint expression. Second, quiescence or dormancy may allow them to evade detection by immune cells over prolonged periods, perhaps even years. Third, to fuel cancer growth, CSCs must reengineer the tumor microenvironment from immunologically "hot" to "cold." They may do this directly and/or indirectly by producing progeny that coordinates this re-engineering (Bonaventura et al. 2019; Duan et al. 2020; Galon and Bruni 2019; Liu and Sun 2021). In the absence of a sufficiently dynamic immunosurveillance system, CSCs can activate and generate tumors that continue to suppress the immune system. Each phase is expected to be dictated by specific molecular dialogs between tumor and immune cells, which could be targeted to shift the balance back toward a more "hot" or immunogenic phenotype.

Interactions between normal and cancer stem cells, their progeny, and the immune system are orchestrated by intra- and extracellular signals (Petrillo et al. 2018; Wu et al. 2018). These occur within the tumorigenic microenvironment, including the tumor parenchyma and related stroma and perivascular niche. As discussed previously, Tregs in the bone marrow provide immune privilege to directly enforce HSC quiescence (Fujisaki et al. 2011; Hirata et al. 2018). This is evidenced by the Treg-dependent persistence of transplanted donor-matched (allogenic) HSCs in non-myeloablative mouse transplant experiments (Fujisaki et al. 2011). Also, pro-inflammatory cytokines, such as TNF, result in upregulation of immune checkpoints including PD-L1, which results in direct suppression of T cells by HSCs (Yamashita and Passegue 2019). Furthermore, evidence for an immune-privileged bone marrow niche as well as HSC suppression of immune cells is provided by the fact that HSC egress from the bone marrow coincides with increased expression of the "don't-eat-me" signal, CD47, which protects extramedullary travels of HSCs by inhibiting innate immune phagocytes much as macrophages and dendritic cells (Jaiswal et al. 2009). Similarly, constitutive overexpression of CD47 in human AML allows immune evasion and correlates with negative outcomes (Majeti et al. 2009). Also, downregulation of human leukocyte antigen (HLA, the human version of MHC) allows for cancer immune escape resulting in relapse following allogenic stem cell transplant therapy (Toffalori et al. 2019). Immunosurveillance by natural killer cells is evaded by human AML LSCs by downregulating NKG2D ligands (Paczulla et al. 2019). These results among others present an overall picture of LSCs avoiding immune detection and elimination by adopting mechanisms that bone marrow HSCs use, especially under stress conditions, to maintain immune privilege.

8.4 Conclusions and Future Directions

Stem cells have unique properties vis-à-vis the immune system. Understanding these properties and their interactions are critical to understanding normal stem cell regulation, cancer development, regenerative medicine, and immunotherapy. Stem cells are not inert bystanders but are governed by, and in turn govern, their microenvironment. Future research will continue to uncover the unique role that various immune cells play in the healthy versus tumorigenic microenvironment. How normal stem cells enforce and regulate their states of immune privilege remains only poorly understood. Efforts in regenerative medicine using non-self cell sources will require improved knowledge of stem cell immune privilege, how it is regulated, and how it might be manipulated. On the other hand, the hijacking of immune privilege by cancer must be understood to prevent immune escape and failure of immunotherapy. While immunotherapy represents the greatest promise for anticancer treatments in decades, its ultimate success may depend on a comprehensive understanding of stem cell immune privilege.

Current anticancer treatment continues to rely on broadly toxic drugs. Multiple chemotherapy agents are administered to eliminate the vast majority of overt cancerous cells and achieve remission. Unfortunately, rare tumor-initiating cells or cancer/leukemia stem cells remain therapy resistant. They achieve this resistance by retention or hijack of the normal properties of stem cells. These properties (i.e., "stemness") provide a combination of perpetual tissue maintenance with self-renewal. In addition to these classic properties, an emerging stemness state includes immune privilege. Indeed, mesenchymal stem cells exhibit immune privilege to the point where intraspecies transplantation is possible (Salem and Thiemermann 2010). Although somewhat less impressive, HSCs also exhibit properties of immune evasion and occupy immune-privileged microenvironments of their own construction. The quintessential adaptive immune checkpoint PD-L1 is expressed on HSCs and increases the capacity for allogeneic HSC transplant (Zheng et al. 2011). Furthermore, HSCs are thought to express low levels of immunosuppressors, which reinforce immune privilege during times of stress (Zheng et al. 2011).

The intersection of stem cell regulation and tumorigenesis is particularly well-illustrated by the Wnt and PI3K pathways, especially with regard to stem cells. The Wnt and PI3K pathways cooperatively interact to drive stem cell self-renewal and, when dysregulated, the transformation of HSCs into LSCs (Perry et al. 2011, 2020). Not coincidentally, these signaling pathways also orchestrate a critical role in immune escape (Sharma et al. 2017). Interestingly, LSCs have been found to express multiple immune checkpoint genes, which were regulated by Wnt/PI3K cooperative signaling (Perry et al. 2020). This signaling, and the immune checkpoint genes driven by its expression, can be inhibited by repurposing a conventional chemotherapy drug as a targeted agent against Wnt-activated PI3K signaling. Interestingly, conventional chemotherapy and even certain targeted anticancer agents can have unappreciated but beneficial immunological side effects (Galluzzi et al. 2015). For example, they can improve immunogenicity of cancer cells and reduce immunosuppression in tumor microenvironments. Considering that cancer is increasingly recognized not simply as a cell-autonomous disease but as a failure of anticancer immunosurveillance, it is interesting to speculate that at least some of the success of current anticancer treatments may be due to unrecognized influences on the immune system. Furthermore, this influence may be particularly relevant to stem cells and their unique immune properties. Clearly, there is a need to better interrogate not only the existing treatments but to develop new ones that account for the crucial role of anticancer immunity, immunosurveillance, and escape, particularly as it relates to stem cells.

References

Abkowitz JL, Linenberger ML, Newton MA, Shelton GH, Ott RL, Guttorp P (1990) Evidence for the maintenance of hematopoiesis in a large animal by the sequential activation of stem-cell clones. Proc Natl Acad Sci U S A 87(22):9062–9066

Barkal AA, Weiskopf K, Kao KS, Gordon SR, Rosental B, Yiu YY et al (2018) Engagement of MHC class I by the inhibitory receptor LILRB1 suppresses macrophages and is a target of cancer immunotherapy. Nat Immunol 19(1):76–84

Baron MH (2003) Embryonic origins of mammalian hematopoiesis. Exp Hematol 31:1160–1169

Becker MW, Jordan CT (2011) Leukemia stemness signatures step toward the clinic. Cell Stem Cell 9(3): 185–186

Beerman I, Bhattacharya D, Zandi S, Sigvardsson M, Weissman IL, Bryder D et al (2010) Functionally distinct hematopoietic stem cells modulate hematopoietic lineage potential during aging by a mechanism of clonal expansion. Proc Natl Acad Sci U S A 107(12): 5465–5470

Benveniste P, Frelin C, Janmohamed S, Barbara M, Herrington R, Hyam D et al (2010) Intermediate-term hematopoietic stem cells with extended but time-limited reconstitution potential. Cell Stem Cell 6(1): 48–58

Benz C, Copley MR, Kent DG, Wohrer S, Cortes A, Aghaeepour N et al (2012) Hematopoietic stem cell subtypes expand differentially during development and display distinct lymphopoietic programs. Cell Stem Cell 10(3):273–283

Bonaventura P, Shekarian T, Alcazer V, Valladeau-Guilemond J, Valsesia-Wittmann S, Amigorena S et al (2019) Cold tumors: a therapeutic challenge for immunotherapy. Front Immunol 10:168

Christopher MJ, Petti AA, Rettig MP, Miller CA, Chendamarai E, Duncavage EJ et al (2018) Immune escape of relapsed AML cells after allogeneic transplantation. N Engl J Med 379(24):2330–2341

Cobbold SP, Adams E, Graca L, Daley S, Yates S, Paterson A et al (2006) Immune privilege induced by regulatory T cells in transplantation tolerance. Immunol Rev 213:239–255

Copley MR, Beer PA, Eaves CJ (2012) Hematopoietic stem cell heterogeneity takes center stage. Cell Stem Cell 10(6):690–697

Crisan M, Dzierzak E (2016) The many faces of hematopoietic stem cell heterogeneity. Development 143(24):4571–4581

Duan Q, Zhang H, Zheng J, Zhang L (2020) Turning cold into hot: firing up the tumor microenvironment. Trends Cancer 6(7):605–618

Dunn GP, Bruce AT, Ikeda H, Old LJ, Schreiber RD (2002) Cancer immunoediting: from immunosurveillance to tumor escape. Nat Immunol 3(11):991–998

Essers MA, Offner S, Blanco-Bose WE, Waibler Z, Kalinke U, Duchosal MA et al (2009) IFNalpha activates dormant haematopoietic stem cells in vivo. Nature 458:904

Finn OJ (2018) A Believer's overview of cancer immunosurveillance and immunotherapy. J Immunol 200(2):385–391

Fitchen JH, Foon KA, Cline MJ (1981) The antigenic characteristics of hematopoietic stem cells. N Engl J Med 305(1):17–25

Fordham SE, Cole M, Irving JA, Allan JM (2015) Cytarabine preferentially induces mutation at specific sequences in the genome which are identifiable in relapsed acute myeloid leukaemia. Leukemia 29(2): 491–494

Fujisaki J, Wu J, Carlson AL, Silberstein L, Putheti P, Larocca R et al (2011) In vivo imaging of treg cells providing immune privilege to the haematopoietic stem-cell niche. Nature 474(7350):216–219

Galluzzi L, Buque A, Kepp O, Zitvogel L, Kroemer G (2015) Immunological effects of conventional chemotherapy and targeted anticancer agents. Cancer Cell 28(6):690–714

Galluzzi L, Spranger S, Fuchs E, Lopez-Soto A (2019) WNT signaling in cancer immunosurveillance. Trends Cell Biol 29(1):44–65

Galon J, Bruni D (2019) Approaches to treat immune hot, altered and cold tumours with combination immunotherapies. Nat Rev Drug Discov 18(3): 197–218

Genovese G, Kahler AK, Handsaker RE, Lindberg J, Rose SA, Bakhoum SF et al (2014) Clonal hematopoiesis and blood-cancer risk inferred from blood DNA sequence. N Engl J Med 371(26):2477–2487

Goodnow CC, Sprent J, de St F, Groth B, Vinuesa CG (2005) Cellular and genetic mechanisms of self tolerance and autoimmunity. Nature 435(7042):590–597

Guezguez B, Campbell CJ, Boyd AL, Karanu F, Casado FL, Di Cresce C et al (2013) Regional localization within the bone marrow influences the functional capacity of human HSCs. Cell Stem Cell 13(2): 175–189

Havel JJ, Chowell D, Chan TA (2019) The evolving landscape of biomarkers for checkpoint inhibitor immunotherapy. Nat Rev Cancer 19(3):133–150

Hernandez-Malmierca P, Vonficht D, Schnell A, Uckelmann HJ, Bollhagen A, Mahmoud MAA et al (2022) Antigen presentation safeguards the integrity of the hematopoietic stem cell pool. Cell Stem Cell 29(5): 760–75.e10

Hirata Y, Furuhashi K, Ishii H, Li HW, Pinho S, Ding L et al (2018) CD150(high) bone marrow Tregs maintain hematopoietic stem cell quiescence and immune privilege via adenosine. Cell Stem Cell 22(3):445–53.e5

Ho TT, Warr MR, Adelman ER, Lansinger OM, Flach J, Verovskaya EV et al (2017) Autophagy maintains the metabolism and function of young and old stem cells. Nature 543(7644):205–210

Ichiryu N, Fairchild PJ (2013) Immune privilege of stem cells. Methods Mol Biol 1029:1–16

Ivanova NB, Dimos JT, Schaniel C, Hackney JA, Moore KA, Ramalho-Santos M et al (2003) Response to comments on "'Stemness': transcriptional profiling of embryonic and adult stem cells" and "a stem cell molecular signature". Science 302(5644):393

Jaiswal S, Jamieson CH, Pang WW, Park CY, Chao MP, Majeti R et al (2009) CD47 is upregulated on circulating hematopoietic stem cells and leukemia cells to avoid phagocytosis. Cell 138(2):271–285

Jaiswal S, Fontanillas P, Flannick J, Manning A, Grauman PV, Mar BG et al (2014) Age-related clonal hematopoiesis associated with adverse outcomes. N Engl J Med 371(26):2488–2498

Jakubzick CV, Randolph GJ, Henson PM (2017) Monocyte differentiation and antigen-presenting functions. Nat Rev Immunol 17(6):349–362

Jurewicz MM, Stern LJ (2019) Class II MHC antigen processing in immune tolerance and inflammation. Immunogenetics 71(3):171–187

Kalavska K, Kucerova L, Schmidtova S, Chovanec M, Mego M (2020) Cancer stem cell niche and immune-active tumor microenvironment in testicular germ cell tumors. Adv Exp Med Biol 1226:111–121

Kambayashi T, Laufer TM (2014) Atypical MHC class II-expressing antigen-presenting cells: can anything replace a dendritic cell? Nat Rev Immunol 14(11): 719–730

Lapidot T, Sirard C, Vormoor J, Murdoch B, Hoang T, Caceres-Cortes J et al (1994) A cell initiating human acute myeloid leukaemia after transplantation into SCID mice. Nature 367:645–648

Lapidot T, Grunberger T, Vormoor J, Estrov Z, Kollet O, Bunin N et al (1996) Identification of human juvenile chronic myelogenous leukemia stem cells capable of initiating the disease in primary and secondary SCID mice. Blood 88:2655–2664

Lensch MW, Daley GQ (2004) Origins of mammalian hematopoiesis: in vivo paradigms and in vitro models. Curr Top Dev Biol 60:127–196

Lesokhin AM, Callahan MK, Postow MA, Wolchok JD (2015) On being less tolerant: enhanced cancer immunosurveillance enabled by targeting checkpoints and agonists of T cell activation. Sci Transl Med 7(280):280sr1

Li L, Clevers H (2010) Coexistence of quiescent and active adult stem cells in mammals. Science 327(5965):542–545

Liu YT, Sun ZJ (2021) Turning cold tumors into hot tumors by improving T-cell infiltration. Theranostics 11(11):5365–5386

Majeti R, Chao MP, Alizadeh AA, Pang WW, Jaiswal S, Gibbs KD Jr et al (2009) CD47 is an adverse prognostic factor and therapeutic antibody target on human acute myeloid leukemia stem cells. Cell 138(2): 286–299

Malta TM, Sokolov A, Gentles AJ, Burzykowski T, Poisson L, Weinstein JN et al (2018) Machine learning identifies stemness features associated with oncogenic dedifferentiation. Cell 173(2):338–54.e15

Merad M, Sathe P, Helft J, Miller J, Mortha A (2013) The dendritic cell lineage: ontogeny and function of dendritic cells and their subsets in the steady state and the inflamed setting. Annu Rev Immunol 31:563–604

Miale TD, Stenke LA, Lindblom JB, Sjogren AM, Reizenstein PG, Uden AM et al (1982) Surface Ia-like expression and MLR-stimulating capacity of human leukemic myeloblasts: implications for immunotherapy and prognosis. Acta Haematol 68(1):3–13

Miao Y, Yang H, Levorse J, Yuan S, Polak L, Sribour M et al (2019) Adaptive immune resistance emerges from tumor-initiating stem cells. Cell 177(5):1172–86.e14

Miyake K, Medina K, Ishihara K, Kimoto M, Auerbach R, Kincade PW (1991) A VCAM-like adhesion molecule on murine bone marrow stromal cells mediates binding of lymphocyte precursors in culture. J Cell Biol 114(3): 557–565

Mrass P, Weninger W (2006) Immune cell migration as a means to control immune privilege: lessons from the CNS and tumors. Immunol Rev 213:195–212

Neefjes J, Jongsma ML, Paul P, Bakke O (2011) Towards a systems understanding of MHC class I and MHC class II antigen presentation. Nat Rev Immunol 11(12):823–836

Newman RA, Delia D, Greaves MF, Navarrete C, Fainboim L, Festenstein H (1983) Differential expression of HLA-DR and DR-linked determinants on human leukemias and lymphoid cells. Eur J Immunol 13(2):172–176

Niederkorn JY (2006) See no evil, hear no evil, do no evil: the lessons of immune privilege. Nat Immunol 7(4): 354–359

Orkin SH, Zon LI (2002) Hematopoiesis and stem cells: plasticity versus developmental heterogeneity. Nat Immunol 3:323–328

Orkin SH, Zon LI (2008) Hematopoiesis: an evolving paradigm for stem cell biology. Cell 132(4):631–644

Paczulla AM, Rothfelder K, Raffel S, Konantz M, Steinbacher J, Wang H et al (2019) Absence of NKG2D ligands defines leukaemia stem cells and mediates their immune evasion. Nature 572(7768): 254–259

Pardal R, Clarke MF, Morrison SJ (2003) Applying the principles of stem-cell biology to cancer. Nat Rev Cancer 3:895–902

Paus R, Nickoloff BJ, Ito T (2005) A 'hairy' privilege. Trends Immunol 26(1):32–40

Perry JM, Li L (2007) Disrupting the stem cell niche: good seeds in bad soil. Cell 129(6):1045–1047

Perry JM, Li L (2020) Overcoming resistance to immunotherapy by teaching old drugs new tricks. Mol Cell Oncol 7(5):1801088

Perry JM, He XC, Sugimura R, Grindley JC, Haug JS, Ding S et al (2011) Cooperation between both Wnt/{beta}-catenin and PTEN/PI3K/Akt signaling promotes primitive hematopoietic stem cell self-renewal and expansion. Genes Dev 25(18):1928–1942

Perry JM, He XC, Sugimura R, Li L (2013) Stem cell dormancy: maintaining a reserved population. In: Meyers RA (ed) Stem cells: from biology to therapy. Current topics in molecular cell biology and molecular medicine. Wiley, John & Sons, Incorporated, Weinheim, pp 119–132

Perry JM, Tao F, Roy A, Lin T, He XC, Chen S et al (2020) Overcoming Wnt-beta-catenin dependent anticancer therapy resistance in leukaemia stem cells. Nat Cell Biol 22(6):689–700

Petrillo C, Thorne LG, Unali G, Schiroli G, Giordano AMS, Piras F et al (2018) Cyclosporine H overcomes innate immune restrictions to improve lentiviral transduction and gene editing in human hematopoietic stem cells. Cell Stem Cell 23(6):820–32.e9

Pinho S, Wei Q, Maryanovich M, Zhang D, Balandran JC, Pierce H et al (2022) VCAM1 confers innate immune tolerance on haematopoietic and leukaemic stem cells. Nat Cell Biol 24(3):290–298

Plaks V, Kong N, Werb Z (2015) The cancer stem cell niche: how essential is the niche in regulating stemness of tumor cells? Cell Stem Cell 16(3):225–238

Ramalho-Santos M, Yoon S, Matsuzaki Y, Mulligan RC, Melton DA (2002) "Stemness": transcriptional profiling of embryonic and adult stem cells. Science 298(5593):597–600

Reya T, Morrison SJ, Clarke MF, Weissman IL (2001) Stem cells, cancer, and cancer stem cells. Nature 414(6859):105–111

Roche PA, Furuta K (2015) The ins and outs of MHC class II-mediated antigen processing and presentation. Nat Rev Immunol 15(4):203–216

Russell JL, van den Engh G (1979) The expression of histocompatibility-2 antigens on hemopoietic stem cells. Tissue Antigens 13(1):45–22

Salem HK, Thiemermann C (2010) Mesenchymal stromal cells: current understanding and clinical status. Stem Cells 28(3):585–596

Sato T, Onai N, Yoshihara H, Arai F, Suda T, Ohteki T (2009) Interferon regulatory factor-2 protects quiescent hematopoietic stem cells from type I interferon-dependent exhaustion. Nat Med 15(6):696–700

Shackleton M, Quintana E, Fearon ER, Morrison SJ (2009) Heterogeneity in cancer: cancer stem cells versus clonal evolution. Cell 138(5):822–829

Sharma P, Hu-Lieskovan S, Wargo JA, Ribas A (2017) Primary, adaptive, and acquired resistance to cancer immunotherapy. Cell 168(4):707–723

Sieff C, Bicknell D, Caine G, Robinson J, Lam G, Greaves MF (1982) Changes in cell surface antigen expression during hemopoietic differentiation. Blood 60(3): 703–713

Simmons PJ, Masinovsky B, Longenecker BM, Berenson R, Torok-Storb B, Gallatin WM (1992) Vascular cell adhesion molecule-1 expressed by bone marrow stromal cells mediates the binding of hematopoietic progenitor cells. Blood 80(2):388–395

Simpson E (2006) A historical perspective on immunological privilege. Immunol Rev 213:12–22

Steinman RM (2007) Dendritic cells: understanding immunogenicity. Eur J Immunol 37(Suppl 1):S53–S60

Streilein JW (2003) Ocular immune privilege: therapeutic opportunities from an experiment of nature. Nat Rev Immunol 3(11):879–889

Szer J, Deeg HJ, Appelbaum FR, Storb R (1985) Failure of autologous marrow reconstitution after cytolytic treatment of marrow with anti-Ia monoclonal antibody. Blood 65(4):819–822

Takizawa H, Fritsch K, Kovtonyuk LV, Saito Y, Yakkala C, Jacobs K et al (2017) Pathogen-induced TLR4-TRIF innate immune signaling in hematopoietic stem cells promotes proliferation but reduces competitive fitness. Cell Stem Cell 21(2):225–40.e5

Toffalori C, Zito L, Gambacorta V, Riba M, Oliveira G, Bucci G et al (2019) Immune signature drives leukemia escape and relapse after hematopoietic cell transplantation. Nat Med 25(4):603–611

Trumpp A, Essers M, Wilson A (2010) Awakening dormant haematopoietic stem cells. Nat Rev Immunol 10(3):201–209

Uchida N, Fleming WH, Alpern EJ, Weissman IL (1993) Heterogeneity of hematopoietic stem cells. Curr Opin Immunol 5:177–184

Ulyanova T, Scott LM, Priestley GV, Jiang Y, Nakamoto B, Koni PA et al (2005) VCAM-1 expression in adult hematopoietic and nonhematopoietic cells is controlled by tissue-inductive signals and reflects their developmental origin. Blood 106(1):86–94

van Galen P, Kreso A, Mbong N, Kent DG, Fitzmaurice T, Chambers JE et al (2014) The unfolded protein response governs integrity of the haematopoietic stem-cell pool during stress. Nature 510(7504): 268–272

Wakkach A, Fournier N, Brun V, Breittmayer JP, Cottrez F, Groux H (2003) Characterization of dendritic cells that induce tolerance and T regulatory 1 cell differentiation in vivo. Immunity 18(5):605–617

Walter D, Lier A, Geiselhart A, Thalheimer FB, Huntscha S, Sobotta MC et al (2015) Exit from dormancy provokes DNA-damage-induced attrition in haematopoietic stem cells. Nature 520(7548):549–552

Wang J, Xie J, Wang D, Han X, Chen M, Shi G et al (2022) CXCR4(high) megakaryocytes regulate host-defense immunity against bacterial pathogens. elife 11:11

Wilson A, Laurenti E, Oser G, van der Wath RC, Blanco-Bose W, Jaworski M et al (2008) Hematopoietic stem cells reversibly switch from dormancy to self-renewal during homeostasis and repair. Cell 135(6):1118–1129

Woo SR, Corrales L, Gajewski TF (2015) Innate immune recognition of cancer. Annu Rev Immunol 33:445–474

Wu X, Dao Thi VL, Huang Y, Billerbeck E, Saha D, Hoffmann HH et al (2018) Intrinsic immunity shapes viral resistance of stem cells. Cell 172(3):423–38.e25

Yamashita M, Passegue E (2019) TNF-alpha coordinates hematopoietic stem cell survival and myeloid regeneration. Cell Stem Cell 25(3):357–72.e7

Zhang H, Rodriguez S, Wang L, Wang S, Serezani H, Kapur R et al (2016) Sepsis induces hematopoietic stem cell exhaustion and myelosuppression through distinct contributions of TRIF and MYD88. Stem Cell Rep 6(6):940–956

Zhao M, Perry JM, Marshall H, Venkatraman A, Qian P, He XC et al (2014) Megakaryocytes maintain homeostatic quiescence and promote post-injury regeneration of hematopoietic stem cells. Nat Med 20(11): 1321–1326

Zheng J, Umikawa M, Zhang S, Huynh H, Silvany R, Chen BP et al (2011) Ex vivo expanded hematopoietic stem cells overcome the MHC barrier in allogeneic transplantation. Cell Stem Cell 9(2):119–130

Zitvogel L, Ayyoub M, Routy B, Kroemer G (2016) Microbiome and anticancer immunosurveillance. Cell 165(2):276–287

Learning from Zebrafish Hematopoiesis 9

Mei Wu, Jin Xu, Yiyue Zhang, and Zilong Wen

Abstract

Hematopoiesis is a complex process that tightly regulates the generation, proliferation, differentiation, and maintenance of hematopoietic cells. Disruptions in hematopoiesis can lead to various diseases affecting both hematopoietic and non-hematopoietic systems, such as leukemia, anemia, thrombocytopenia, rheumatoid arthritis, and chronic granuloma. The zebrafish serves as a powerful vertebrate model for studying hematopoiesis, offering valuable insights into both hematopoietic regulation and hematopoietic diseases. In this chapter, we present a comprehensive overview of zebrafish hematopoiesis, highlighting its distinctive characteristics in hematopoietic processes. We discuss the ontogeny and modulation of both primitive and definitive hematopoiesis, as well as the microenvironment that supports hematopoietic stem/progenitor cells. Additionally, we explore the utility of zebrafish as a disease model and its potential in drug discovery, which not only advances our understanding of the regulatory mechanisms underlying hematopoiesis but also facilitates the exploration of novel therapeutic strategies for hematopoietic diseases.

Keywords

Zebrafish · Hematopoiesis · Hematopoietic diseases · Hematopoietic stem/progenitor cells · Disease models

9.1 Unique Advantages of Zebrafish for the Research of Hematopoiesis

Zebrafish, a small aquarium fish, has flourished in many research fields of biology since its introduction into laboratories about 40 years ago. Hematopoiesis is one of the fields that benefit most from the studies in zebrafish. Zebrafish provides opportunities to study hematopoiesis from a new perspective owing to its unique biology properties. First, zebrafish can produce hundreds of embryos per mating. Its body is much smaller, its sexual maturation only takes 3 months, and its raising cost is much lower compared with other vertebrate model animals. One could raise tens of thousands of zebrafish in limited spaces. These

M. Wu (✉)
Affiliated Hospital of Guangdong Medical University and Key Laboratory of Zebrafish Model for Development and Disease of Guangdong Medical University, Zhanjiang, Guangdong, China

J. Xu (✉) · Y. Zhang (✉)
South China University of Technology, School of Medicine, Guangzhou, Guangdong, China
e-mail: jinxu@scut.edu.cn; mczhangyy@scut.edu.cn

Z. Wen (✉)
Southern University of Science and Technology, School of Life Sciences, Shenzhen, Guangdong, China
e-mail: wenzl@sustech.edu.cn

M. Zhao, P. Qian (eds.), *Hematopoietic Stem Cells*, Advances in Experimental Medicine and Biology 1442, https://doi.org/10.1007/978-981-99-7471-9_9

properties make large-scale experiments such as genetic screens and chemical screens more feasible in zebrafish than in other vertebrates. Indeed, many such kinds of experiments have been carried out in zebrafish and have identified novel genes or chemicals affecting hematopoiesis (Weinstein et al. 1996; Ransom et al. 1996; Chen et al. 2014; Zon and Peterson 2010; North et al. 2007; Li et al. 2015). Second, zebrafish fertilizes externally and develops quickly. The primitive and definitive hematopoiesis emerges at about 11-hour and 1-day post fertilization, respectively. Because zebrafish embryos are transparent, the hematopoietic process could be directly visualized in living embryos. The external development of the embryos makes experimental manipulations such as DNA/RNA microinjection, chimeric transplantation, and chemical treatment much easier in zebrafish than in mice. Furthermore, transparent adult zebrafish has been created and has provided a valuable tool for imaging normal and disordered hematopoiesis in adulthood (White et al. 2008). Imaging techniques in high vertebrates have advanced tremendously during recent years and are promising to promote hematopoietic studies in the future. However, live imaging with zebrafish is easy to handle and, more importantly, is noninvasive. These properties make zebrafish indispensable for the study of hematopoiesis. Besides, state-of-the-art biotechniques, such as light-induced gene manipulation, have been introduced in zebrafish and further enhanced the unique advantages of zebrafish (Halloran et al. 2000; Deguchi et al. 2009; Kamei et al. 2009).

9.2 Initiation of the Primitive Hematopoiesis in Zebrafish

The hematopoietic programs are highly conserved in zebrafish. Like in other vertebrates, hematopoiesis in zebrafish occurs in multiple waves. The first wave, namely, primitive hematopoiesis, produces mainly erythroid and myeloid lineages. For a long time, primitive erythrocytes and endothelium have been considered to arise from common mesoderm precursors: the hemangioblasts (Choi et al. 1998; Nishikawa et al. 1998). In vitro experiments have identified hemangioblasts in mouse embryos and human embryonic stem cells (Huber et al. 2004; Kennedy et al. 2007). However, the direct proof of hemangioblasts in vivo, although has been shown in *Drosophila*, is still debatable in vertebrates (Mandal et al. 2004). Some in vivo lineage tracing studies have challenged the existence of hemangioblasts in mice (Ueno and Weissman 2006; Padron-Barthe et al. 2014). In zebrafish, both hematopoietic and endothelial progenitors emerge from the anterior and posterior lateral mesoderm (ALM and PLM). They locate close to each other and share a series of transcription factors such as *tal1*, *etv2*, and *lmo2* (Frame et al. 2017). Furthermore, the zebrafish *cloche* mutant, which is mutated in the transcription factor *npas4l*, is deficient in both hematopoietic and endothelial lineages (Stainier et al. 1995; Reischauer et al. 2016). These results suggest the existence of hemangioblasts in zebrafish. Taking advantage of the unique properties of zebrafish, Vogeli and his colleagues labeled individual cells at late blastula/gastrula stages with photo-activated fluorescein and performed linage tracing. This single-cell-resolution fate mapping suggests that a portion of red blood cells and endothelium are derived from common progenitors, namely, hemangioblasts (Vogeli et al. 2006). A later fate-mapping study, which labeled individual cells by direct single-cell injection of fluorescent dye, suggested that primitive myeloid cells could also be generated from hemangioblasts (Warga et al. 2009). Therefore, zebrafish is the first vertebrate model animal to demonstrate the presence of hemangioblasts in vivo at single-cell resolution.

At the early segmentation stages of zebrafish development, the ALM and PLM form the rostral blood island (RBI) and intermediate cell mass (ICM), respectively. Primitive erythropoiesis is evidenced by *gata1* expression in the ICM, whereas primitive myeloid cells are determined by *pu.1* expression in the RBI region (Davidson and Zon 2004; Detrich 3rd et al. 1995; Bennett et al. 2001; Lieschke et al. 2002). The antagonism of *gata1* and *pu.1* to each other determines erythroid versus myeloid fate choice (Rhodes et al. 2005; Galloway et al. 2005). The *pu.1*-positive myeloid progenitors further differentiate to become primitive macrophages and neutrophils. The adoption of macrophage or neutrophil fate is determined by several master regulators. *pu.1* itself affects macrophage versus neutrophil fate choice. A high level of *pu.1* promotes macrophage fate choice, whereas a relatively lower level of *pu.1* encourages neutrophil lineages (Jin et al. 2012). In contrast to *pu.1*, myeloid fate determinant *runx1* favors neutrophils than macrophages. Further study reveals that *pu.1* induces *runx1* expression, whereas *runx1*, in turn, represses *pu.1* to balance macrophage versus neutrophil fate commitment (Jin et al. 2012). Another master regulator, *irf8*, works downstream of *pu.1* to promote the macrophage lineage. In the absence of *irf8*, myeloid progenitors adopt neutrophil instead of macrophage fate (Li et al. 2011). In the ICM region, several genes, including *klf17* and *gfi1aa*, are important to promote the erythroid lineage and prevent myeloid gene expression (Kitaguchi et al. 2009; Wei et al. 2008). Furthermore, the sumoylation status of C/ebpα also affects the erythroid versus myeloid differentiation (Yuan et al. 2011).

9.3 The Emergence of the Definitive Hematopoiesis in Zebrafish

Soon after the primitive hematopoiesis, an intermediate wave of hematopoiesis emerges in the posterior blood island (PBI) in zebrafish. Similar to mice, this wave of hematopoiesis generates erythro-myeloid hematopoietic progenitors (EMPs), which produce erythroid and myeloid cells (Bertrand et al. 2007). EMPs in zebrafish arise directly from the PBI. Although EMPs in mice are suggested to emerge from endothelial cells via endothelial-hematopoietic transition (EHT), this has not been demonstrated in zebrafish (Kasaai et al. 2017; McGrath et al. 2015; Frame et al. 2016; Chen et al. 2011). Whether EMPs in zebrafish are generated by endothelial cells or not remains to be explored. Unlike mice EMPs, which contribute the majority of tissue-resident macrophages in adulthood, fish EMPs mainly generate erythroid and myeloid cells at larval stages and only contribute limited tissue-resident macrophages in eyes of adult zebrafish (Kasaai et al. 2017; Gomez Perdiguero et al. 2015; Epelman et al. 2014; Ginhoux et al. 2010; Schulz et al. 2012; Hoeffel et al. 2015; Xu et al. 2015; He et al. 2018). Although EMPs possess only a limited ability of proliferation and maintenance, they sometimes are still classified as the definitive hematopoiesis (Frame et al. 2017; Gore et al. 2018).

The conventional definitive hematopoiesis, which gives rise to hematopoietic stem cells (HSCs), initiates at the ventral wall of the dorsal aorta (VDA), the equivalent of mammalian aorta-gonad-mesonephros (AGM) in zebrafish, at around 26-hour post fertilization (Kissa and Herbomel 2010). For a long time, it has been debatable whether HSCs were derived from vascular endothelial cells or mesenchyme (Dieterlen-Lievre et al. 2006; Bertrand et al. 2005; Yoshimoto and Yoder 2009). This question is very difficult to answer in mammals because noninvasive observation without disturbing normal hematopoiesis in high vertebrates is technically challenging. The first in vivo live imaging demonstrating that hematopoietic stem/progenitor cells (HSPCs) are directly derived from endothelial cells was reported by several groups in zebrafish (Kissa and Herbomel 2010; Bertrand et al. 2010; Lam et al. 2010). In the meantime, ex vivo live imaging in mice suggests the same conclusion (Boisset et al. 2010). In zebrafish, it is found that endothelial cells in VDA bend their bodies toward the ventral space, contract to bring their neighbor endothelial cells together, and detach themselves off the vessel from 26- to 60-hour post fertilization (Kissa and Herbomel 2010). These budding-off cells turn on a series of hematopoietic markers including *cd41* and *cmyb*, which distinguish them from other endothelial cells (Kissa and Herbomel 2010; Bertrand et al. 2010). The whole process from endothelium to become hematopoietic cells is named endothelial-hematopoietic transition (EHT), and these special endothelial cells are called hemogenic endothelial cells (Kissa and Herbomel 2010). Lineage tracing study further shows that adult hematopoiesis can be efficiently marked by *Tg(kdrl:Cre)*, a vessel-specific *cre* transgenic fish, suggesting the endothelial origin of HSCs (Bertrand et al. 2010). Therefore, these budding-off hemogenic endothelial cells are considered HSPCs. A Brainbow-based multicolor Zebrabow system has been applied to quantify the HSC number. This system distinguishes individual hematopoietic progenitors according to their unique "color hue," which is formed by a random combination of different fluorescent reporters in each cell. The result shows that about 20–30 genuine HSCs are generated from aortic endothelium per fish embryo (Henninger et al. 2017). Unlike in mice, where HSPCs bud toward the artery lumen and can go into circulation directly, HSPCs in zebrafish bud toward the subaortic space and do not enter circulation directly. Instead, they travel downward in the subaortic space and finally penetrate the vein. Then they move along with the circulation to seed the caudal hematopoietic tissue (CHT) (Kissa et al. 2008; Murayama et al. 2006; Jin et al. 2007). Interestingly, recent studies have shown that EHT exists not only in VDA but also in CHT (Zhan et al. 2018; Tian et al. 2017). HSPCs in CHT further undergo differentiation and proliferation. Multiple hematopoietic lineages including erythrocytes and myeloid cells were generated locally. HSPCs also migrate from CHT to the thymus to give rise to T cells. Whether HSPCs could directly immigrate to the thymus without passing through CHT remains an open question. CHT is only a transient harbor for HSPCs. The ultimate destination of HSPCs is the kidney, the equivalent of mammalian bone marrow, where hematopoiesis is sustained throughout the whole life of zebrafish (Murayama et al. 2006; Jin et al. 2007).

Because limited HSC markers are characterized in zebrafish, it is difficult to determine whether all hemogenic endothelial cells represent HSCs or a mixture of stem cells and progenitors. Recently, a new lineage tracing method with high spatial-temporal resolution has been introduced to answer this question. This method takes advantage of infrared laser-mediated local heat shock to specifically activate CreER expression in cells of targeted anatomic positions at the desired time point (Deguchi et al. 2009; Kamei et al. 2009; Xu et al. 2015; He et al. 2018; Tian et al. 2017). These activated cells are permanently labeled by fluorescent reporter genes, and their fates are traced. With this method, it has been shown that genuine HSCs are only generated in the VDA but not in the PBI region in zebrafish. Furthermore, HSCs are more enriched in the anterior part of VDA than in the posterior part. Non-HSC progenitors are generated both in the VDA and in the PBI. These progenitors give rise to hematopoietic lineages including T cells and macrophages during the early life of zebrafish. However, they are unable to last to adulthood (He et al. 2018; Tian et al. 2017). Another important finding with this method is that most, if not all, adult tissue-resident macrophages including microglia are derived from HSCs in zebrafish (Xu et al. 2015; He et al. 2018). This conclusion is different from mice in which adult tissue-resident macrophages are reported mainly generated by EMPs (Gomez Perdiguero et al. 2015; Schulz et al. 2012; Hoeffel et al. 2015; Ginhoux and Guilliams 2016; Sheng et al. 2015; Kierdorf et al. 2015). Recently, refinement has been made for this method so that labeling at the single-cell resolution has been achieved. With the improved method, the heterogeneity of hemogenic endothelial cells in CHT has been further revealed. It has been shown that only myeloid and myeloid-T cell bipotential progenitors but not T cell-only progenitors exist in the PBI region (He et al. 2020). It is conceivable that this state-of-the-art cell tracing technique will reveal more secrets of hematopoiesis in the future.

9.4 Hematopoietic Niches in Zebrafish

Intrinsic factors such as the Notch-Runx1 pathway directly control HSPC formation in zebrafish, whereas extrinsic cues from various hematopoietic niches are essential for activating intrinsic factors (Burns et al. 2005; Nik et al. 2017; Wattrus and Zon 2018). During somitogenesis stages, cells in bilateral PLM migrate medially and converge to form the dorsal aorta (Davidson and Zon 2004). These PLM cells contact the somites intimately. The non-canonical Wnt16 signaling activates the expression of the Notch ligands *dlc* and *dld* in somite cells (Clements et al. 2011). On the one hand, *dlc* and *dld* function upstream of Notch3, which functions within the somite to regulate HSPC formation via non-cell-autonomous effects. On the other hand, *dlc* and *dld* work upstream of the PLM cell-expressed Notch receptor(s), probably Notch1a and Notch1b. The PLM cell Notch signaling controls HSPC development cell-autonomously (Kim et al. 2014; Kobayashi et al. 2014). How would the Wnt signaling relay to Notch pathway remains an open question. It has been shown that Fgfr4, the receptor for FGF signaling, forms the bridge between Wnt16 and Dlc, but not Dld (Lee et al. 2014). The transduction of Notch signaling between somites and PLM cells requires additional factors. Jam1a and Jam2a are junctional adhesion molecules located on migrating PLM cells and somatic cells, respectively. Their interaction is essential for Notch activation in PLM cells. Disruption of either Jam1a or Jam2a dampens HSPC specification in the VDA (Kobayashi et al. 2014). Another Wnt signaling factor, Wnt9a, is also released from somites. It signals through Fzd9b with EGFR as a cofactor to provoke β-catenin-dependent Wnt signaling in HSPCs and expand HSPCs in the VDA (Grainger et al. 2016; Grainger et al. 2019). In addition to Wnt, VegfA is released from somites and specifies hemogenic endothelium via TGFβ signaling (Monteiro et al. 2016; Leung et al. 2013). Somites not only relay signals but also contribute to endothelial cells or vascular associated cells at the dorsal aorta directly. These somite-derived cells, although do not contribute to HSPCs, can induce HSPCs by releasing Cxcl12b (Nguyen et al. 2014). Interestingly, somite-origin definitive HSCs have been reported in zebrafish recently (Qiu et al. 2016). The relationship between somite and hematopoiesis remains to be explored.

After the convergence of PLM cells, the dorsal aorta forms. Blood flow, which initiates soon after the connection of main vessels, has been demonstrated an important trigger for HSPC formation (North et al. 2009). Blood flow could be sensed by primary cilia on endothelial cells (Liu et al. 2019; Goetz et al. 2014). Yap, *klf2a*-NO, and adenosine signaling pathways are downstream of blood flow to control HSPC programming (Lundin et al. 2020; Wang et al. 2011; Jing et al. 2015). Interestingly, one recent report has shown a counterintuitive role of blood flow. The authors found that impaired mechanosensing of the endothelium by *pkd2* depletion or reduced blood flow could increase endothelial cell extrusion in the VDA (Campinho et al. 2020). Therefore, the exact mechanisms of blood flow-induced HSPC formation remain to be explored. In addition to blood flow, many other factors have been reported to regulate HSPC development in the VDA. FGF signaling modulates HSPC specification by regulating BMP signaling around the aortic microenvironment (Pouget et al. 2014). Cholesterol and vitamin D metabolism, docosahexaenoic acid (DHA) supply, estrogen and prostaglandin (PG) E2 signaling, TGFβ signaling input from the notochord, inflammatory signaling, Gcsf signaling, and 7α,25-OHC-mediated Gpr183 activation are all involved in HSPC emergence (North et al. 2007; Monteiro et al. 2016; Gu et al. 2019; Zhang et al. 2015; Carroll et al. 2014; Cortes et al. 2016; Cortes et al. 2015; Liu et al. 2018; Li et al. 2014; He et al. 2015; Sawamiphak et al. 2014; Frame et al. 2020; Espin-Palazon et al. 2014; Stachura et al. 2013). The close association between the dorsal aorta and neural crest cells is necessary for HSPC formation (Damm and Clements 2017). The HSPC specification signal, presumably released from neural crest cells, remains to be explored.

After the formation of HSPCs in the VDA, they travel through the stroma under the aorta, enter circulation via the vein, and lodge in the CHT. Macrophages contact HSPCs around the ventral aorta and facilitate HSPC mobilization by degrading extracellular matrix (ECM) (Travnickova et al. 2015). Besides, VCAM-1^+ macrophages facilitate HSPC retention in the CHT via an ITGA4-dependent manner (Li et al. 2018). Interestingly, definitive hematopoiesis in the CHT appears not affected when primitive macrophages are abrogated by *cebpα* mutation (Yuan et al. 2019). Therefore, it is worthy to explore whether there is a compensatory mechanism for HSPC migration/homing in the absence of macrophages. To prosper in the CHT, the newly arrived HSPCs require supports from the surrounding niche. Upon the arrival of HSPCs, CHT endothelial cells remodel to surround HSPCs (Xue et al. 2017). Many cytokines, including *kitlgb*, *cxcl8*, and *ccl25b*, are expressed by endothelial cells and promote the lodgment, expansion, and differentiation of HSPCs (Nik et al. 2017; Wattrus and Zon 2018). *Kitlgb* works downstream of the endothelial-enriched *tfec* transcription factor to promote CHT hematopoiesis (Stachura et al. 2013). Endothelial *cxcl8/cxcr1* signaling positively regulates HSPC colonization possibly by the induction of *cxcl12a* expression in the CHT (Stachura et al. 2013). The endothelial-enriched *klf6a* transcription factor regulates *ccl25b* and *Cxcl8b* expression. Knockdown of *ccl25b* impairs hematopoiesis in the CHT, and replenishment of *ccl25b* recovers HSPC defects in *klf6a* morphants (Stachura et al. 2013). In addition to endothelial cells, the stromal cells, which arise from the caudal somites via an epithelial-mesenchymal transition (EMT), are necessary for supporting the CHT hematopoiesis (Murayama et al. 2015).

HSPCs finally reach the kidney after 4 days post fertilization. Interestingly, a population of melanocytes, the pigment cells with black color, locates just above the head kidney where it accommodates HSPCs in zebrafish. These melanocytes form a cover to protect HSPCs from UV irradiation. This population of melanocytes exists in both teleost fish and tadpole suggesting their conserved role during evolution (Kapp et al. 2018).

9.5 Forward and Reverse Genetic Studies of Zebrafish Hematopoiesis

To investigate the genes that regulate hematopoiesis using zebrafish, the forward and reverse genetic approaches provide great advancement for insights into the function of the genes. The forward genetic approach is based on the hematopoietic phenotype to identify the genes involved in the process, whereas the reverse genetic approach is modifying the interesting genes to reveal the effect on hematopoiesis (Lawson and Wolfe 2011). For the forward genetic screen, the chemical mutagen N-ethyl-N-nitrosourea (ENU) is the most frequently applied in zebrafish, which has generated a series of mutations that are involved in various biological processes. In the past decades, several groups carried out the ENU large-scale genetic screens and identified a large number of genes involved in hematopoiesis (Weinstein et al. 1996; Ransom et al. 1996; Trede et al. 2008; Wang et al. 2012). Christiane Nüsslein-Volhard group and Wolfgang Driever group were the first to classify the zebrafish blood mutant and isolated about 25 mutants that affected blood cell generation, differentiation, and function (Weinstein et al. 1996; Ransom et al. 1996). Then through positional cloning, they identified the causative genes of those processes, such as *alas2*, *urod*, and *sptb* (Liao et al. 2000; Brownlie et al. 1998; Wang et al. 1998), which provides great support to discover unknown genes and new functions of known genes in hematopoiesis. In addition, through ENU screening, studies have not only identified new genes related to hematopoiesis but also revealed new roles of some known genes in hematopoiesis, including the *udu*, *rumba*, *samba*, *c-myb*, *cebpa*, *alas1*, *slc20a1b*, and *sart3* (Liu et al. 2007; Du et al. 2011; Zhang et al. 2011; Dai et al. 2016; Lian et al. 2018; Chen et al. 2021; Zhao et al. 2021), which provide important insights in erythropoiesis, myelopoiesis, and HSPC development. Besides, insertion mutagens including pseudotyped retrovirus and transposons are also applied in zebrafish to link specific phenotype to the genotype. Nancy Hopkins group pioneered the insertional mutagenesis strategy with retroviral vector in zebrafish and isolated about 385 mutant genes, which were essential for early development (Allende et al. 1996; Gaiano et al. 1996; Amsterdam et al. 1999; Golling et al. 2002; Amsterdam et al. 2004). Subsequently, Leonard I. Zon group utilized the insertional mutant lines from Nancy group to screen the HSC marker *cmyb* and found nine genes (e.g., *tbx16*, *plcr1*, *hdac1*) were essential for HSC emergence. Through the screening, a new hierarchical network required for HSC emergence was clarified (Burns et al. 2009). For the transposon-mediated mutations, the *Sleeping Beauty* and *Tol2* transposon systems are widely applied in zebrafish. The *Sleeping Beauty* in zebrafish is induced by Stephen C. Ekker group, whose lab demonstrates that it could generate not only ubiquitous and tissue-specific transgenic animals but also insertional mutations (Davidson et al. 2003). Transposable element *Tol2* is first identified in medaka fish (Koga et al. 1996). Then the *Tol2* element is found to be excised, and the high-efficiency chromosomal integration *Tol2* transposon system is established in zebrafish (Kawakami et al. 1998; Kawakami et al. 2004). Through the system, the *Tol2*-based gene trap and enhancer trap were developed (Kawakami et al. 2004; Clark et al. 2004; Parinov et al. 2004). Using the *Tol2*-mediated enhancer trap screen, the *gfi1aa:GFP* transgenic line is screened, and the expression in embryonic hematopoiesis is identified (Wei et al. 2008). Through the *Tol2*-mediated gene trap screen, *fli1a* and *fli1b* mutant lines are generated and demonstrate *etv2* and *fli1b* but not *fli1a* act redundantly in regulation of vasculogenesis and angiogenesis (Craig et al. 2015). Overall, the forward genetic approach provides vital basis in analyzing the gene regulation of early hematopoiesis.

As forward genetic screening is not likely to identify all mutations related to hematopoiesis, it is necessary to utilize reverse genetic approaches to define the gene function in hematopoiesis. To date, several reverse genetic tools have been applied in zebrafish, such as transgenics, morpholino oligonucleotide (MO), targeting induced local lesions in genomes (TILLING), zinc finger nucleases (ZFNs), transcription activator-like effector nucleases (TALENs), and clustered regularly interspersed short palindromic repeats/CRISPR-associated protein 9 (CRISPR/Cas9) (Lawson et al. 2001; Feng et al. 2007; Le et al. 2007; Nasevicius and Ekker 2000; Wienholds et al. 2002; Doyon et al. 2008; Sander et al. 2011; Huang et al. 2011; Chang et al. 2013; Hwang et al. 2013). Transgenic mediated gene alteration is widely achieved by the Gal4/UAS system and the Cre/loxp system. For example, the activation of the Notch signaling using the *hsp70:Gal4;uas:NICD* system has revealed that Notch signaling is sufficient for expanding the definitive HSCs (Burns et al. 2005). *hsp-cre*-induced KRASG12D activation in zebrafish developed myeloproliferative disorder and other tumors (Le et al. 2007). Although the transgenic is convenient to overexpress target genes, it is not applicable to knock down or knock out the gene. Using MO, it is able to knock down gene expression and identify gene functions in hematopoiesis, such as Mat3, a subunit of the nucleosome remodeling and deacetylase complexes, which knockdown caused primitive hematopoiesis defect (Li et al. 2009). Moreover, MO was applied in the genetic screen, one of which, chromatin factors involved in embryonic hematopoiesis were analyzed (Huang et al. 2013). However, MO-mediated gene knockdown but not knockout shows short effective duration and off-target effect (Robu et al. 2007), and it could not assess the compensatory effect of some gene mutations (Rossi et al. 2015). Thus, the germline mutation approaches are indispensable to decipher the gene function. TILLING was the first gene knockout approach in zebrafish and identified mutants such as *rag1*, *runx1*, and *pu.1* that are involved in hematopoiesis (Jin et al. 2012; Wienholds et al. 2002; Jin et al. 2009). With the development of engineered endonucleases, ZFNs and TALEN knockout technology were applied to edit the target genes in zebrafish (Doyon et al. 2008; Sander et al. 2011; Huang et al. 2011; Meng et al. 2008). Using ZFNs, the miR-142a and miR-142b were mutated which showed their mature miRNA miR-142b-3p was essential for neutrophil development in zebrafish (Fan et al. 2014). However, the mutagenicity of ZFNs was less efficient compared with TALEN (Chen et al. 2013); therefore, candidate gene mutations associated with hematopoiesis were frequently edited by TALEN since it was developed. Using TALEN, the novel functions of a series of genes in hematopoiesis were revealed, such as *idh1*, *zap70*, *bcas2*, *arl4aa*, *dhx15*, and *rb1* (Cai et al. 2021; Shi et al. 2015; Guo et al. 2020; Moore et al. 2016; Yu et al. 2019; Zhang et al. 2018); meanwhile, some of the mutations established related disease models, such as *rps19*, *mpl*, and *asxl1* (Zhang et al. 2014; Lin et al. 2017; Fang et al. 2021). Although ZFNs and TALEN have successfully generated mutations in zebrafish, their manipulation procedures are complicated; nowadays, CRISPR/Cas9 has emerged as an excellent approach to gene editing (Varshney et al. 2015). In recent years, an increasing number of genes are edited by CRISPR/Cas9, and their novel roles in hematopoiesis are illustrated. For example, *tfec* mutated by CRISPR/Cas9 showed HSC expansion and maintenance defects (Mahony et al. 2016). CRISPR/Cas9-mediated mutation of *ptprj* showed the thrombocytopenia and revealed the important roles of *ptprj* in thrombocyte development (Marconi et al. 2019). *rpl18* mutant generated by CRISPR/Cas9 phenocopied the erythroid defect of DBA (Chen et al. 2020). These studies suggest CRISPR/Cas9 is a bona fide powerful system for hematopoietic gene editing. Moreover, several studies have shown that CRISPR/Cas9 is a powerful screening strategy to uncover the genes that affect different biological processes in zebrafish (Shah et al. 2015; Wu et al. 2018; Keatinge et al. 2021; Kroll et al. 2021; Parvez et al. 2021; Shin et al. 2021). Three genes, *atad5a*, *ddb1*, and *pcna*, are screened from 32 DNA repair-related genes and found to be critical for HSPC development (Shin et al. 2021). However, although CRISPR/Cas9 was thought as high efficiency in the gene editing of double-stranded DNA, efficient and accurate CRISPR/Cas9-mediated knock-in approaches are still to be optimized in zebrafish, and only several hematopoietic associated knock-in lines have been established (Solman et al. 2022; Chestnut and Sumanas 2020). Moreover, a recent study has shown that TALEN exhibits more efficiency than CRISPR/Cas9 in editing the target sites of heterochromatin (Jain et al. 2021), suggesting the requirement of multiple technologies collaboration to comprehensively understand the gene function. With the repaid development of gene editing, we will have a comprehensive understanding of the hematopoietic system.

9.6 Zebrafish Models of Human Hematopoietic Disorders and Drug Discovery

As the hematopoiesis of zebrafish and humans is highly conserved, zebrafish have been widely used to study hematopoietic diseases. Using the genetic approaches mentioned above, researchers have established a large number of human hematopoietic disorder models, mainly including malignant hematopoiesis and inherited bone marrow failure syndromes. The most studied malignant hematopoietic diseases in zebrafish mainly consist of leukemia, myelodysplastic syndromes (MDS), and myeloproliferative neoplasms (MPN). The first reported leukemia in zebrafish is to overexpress the human *RUNX1-CBF2T1 (AML1-ETO)* cDNA to identify the leukemogenesis role of Runx1, and the authors find the overexpression reproduces the phenotype of *RUNX1-CBF2T1* knock-in mice, suggesting the zebrafish can be a model system to study acute myeloid leukemia (AML) (Kalev-Zylinska et al. 2002). Subsequently, numerous leukemia models are developed to investigate leukemogenesis in different types of leukemia. Langenau and colleagues established the first leukemia model by using the zebrafish *rag2* promoter to drive the murine *c-Myc* with a green fluorescent protein tag (Langenau et al. 2003). However, the transgenic-induced T cell acute lymphoblastic leukemia (T-ALL) model was predisposed to death before sexual maturity, so it was exclusively maintained by in vitro fertilization. Soon after, several groups introduce the Cre/loxp system to conditionally control the expression of c-MYC (Feng et al. 2007; Langenau et al. 2005; Gutierrez et al. 2011), which not only improves our understanding of the mechanism but also the disease progression of MYC-induced T-ALL. NOTCH1 is another common mutation found in more than 50% of T-ALL patients (Weng et al. 2004), clarifying the pathogenesis of NOTCH-associated T-ALL has important implications for the understanding of the disease. Chen et al. generate the NOTCH1-induced zebrafish T-ALL model and find NOTCH1 and Bcl2 synergistically induce leukemogenesis (Chen et al. 2007). Moreover, using the *rag2-notch1a*^{ICD} zebrafish T-ALL model, Blackburn et al. demonstrate activation of NOTCH1 only increases the number of pre-malignant T cells but has no effect on the frequency of leukemia-propagating cells (Blackburn et al. 2012). In addition to the T-ALL models, researchers also establish the B-ALL models, such as the *TEL-AML1* and the *hMYC;GFP*^{lo} transgenic lines (Sabaawy et al. 2006; Borga et al. 2019). Compared with ALL, more myeloid neoplasm models have been established based on myeloid neoplasm common mutations. Yeh and colleagues utilize the zebrafish *hsp70* promoter to drive the human *AML1-ETO* and generate inducible AML1-ETO-related leukemia (Yeh et al. 2008). Zhuravleva et al. overexpress the *MOZ/TIF2* fusion gene, which is controlled by myeloid-specific promoter *spi1*, and ~1% of the model develops to AML (Zhuravleva et al. 2008). Similarly, He et al. utilize the *spi1* promoter to drive the *FLT3-ITD* or *FLT3-TKD* fusion gene and establish the associated zebrafish AML models (He et al. 2014). Besides the AML, the models for studying myeloid neoplasm MDS and MPN are as well extensively developed. Mutation of mitochondrial matrix chaperone *hspa9b* induces blood cell apoptosis via excess ROS that result in MDS-like ineffective hematopoiesis, suggesting *hspa9b* deficiency and mitochondrial dysfunction can be involved in the progress of MDS (Craven et al. 2005). However, the mutant is embryonic lethality, whereas MDS occurs primarily in elder patients. Sun et al. utilize the Pu.1 activity insufficient mutant *pu.1*^{G242D}, and Gjini et al. generate *tet2*$^{m/m}$ mutant, both of which survive to adulthood and exhibit the phenotype of MDS (Sun et al. 2013; Gjini et al. 2015). In addition, Liu et al. find the *cmyb-eGFP* transgenic fish exhibit abnormal granulocyte expansion and display the MDS-like phenotype in adult zebrafish due to *cmyb* hyperactivity (Liu et al. 2017). Compared with the ineffective hematopoiesis of MDS, MPN is characterized by anomaly proliferation of one or more myeloid lineages. Le and colleagues develop the inducible KRASG12D transgenic line that shows the kidney marrow (KM) myelomonocyte and precursors increased (Le et al. 2007). Like the hemogram of KRASG12D induced, *irf8* mutants show the myeloid expansion as well; moreover, Mertk activation is showed to participate in the *irf8*-related leukemogenesis (Zhao et al. 2018). Apart from the myeloid cell expansion, *JAK2*^{V581F} is overexpressed in zebrafish embryos that show significantly erythroid increased, which phenocopied the human polycythemia vera (Ma et al. 2009). Overexpressing the CALR-del52 mutation mRNA displays thrombocytosis that is consistent with the human essential thrombocythemia (Lim et al. 2016). Researchers have established a series of MDS (*hspa9b*^{G492E}, *pu.1*^{G242D}, *tet2*$^{m/m}$, *cmyb*^{hyper}) and MPN models (*kRAS*^{G12D}, *irf8*$^{-/-}$, *jak2a*^{v581F}, *CALR-del52*) that provide invaluable implication for understanding the pathogenesis of MDS and MPN.

Besides malignant hematopoiesis models, the zebrafish models for inherited bone marrow failure syndromes (IBMFS) are developed for clarifying their pathogenesis and relationships with leukemogenesis. IBMFS is a group of rare diseases characterized by hematopoietic failure with cancer predisposition, mainly including Fanconi anemia (FA), dyskeratosis congenita (DC), Diamond-Blackfan anemia (DBA), Shwachman-Diamond syndrome (SDS), severe congenital neutropenia (SCN), and congenital amegakaryocytic thrombocytopenia (CAMT). For the FA, there are 22 causative genes identified so far. Except for the unidentified FANCS/BRCA1 ortholog in zebrafish and three newly identified FA genes (*FANCU*, *FANCV*, *FANCW*) in mice, researchers have established models for all other FA genes in zebrafish (Guitton-Sert et al. 2021). Liu et al. generate the first FA zebrafish model through the MO knockdown, and Fancd2-deficient zebrafish are sensitive to DNA damage and exhibit most of the phenotypes of childhood FA; in addition, the authors found that apoptosis in fancd2-deficient zebrafish is p53-dependent (Liu et al. 2003). Subsequently, stable heritable FA-associated mutants *brca2*^{Q658X}, *brca2*$^{-/-}$, and *rad51*^{sa23805} are established, whereas the *brca2*^{Q658X} and *brca2*$^{-/-}$ mutants show no hematopoietic phenotype but exhibit genome instability. Moreover, the two models closely mimic the gonad tumors in humans (Shive et al. 2010; Rodriguez-Mari et al. 2011). In contrast, the *rad51*^{sa23805} mutants recapitulate the FA phenotypes such as microphthalmia, DNA damage sensitivity, and KM cytopenia, suggesting the model is valuable for uncovering the hematopoietic pathway of FA (Botthof et al. 2017). After that, 17 FA genes are knockout using the CRISPR/Cas9, which provides huge research resources for revealing the pathogenesis of FA-related diseases (Ramanagoudr-Bhojappa et al. 2018). For DC, in terms of hematopoiesis dysfunction, it is similar to FA that exhibits aplastic anemia and leukemia susceptibility. But FA is characterized by the DNA repair defect, and DC is accompanied by the telomere maintenance impaired. Notably, no zebrafish DC-associated model has yet fully mimicked the DC patient phenotypes. Pereboom and colleagues generated the first zebrafish DC model *nop10*^{hi2578} by the insertional mutagenesis screen. However, the *nop10* mutation does not affect telomere function but results in HSC deficiency, which is dependent on p53 (Pereboom et al. 2011). Similarly, the other two zebrafish DC models *dkc*^{KD} and *nola1*^{KO} show hematopoietic defects and no telomere maintenance defect (Zhang et al. 2012). Different from the impaired hematopoiesis caused by the above three gene defects, *tert*$^{-/-}$ zebrafish mutants display tissue atrophy, premature death, and shortened telomeres (Anchelin et al. 2013; Henriques et al. 2013). These models suggest that different mutants can be used to reveal different DC-related pathogenesis, respectively. Unlike FA and DC, which are often accompanied by pancytopenia, DBA usually presents with erythroid hypoplasia. Genetically, DBA is often associated with ribosomal protein (RP) gene mutations. So far, 20 RP gene mutations have been found to be involved in DBA (Da Costa et al. 2020). In zebrafish, approximately half of DAB-related genes are knockdown or knockout, and defects in these genes mimic DBA phenotypes such as severe anemia, developmental defects, and cancer susceptibility (Oyarbide et al. 2019). Danilova et al. knock down the first DAB-associated gene *rps19* in zebrafish. The *rps19* deficiency exhibits impaired erythropoiesis, and developmental abnormality resembles the DBA (Danilova et al. 2008). Mechanistically, the developmental defects but not anemia are p53 dependent (Torihara et al. 2011). Since the model of *rps19* deficiency was established, other DBA-related gene models have been created to reveal the pathogenesis of DAB and DAB-related drug screening. Consistent with DBA affecting ribosome biogenesis, SDS also causes defects in ribosome biogenesis. In addition, SDS patients exhibit neutropenia and exocrine pancreatic insufficiency, and most of them have *SBDS* mutations (Woloszynek et al. 2004). Venkatasubramani and Mayer knock down *sbds* in zebrafish and find that *sbds* deficiency exhibits partial loss of exocrine pancreas and impaired granulocyte migration, which resembles the SDS phenotypes (Venkatasubramani and Mayer 2008). Subsequently, the other two groups describe the similar phenotypes by knockdown or knockout *sbds* (Provost et al. 2012; Oyarbide et al. 2020). Moreover, Leach group demonstrates that the exocrine pancreas defect is p53 independent, and the Corey group shows that the p53 is activated in *sbds*$^{nu132/nu132}$ mutant and the p53 loss can partially alleviate the mutant lethality. Interestingly, *srp54* deficiency displays the neutrophilia and exocrine pancreas in zebrafish, revealing the relevance between SDS and human *SRP54* mutation in vivo (Carapito et al. 2017). Moreover, patients with *SRP54* mutation are linked to SDS as well as SCN (Bellanne-Chantelot et al. 2018). SCN and SDS have a similar hematopoietic defect with low absolute neutrophil counts, whereas the SCN does not have the extra-hematopoietic manifestation of exocrine pancreatic insufficiency. In the past decades, several genes involved in SCN have been identified, such as *ELANE*, *CSF3R*, WAS, *VPS45*, *HAX1*, and *GFI1* (Skokowa et al. 2017). However, only a subset of zebrafish SCN models has been established so far, including *csf3r*^{KO}, *vps45*^{KD}, and *hax1*^{KD} (Vilboux et al. 2013; Pazhakh et al. 2017; Basheer et al. 2019; Doll et al. 2021). All three models exhibit neutrophil development defects with the reduced neutrophil number, providing new platforms for SCN-associated drug screening. CAMT is another IBMFS characterized by severe reduction of a single hematopoietic lineage, which is mainly caused by *MPL* or *TPO* mutations and manifests as severe thrombocytopenia due to megakaryocytopoiesis failure. The *mpl* or *tpo* knockout zebrafish exhibit a severe reduction of thrombocytes similar to thrombocytopenia in CAMT patients (Lin et al. 2017; Yang et al. 2022), and the mutants combined with the thrombocyte-specific transgenic line *mpl:eGFP* could serve as a powerful tool for identifying the pathogenesis of CAMT and thrombopoietic drug screening.

Zebrafish not only can reveal the pathogenesis of hematopoietic diseases but can also be a valuable platform for drug discovery. To date, drug screening using zebrafish for the discovery of therapeutic drugs for hematopoietic-related diseases has become increasingly widespread, and a series of drugs have been identified. The compounds for drug screening are mainly divided into two categories: one is the new effects or new targets of the known compounds and the other is the novel compounds. For the former, the most high-profile drug is ProHema, a derivative of prostaglandin E2 that is found to promote HSC homeostasis in a zebrafish screen for bioactive compounds that modulate HSC formation. The group demonstrates that ProHema can expand not only zebrafish HSC numbers but also murine bone marrow HSCs, suggesting that ProHema may be useful for HSC transplantation in patients (North et al. 2007). Based on this study, Fate Therapeutics initiates phase I and II clinical trials of ProHema to evaluate its safety and efficacy in umbilical cord blood (UCB) transplantation in patients with hematologic malignancies. The trials show that ProHema improves UCB transplant efficiency and reduces the incidence of graft-versus-host disease in patients (Cully 2019). In addition to the ProHema, the same research group develops a zebrafish KM competitive transplantation-based screen and identified another two known compounds, 11,12-epoxyeicosatrienoic acids and 14,15-epoxyeicosatrienoic acids, which improve HSPC engraftment and may aid in bone marrow transplantation in patients (Li et al. 2015). Through screening known chemical compounds in zebrafish, researchers not only identify drugs that enhance HSC transplantability but also discover potential drugs to treat leukemia. Using the *hsp70:AML1-ETO* transgenic line, Yeh et al. screen chemical modifiers that regulate AML1-ETO function and validate two compounds, nimesulide and dicumarol, that reverse heat-treated *hsp70:AML1-ETO* embryo phenotypes. Moreover, as nimesulide is a cyclooxygenase-2-inhibitor, analysis of the role reveals that the COX-2 and β-catenin pathways are critical for the function of AML1-ETO, providing important insights for us to understand the effects of oncogenes on hematopoietic differentiation and discover new drug targets (Yeh et al. 2009). Besides drug screening efforts on AML, the phenothiazine antipsychotic drug perphenazine is found to inhibit T-ALL cell growth when screening small molecules against human and zebrafish T-ALL cells. By analyzing how perphenazine and its derivatives exert anti-leukemia effects, the protein phosphatase PP2A is identified as one of the targets of phenothiazine, which introduces PP2A activators into preclinical drugs for T-ALL treatment (Gutierrez et al. 2014). In addition to single-drug treatment for leukemias, researchers have also focused drug screening strategies on combination therapies of different drugs, especially for relapse and refractory leukemias. Through an ex vivo drug screening based on cultured zebrafish ALL cells, the combination of dasatinib and temsirolimus is found to be an effective therapy for human T-ALLs, including the refractory dexamethasone-resistant T-ALL (Laukkanen et al. 2022). Screening known drugs not only uncovers potential treatments for leukemias but also DBA. Through compound screening of *rps29-/-* mutant zebrafish, several calmodulin inhibitors are found to rescue the *rps29-/-* deficiency. Interestingly, one of the FDA-approved antipsychotics, the calmodulin inhibitor trifluoperazine (TFP), effectively reversed the *rps29-/-* mutant phenotype.

Importantly, TFP has been shown to improve erythropoiesis both in mouse and human DBA models (Taylor et al. 2020). Moreover, the TFP has been performed in a phase I clinical trial (NCT03966053) to assess its tolerability in DBA treatment. From the above, researchers have identified several new applications of drugs in hematopoietic diseases by screening known compounds. Compared to known compounds, new compound library screening is also an important approach to discovering drugs that are effective for hematopoietic diseases. Most representatively, in a screen of 26,400 small molecules, Lenaldekar (LDK) is identified as selectively eliminating immature T cells in zebrafish. It was then shown that LDK treatment not only efficaciously inhibits the tumor progression in zebrafish and mouse T-ALL models but also inhibits the proliferation of patient-derived primary leukemia cells, suggesting that LDK may be a potential therapy for a subset of leukemia patients (Ridges et al. 2012). In addition to a direct drug screening using zebrafish, ex vivo screening or synthetic drugs can also be first validated in zebrafish. The benzodiazepine drug Ro5-3335 is identified from screening for inhibitors of RUNX1-CBFβ interaction and is shown to inhibit definitive hematopoiesis in zebrafish that is similar to the hematopoietic defects in *runx1*-truncated mutants in zebrafish. Then the drug inhibition of the RUNX1-CBFβ interaction is validated in human leukemia cells, zebrafish leukemia models, and mouse leukemia models, thus providing a promising therapeutic agent for RUNX1-CBFβ-related leukemias (Cunningham et al. 2012). Furthermore, a chemical synthesized small molecule inhibitor 5b targeting the C-terminal domain of Hsp90 is recently found to kill human leukemia cells in the zebrafish xenotransplantation model (Bhatia et al. 2022). From all the abovementioned compounds, the high-throughput drug screenings of zebrafish play a great insight in promoting the search for new therapeutic drugs for blood diseases.

9.7 Conclusions and Perspective

Comparable primitive and definitive hematopoiesis, hematopoiesisrelated regulators and pathways suggest that the hematopoiesis in zebrafish and mammals are highly conserved. Owing to these, zebrafish serve as an important complement for elucidating mammalian hematopoiesis as follows. On the one hand, using zebrafish to illuminate the regulatory mechanism of hematopoiesis and hematopoietic microenvironment, such as the roles of transcription factors, epigenetic regulators, and signal pathways in these processes, provides important insights for our understanding of hematopoiesis. On the other hand, the establishment of hematopoietic disorder models and new drug discoveries in zebrafish provide us with new ideas for revealing the pathogenesis of hematopoietic diseases and developing therapeutic strategies. Although the zebrafish has conducted great progress in hematopoiesis, some open questions remained. Hematopoietic cells, especially leukemia cells, are a class of heterogeneous cells. Although single-cell sequencing has revealed the cellular atlas of some hematopoietic stages and identified some regulatory factors, the atlas of hematopoietic cells at various stages in both physiological and pathological states is still largely unknown. Besides this, the cell origins and interactions, especially the regulatory pathways and mechanisms between hematopoietic cells and the hematopoietic microenvironment, remain to be resolved. Moreover, several hematopoietic disorder models have not been established in zebrafish such as primary myelofibrosis and chronic lymphocytic leukemia. In addition, there are still some limitations that restrict the depth of hematopoietic research. Most zebrafish proteins lack reliable antibodies so that we are unable to assess gene functions at protein levels. Due to the lack of cell-specific surface markers like in mice, isolation of specific zebrafish hematopoietic cells usually relies on transgenic lines. Moreover, in contrast to the mammal, the zebrafish genome exhibits duplications that we need to carefully dissect the difference between different paralogs, and some mammalian homologous genes have not been found in zebrafish, so it is difficult to study the relevant regulatory processes. Despite these questions, several technological advances have opened new avenues for deciphering hematopoiesis. The first is the CRISPR/Cas9-mediated gene-editing technologies. The versatile technologies allow us to disrupt and modify genes more conveniently and to verify the newly screened gene functions more quickly. Excitingly, several novel regulators involved in cardiac development were recently identified using a Multiplexed Intermixed CRISPR Droplets platform on zebrafish (Parvez et al. 2021), which may also have the potential to be applied in hematopoietic-related gene screening. Secondly, with the development of live imaging, such as using IR-LEGO heat-shock microscopy and light-sheet fluorescence microscopy, the origin of several hematopoietic-related cells such as microglia, T cell, and vascular endothelial cell has been identified (Xu et al. 2015; Tian et al. 2017; Pang et al. 2020). The origins of other hematopoietic cells will be gradually traced and discovered. Lastly, the combination of various sequencing technologies, including transcriptome, proteome, and epigenome, will deepen our understanding of hematopoietic regulation. In the future, the limitations will be gradually solved through new technologies, and new regulatory mechanisms of developmental and pathological hematopoiesis and new therapeutic drugs for hematopoietic diseases will be continuously discovered using zebrafish, which will eventually be applied in clinical treatment.

References

Allende ML et al (1996) Insertional mutagenesis in zebrafish identifies two novel genes, pescadillo and dead eye, essential for embryonic development. Genes Dev 10:3141–3155. https://doi.org/10.1101/gad.10.24.3141

Amsterdam A et al (1999) A large-scale insertional mutagenesis screen in zebrafish. Genes Dev 13:2713–2724. https://doi.org/10.1101/gad.13.20.2713

Amsterdam A et al (2004) Identification of 315 genes essential for early zebrafish development. Proc Natl Acad Sci U S A 101:12792–12797. https://doi.org/10.1073/pnas.0403929101

Anchelin M et al (2013) Premature aging in telomerase-deficient zebrafish. Disease Models & Mechanisms 6: 1101–1112. https://doi.org/10.1242/dmm.011635

Basheer F, Rasighaemi P, Liongue C, Ward AC (2019) Zebrafish granulocyte colony-stimulating factor receptor maintains neutrophil number and function throughout the life span. Infect Immun 87. https://doi.org/10.1128/IAI.00793-18

Bellanne-Chantelot C et al (2018) Mutations in the SRP54 gene cause severe congenital neutropenia as well as Shwachman-Diamond-like syndrome. Blood 132: 1318–1331. https://doi.org/10.1182/blood-2017-12-820308

Bennett CM et al (2001) Myelopoiesis in the zebrafish. Danio rerio. Blood 98:643–651. https://doi.org/10.1182/blood.v98.3.643

Bertrand JY et al (2005) Characterization of purified intraembryonic hematopoietic stem cells as a tool to define their site of origin. Proc Natl Acad Sci U S A 102:134–139. https://doi.org/10.1073/pnas.0402270102

Bertrand JY et al (2007) Definitive hematopoiesis initiates through a committed erythromyeloid progenitor in the zebrafish embryo. Development 134:4147–4156. https://doi.org/10.1242/dev.012385

Bertrand JY et al (2010) Haematopoietic stem cells derive directly from aortic endothelium during development. Nature 464:108–111. https://doi.org/10.1038/nature08738nature08738

Bhatia S et al (2022) Development of a first-in-class small-molecule inhibitor of the C-terminal Hsp90 dimerization. ACS Cent Sci 8:636–655. https://doi.org/10.1021/acscentsci.2c00013

Blackburn JS et al (2012) Notch signaling expands a pre-malignant pool of T-cell acute lymphoblastic leukemia clones without affecting leukemia-propagating cell frequency. Leukemia 26:2069–2078. https://doi.org/10.1038/leu.2012.116

Boisset JC et al (2010) In vivo imaging of haematopoietic cells emerging from the mouse aortic endothelium. Nature 464:116–120. https://doi.org/10.1038/nature08764nature08764

Borga C et al (2019) Simultaneous B and T cell acute lymphoblastic leukemias in zebrafish driven by transgenic MYC: implications for oncogenesis and lymphopoiesis. Leukemia 33:333–347. https://doi.org/10.1038/s41375-018-0226-6

Botthof JG, Bielczyk-Maczynska E, Ferreira L, Cvejic A (2017) Loss of the homologous recombination gene rad51 leads to Fanconi anemia-like symptoms in zebrafish. Proc Natl Acad Sci U S A 114:E4452–E4461. https://doi.org/10.1073/pnas.1620631114

Brownlie A et al (1998) Positional cloning of the zebrafish sauternes gene: a model for congenital sideroblastic anaemia. Nat Genet 20:244–250. https://doi.org/10.1038/3049

Burns CE, Traver D, Mayhall E, Shepard JL, Zon LI (2005) Hematopoietic stem cell fate is established by the Notch-Runx pathway. Genes Dev 19:2331–2342. https://doi.org/10.1101/gad.1337005

Burns CE et al (2009) A genetic screen in zebrafish defines a hierarchical network of pathways required for hematopoietic stem cell emergence. Blood 113:5776–5782. https://doi.org/10.1182/blood-2008-12-193607

Cai Y et al (2021) Dhx15 regulates zebrafish definitive hematopoiesis through the unfolded protein response pathway. Cancer Sci 112:3884–3894. https://doi.org/10.1111/cas.15002

Campinho P, Lamperti P, Boselli F, Vilfan A, Vermot J (2020) Blood flow limits endothelial cell extrusion in the zebrafish dorsal aorta. Cell Rep 31:107505. https://doi.org/10.1016/j.celrep.2020.03.069

Carapito R et al (2017) Mutations in signal recognition particle SRP54 cause syndromic neutropenia with Shwachman-Diamond-like features. J Clin Invest 127:4090–4103. https://doi.org/10.1172/JCI92876

Carroll KJ et al (2014) Estrogen defines the dorsal-ventral limit of VEGF regulation to specify the location of the hemogenic endothelial niche. Dev Cell 29:437–453. https://doi.org/10.1016/j.devcel.2014.04.012S1534-5807(14)00232-9

Chang N et al (2013) Genome editing with RNA-guided Cas9 nuclease in zebrafish embryos. Cell Res 23:465–472. https://doi.org/10.1038/cr.2013.45

Chen C, Lu MJ, Lin S, Qin W (2020) The nuclear gene rpl18 regulates erythroid maturation via JAK2-STAT3 signaling in zebrafish model of Diamond-Blackfan anemia. Cell Death Dis 11, . ARTN: 135. https://doi.org/10.1038/s41419-020-2331-5

Chen EY et al (2014) Glycogen synthase kinase 3 inhibitors induce the canonical WNT/beta-catenin pathway to suppress growth and self-renewal in embryonal rhabdomyosarcoma. Proc Natl Acad Sci U S A 111:5349–5354. https://doi.org/10.1073/pnas.13177311111317731111

Chen J et al (2007) NOTCH1-induced T-cell leukemia in transgenic zebrafish. Leukemia 21:462–471. https://doi.org/10.1038/sj.leu.2404546

Chen JK et al (2021) Slc20a1b is essential for hematopoietic stem/progenitor cell expansion in zebrafish. Sci China Life Sci 64:2186–2201. https://doi.org/10.1007/s11427-020-1878-8

Chen MJ et al (2011) Erythroid/myeloid progenitors and hematopoietic stem cells originate from distinct populations of endothelial cells. Cell Stem Cell 9: 541–552. https://doi.org/10.1016/j.stem.2011.10.003S1934-5909(11)00483-8

Chen SJ et al (2013) A large-scale in vivo analysis reveals that TALENs are significantly more mutagenic than ZFNs generated using context-dependent assembly. Nucleic Acids Res 41:2769–2778. https://doi.org/10.1093/nar/gks1356

Chestnut B, Sumanas S (2020) Zebrafish etv2 knock-in line labels vascular endothelial and blood progenitor cells. Dev Dyn 249:245–261. https://doi.org/10.1002/dvdy.130

Choi K, Kennedy M, Kazarov A, Papadimitriou JC, Keller G (1998) A common precursor for hematopoietic and endothelial cells. Development 125:725–732

Clark KJ, Geurts AM, Bell JB, Hackett PB (2004) Transposon vectors for gene-trap insertional mutagenesis in vertebrates. Genesis 39:225–233. https://doi.org/10.1002/gene.20049

Clements WK et al (2011) A somitic Wnt16/Notch pathway specifies haematopoietic stem cells. Nature 474: 220–224. https://doi.org/10.1038/nature10107nature10107

Cortes M et al (2015) Accumulation of the vitamin D precursor cholecalciferol antagonizes hedgehog signaling to impair hemogenic endothelium formation. Stem Cell Reports 5:471–479. https://doi.org/10.1016/j.stemcr.2015.08.010S2213-6711(15)00244-1

Cortes M et al (2016) Developmental vitamin D availability impacts hematopoietic stem cell production. Cell Rep 17:458–468. https://doi.org/10.1016/j.celrep.2016.09.012

Craig MP et al (2015) Etv2 and fli1b function together as key regulators of vasculogenesis and angiogenesis. Arterioscler Thromb Vasc Biol 35:865–876. https://doi.org/10.1161/ATVBAHA.114.304768

Craven SE, French D, Ye W, de Sauvage F, Rosenthal A (2005) Loss of Hspa9b in zebrafish recapitulates the ineffective hematopoiesis of the myelodysplastic syndrome. Blood 105:3528–3534. https://doi.org/10.1182/blood-2004-03-1089

Cully M (2019) Zebrafish earn their drug discovery stripes. Nat Rev Drug Discov 18:811–813. https://doi.org/10.1038/d41573-019-00165-x

Cunningham L et al (2012) Identification of benzodiazepine Ro5-3335 as an inhibitor of CBF leukemia through quantitative high throughput screen against RUNX1-CBFbeta interaction. Proc Natl Acad Sci U S A 109:14592–14597. https://doi.org/10.1073/pnas.1200037109

Da Costa L, Leblanc T, Mohandas N (2020) Diamond-Blackfan anemia. Blood 136:1262–1273. https://doi.org/10.1182/blood.2019000947

Dai Y et al (2016) Cebpalpha is essential for the embryonic myeloid progenitor and neutrophil maintenance in zebrafish. J Genet Genomics 43:593–600. https://doi.org/10.1016/j.jgg.2016.09.001

Damm EW, Clements WK (2017) Pdgf signalling guides neural crest contribution to the haematopoietic stem cell specification niche. Nat Cell Biol 19:457–467. https://doi.org/10.1038/ncb3508ncb3508

Danilova N, Sakamoto KM, Lin S (2008) Ribosomal protein S19 deficiency in zebrafish leads to developmental abnormalities and defective erythropoiesis through activation of p53 protein family. Blood 112: 5228–5237. https://doi.org/10.1182/blood-2008-01-132290

Davidson AE et al (2003) Efficient gene delivery and gene expression in zebrafish using the Sleeping Beauty

transposon. Dev Biol 263:191–202. https://doi.org/10.1016/j.ydbio.2003.07.013
Davidson AJ, Zon LI (2004) The 'definitive' (and 'primitive') guide to zebrafish hematopoiesis. Oncogene 23: 7233–7246. https://doi.org/10.1038/sj.onc.12079431207943
Deguchi T et al (2009) Infrared laser-mediated local gene induction in medaka, zebrafish and Arabidopsis thaliana. Dev Growth Differ 51:769–775. https://doi.org/10.1111/j.1440-169X.2009.01135.xDGD1135
Detrich HW 3rd et al (1995) Intraembryonic hematopoietic cell migration during vertebrate development. Proc Natl Acad Sci U S A 92:10713–10717. https://doi.org/10.1073/pnas.92.23.10713
Dieterlen-Lievre F, Pouget C, Bollerot K, Jaffredo T (2006) Are intra-aortic hemopoietic cells derived from endothelial cells during ontogeny? Trends Cardiovasc Med 16:128–139. https://doi.org/10.1016/j.tcm.2006.02.005
Doll L et al (2021) A zebrafish model for HAX1-associated congenital neutropenia. Haematologica 106:1311–1320. https://doi.org/10.3324/haematol.2019.240200
Doyon Y et al (2008) Heritable targeted gene disruption in zebrafish using designed zinc-finger nucleases. Nat Biotechnol 26:702–708. https://doi.org/10.1038/nbt1409
Du L et al (2011) Rumba and Haus3 are essential factors for the maintenance of hematopoietic stem/progenitor cells during zebrafish hematopoiesis. Development 138:619–629. https://doi.org/10.1242/dev.054536
Epelman S et al (2014) Embryonic and adult-derived resident cardiac macrophages are maintained through distinct mechanisms at steady state and during inflammation. Immunity 40:91–104. https://doi.org/10.1016/j.immuni.2013.11.019S1074-7613(13)00558-X
Espin-Palazon R et al (2014) Proinflammatory signaling regulates hematopoietic stem cell emergence. Cell 159: 1070–1085. https://doi.org/10.1016/j.cell.2014.10.031
Fan HB et al (2014) miR-142-3p acts as an essential modulator of neutrophil development in zebrafish. Blood 124:1320–1330. https://doi.org/10.1182/blood-2013-12-545012
Fang X et al (2021) Asxl1 C-terminal mutation perturbs neutrophil differentiation in zebrafish. Leukemia 35: 2299–2310. https://doi.org/10.1038/s41375-021-01121-8
Feng H et al (2007) Heat-shock induction of T-cell lymphoma/leukaemia in conditional Cre/lox-regulated transgenic zebrafish. Br J Haematol 138:169–175. https://doi.org/10.1111/j.1365-2141.2007.06625.x
Frame JM, Fegan KH, Conway SJ, McGrath KE, Palis J (2016) Definitive hematopoiesis in the yolk sac emerges from Wnt-responsive hemogenic endothelium independently of circulation and arterial identity. Stem Cells 34:431–444. https://doi.org/10.1002/stem.2213
Frame JM, Lim SE, North TE (2017) Hematopoietic stem cell development: Using the zebrafish to identify extrinsic and intrinsic mechanisms regulating hematopoiesis. Methods Cell Biol 138:165–192, . S0091-679X(16)30140-6. https://doi.org/10.1016/bs.mcb.2016.08.004
Frame JM et al (2020) Metabolic regulation of inflammasome activity controls embryonic hematopoietic stem and progenitor cell production. Dev Cell. https://doi.org/10.1016/j.devcel.2020.07.015
Gaiano N et al (1996) Insertional mutagenesis and rapid cloning of essential genes in zebrafish. Nature 383: 829–832. https://doi.org/10.1038/383829a0
Galloway JL, Wingert RA, Thisse C, Thisse B, Zon LI (2005) Loss of gata1 but not gata2 converts erythropoiesis to myelopoiesis in zebrafish embryos. Dev Cell 8:109–116. https://doi.org/10.1016/j.devcel.2004.12.001
Ginhoux F, Guilliams M (2016) Tissue-resident macrophage ontogeny and homeostasis. Immunity 44:439–449. https://doi.org/10.1016/j.immuni.2016.02.024S1074-7613(16)30063-2
Ginhoux F et al (2010) Fate mapping analysis reveals that adult microglia derive from primitive macrophages. Science 330:841–845. https://doi.org/10.1126/science.1194637
Gjini E et al (2015) A zebrafish model of myelodysplastic syndrome produced through tet2 genomic editing. Molecular and Cellular Biology 35:789–804. https://doi.org/10.1128/Mcb.00971-14
Goetz JG et al (2014) Endothelial cilia mediate low flow sensing during zebrafish vascular development. Cell Rep 6:799–808. https://doi.org/10.1016/j.celrep.2014.01.032S2211-1247(14)00066-7
Golling G et al (2002) Insertional mutagenesis in zebrafish rapidly identifies genes essential for early vertebrate development. Nat Genet 31:135–140. https://doi.org/10.1038/ng896
Gomez Perdiguero E et al (2015) Tissue-resident macrophages originate from yolk-sac-derived erythro-myeloid progenitors. Nature 518:547–551. https://doi.org/10.1038/nature13989nature13989
Gore AV, Pillay LM, Venero Galanternik M, Weinstein BM (2018) The zebrafish: a fintastic model for hematopoietic development and disease. Wiley Interdiscip Rev Dev Biol 7:e312. https://doi.org/10.1002/wdev.312
Grainger S et al (2016) Wnt9a is required for the aortic amplification of nascent hematopoietic stem cells. Cell Rep 17:1595–1606. https://doi.org/10.1016/j.celrep.2016.10.027
Grainger S et al (2019) EGFR is required for Wnt9a-Fzd9b signalling specificity in haematopoietic stem cells. Nat Cell Biol 21:721–730. https://doi.org/10.1038/s41556-019-0330-5
Gu Q et al (2019) AIBP-mediated cholesterol efflux instructs hematopoietic stem and progenitor cell fate. Science 363:1085–1088. https://doi.org/10.1126/science.aav1749science.aav1749
Guitton-Sert L, Gao Y, Masson JY (2021) Animal models of Fanconi anemia: a developmental and therapeutic

perspective on a multifaceted disease. Semin Cell Dev Biol 113:113–131. https://doi.org/10.1016/j.semcdb.2020.11.010
Guo Y et al (2020) Function of Arl4aa in the initiation of hematopoiesis in zebrafish by maintaining Golgi complex integrity in hemogenic endothelium. Stem Cell Reports 14:575–589. https://doi.org/10.1016/j.stemcr.2020.02.012
Gutierrez A et al (2011) Pten mediates Myc oncogene dependence in a conditional zebrafish model of T cell acute lymphoblastic leukemia. J Exp Med 208:1595–1603. https://doi.org/10.1084/jem.20101691
Gutierrez A et al (2014) Phenothiazines induce PP2A-mediated apoptosis in T cell acute lymphoblastic leukemia. J Clin Invest 124:644–655. https://doi.org/10.1172/JCI65093
Halloran MC et al (2000) Laser-induced gene expression in specific cells of transgenic zebrafish. Development 127:1953–1960
He BL et al (2014) Functions of flt3 in zebrafish hematopoiesis and its relevance to human acute myeloid leukemia. Blood 123:2518–2529. https://doi.org/10.1182/blood-2013-02-486688
He Q et al (2015) Inflammatory signaling regulates hematopoietic stem and progenitor cell emergence in vertebrates. Blood 125:1098–1106. https://doi.org/10.1182/blood-2014-09-601542
He S et al (2018) Adult zebrafish Langerhans cells arise from hematopoietic stem/progenitor cells. Elife 7: e36131. https://doi.org/10.7554/eLife.36131e36131
He S et al (2020) In vivo single-cell lineage tracing in zebrafish using high-resolution infrared laser-mediated gene induction microscopy. Elife 9:e52024. https://doi.org/10.7554/eLife.52024e52024
Henninger J et al (2017) Clonal fate mapping quantifies the number of haematopoietic stem cells that arise during development. Nat Cell Biol 19:17–27. https://doi.org/10.1038/ncb3444ncb3444
Henriques CM, Carneiro MC, Tenente IM, Jacinto A, Ferreira MG (2013) Telomerase is required for zebrafish lifespan. Plos Genetics 9 . ARTN: e1003214. https://doi.org/10.1371/journal.pgen.1003214
Hoeffel G et al (2015) C-Myb(+) erythro-myeloid progenitor-derived fetal monocytes give rise to adult tissue-resident macrophages. Immunity 42:665–678. https://doi.org/10.1016/j.immuni.2015.03.011S1074-7613(15)00125-9
Huang HT et al (2013) A network of epigenetic regulators guides developmental haematopoiesis in vivo. Nat Cell Biol 15:1516–1525. https://doi.org/10.1038/ncb2870
Huang P et al (2011) Heritable gene targeting in zebrafish using customized TALENs. Nat Biotechnol 29:699–700. https://doi.org/10.1038/nbt.1939
Huber TL, Kouskoff V, Fehling HJ, Palis J, Keller G (2004) Haemangioblast commitment is initiated in the primitive streak of the mouse embryo. Nature 432: 625–630. https://doi.org/10.1038/nature03122
Hwang WY et al (2013) Efficient genome editing in zebrafish using a CRISPR-Cas system. Nat Biotechnol 31:227–229. https://doi.org/10.1038/nbt.2501
Jain S et al (2021) TALEN outperforms Cas9 in editing heterochromatin target sites. Nat Commun 12:606. https://doi.org/10.1038/s41467-020-20672-5
Jin H, Xu J, Wen Z (2007) Migratory path of definitive hematopoietic stem/progenitor cells during zebrafish development. Blood 109:5208–5214. https://doi.org/10.1182/blood-2007-01-069005
Jin H et al (2009) Definitive hematopoietic stem/progenitor cells manifest distinct differentiation output in the zebrafish VDA and PBI. Development 136:647–654. https://doi.org/10.1242/dev.029637
Jin H et al (2012) Runx1 regulates embryonic myeloid fate choice in zebrafish through a negative feedback loop inhibiting Pu.1 expression. Blood 119:5239–5249. https://doi.org/10.1182/blood-2011-12-398362blood-2011-12-398362
Jing L et al (2015) Adenosine signaling promotes hematopoietic stem and progenitor cell emergence. J Exp Med 212:649–663. https://doi.org/10.1084/jem.20141528jem.20141528
Kalev-Zylinska ML et al (2002) Runx1 is required for zebrafish blood and vessel development and expression of a human RUNX1-CBF2T1 transgene advances a model for studies of leukemogenesis. Development 129:2015–2030. https://doi.org/10.1242/dev.129.8.2015
Kamei Y et al (2009) Infrared laser-mediated gene induction in targeted single cells in vivo. Nat Methods 6:79–81. https://doi.org/10.1038/nmeth.1278nmeth.1278
Kapp FG et al (2018) Protection from UV light is an evolutionarily conserved feature of the haematopoietic niche. Nature 558:445–448. https://doi.org/10.1038/s41586-018-0213-010.1038/s41586-018-0213-0
Kasaai B et al (2017) Erythro-myeloid progenitors can differentiate from endothelial cells and modulate embryonic vascular remodeling. Sci Rep 7:43817. https://doi.org/10.1038/srep43817srep43817
Kawakami K, Koga A, Hori H, Shima A (1998) Excision of the tol2 transposable element of the medaka fish, Oryzias latipes, in zebrafish. Danio rerio. Gene 225: 17–22. https://doi.org/10.1016/s0378-1119(98)00537-x
Kawakami K et al (2004) A transposon-mediated gene trap approach identifies developmentally regulated genes in zebrafish. Dev Cell 7:133–144. https://doi.org/10.1016/j.devcel.2004.06.005
Keatinge M et al (2021) CRISPR gRNA phenotypic screening in zebrafish reveals pro-regenerative genes in spinal cord injury. PLoS Genet 17:e1009515. https://doi.org/10.1371/journal.pgen.1009515
Kennedy M, D'Souza SL, Lynch-Kattman M, Schwantz S, Keller G (2007) Development of the hemangioblast defines the onset of hematopoiesis in human ES cell differentiation cultures. Blood 109:2679–2687. https://doi.org/10.1182/blood-2006-09-047704

Kierdorf K, Prinz M, Geissmann F, Gomez Perdiguero E (2015) Development and function of tissue resident macrophages in mice. Seminars Immunol 27:369–378. https://doi.org/10.1016/j.smim.2016.03.017

Kim AD et al (2014) Discrete Notch signaling requirements in the specification of hematopoietic stem cells. EMBO J 33:2363–2373. https://doi.org/10.15252/embj.201488784embj.201488784

Kissa K, Herbomel P (2010) Blood stem cells emerge from aortic endothelium by a novel type of cell transition. Nature 464:112–115. https://doi.org/10.1038/nature08761nature08761

Kissa K et al (2008) Live imaging of emerging hematopoietic stem cells and early thymus colonization. Blood 111:1147–1156. https://doi.org/10.1182/blood-2007-07-099499

Kitaguchi T, Kawakami K, Kawahara A (2009) Transcriptional regulation of a myeloid-lineage specific gene lysozyme C during zebrafish myelopoiesis. Mech Dev 126:314–323. https://doi.org/10.1016/j.mod.2009.02.007S0925-4773(09)00033-1

Kobayashi I et al (2014) Jam1a-Jam2a interactions regulate haematopoietic stem cell fate through Notch signalling. Nature 512:319–323. https://doi.org/10.1038/nature13623nature13623

Koga A, Suzuki M, Inagaki H, Bessho Y, Hori H (1996) Transposable element in fish. Nature 383:30. https://doi.org/10.1038/383030a0

Kroll F et al (2021) A simple and effective F0 knockout method for rapid screening of behaviour and other complex phenotypes. Elife 10. https://doi.org/10.7554/eLife.59683

Lam EY, Hall CJ, Crosier PS, Crosier KE, Flores MV (2010) Live imaging of Runx1 expression in the dorsal aorta tracks the emergence of blood progenitors from endothelial cells. Blood 116:909–914. https://doi.org/10.1182/blood-2010-01-264382blood-2010-01-264382

Langenau DM et al (2003) Myc-induced T cell leukemia in transgenic zebrafish. Science 299:887–890. https://doi.org/10.1126/science.1080280

Langenau DM et al (2005) Cre/lox-regulated transgenic zebrafish model with conditional myc-induced T cell acute lymphoblastic leukemia. Proc Natl Acad Sci U S A 102:6068–6073. https://doi.org/10.1073/pnas.0408708102

Laukkanen S et al (2022) Combination therapies to inhibit LCK tyrosine kinase and mTOR signaling in T-cell acute lymphoblastic leukemia. Blood. https://doi.org/10.1182/blood.2021015106

Lawson ND, Wolfe SA (2011) Forward and reverse genetic approaches for the analysis of vertebrate development in the zebrafish. Dev Cell 21:48–64. https://doi.org/10.1016/j.devcel.2011.06.007

Lawson ND et al (2001) Notch signaling is required for arterial-venous differentiation during embryonic vascular development. Development 128:3675–3683. https://doi.org/10.1242/dev.128.19.3675

Le X et al (2007) Heat shock-inducible Cre/Lox approaches to induce diverse types of tumors and hyperplasia in transgenic zebrafish. Proc Natl Acad Sci U S A 104:9410–9415. https://doi.org/10.1073/pnas.0611302104

Lee Y et al (2014) FGF signalling specifies haematopoietic stem cells through its regulation of somitic Notch signalling. Nat Commun 5:5583. https://doi.org/10.1038/ncomms6583ncomms6583

Leung A et al (2013) Uncoupling VEGFA functions in arteriogenesis and hematopoietic stem cell specification. Dev Cell 24:144–158. https://doi.org/10.1016/j.devcel.2012.12.004S1534-5807(12)00577-1

Li D et al (2018) VCAM-1(+) macrophages guide the homing of HSPCs to a vascular niche. Nature 564:119–124. https://doi.org/10.1038/s41586-018-0709-710.1038/s41586-018-0709-7

Li L, Jin H, Xu J, Shi Y, Wen Z (2011) Irf8 regulates macrophage versus neutrophil fate during zebrafish primitive myelopoiesis. Blood 117:1359–1369. https://doi.org/10.1182/blood-2010-06-290700blood-2010-06-290700

Li P et al (2015) Epoxyeicosatrienoic acids enhance embryonic haematopoiesis and adult marrow engraftment. Nature 523:468–471. https://doi.org/10.1038/nature14569nature14569

Li X, Jia S, Wang S, Wang Y, Meng A (2009) Mta3-NuRD complex is a master regulator for initiation of primitive hematopoiesis in vertebrate embryos. Blood 114:5464–5472. https://doi.org/10.1182/blood-2009-06-227777

Li Y et al (2014) Inflammatory signaling regulates embryonic hematopoietic stem and progenitor cell production. Genes Dev 28:2597–2612. https://doi.org/10.1101/gad.253302.114gad.253302.114

Lian JW et al (2018) Alas1 is essential for neutrophil maturation in zebrafish. Haematologica 103:1785–1795. https://doi.org/10.3324/haematol.2018.194316

Liao EC et al (2000) Hereditary spherocytosis in zebrafish riesling illustrates evolution of erythroid beta-spectrin structure, and function in red cell morphogenesis and membrane stability. Development 127:5123–5132. https://doi.org/10.1242/dev.127.23.5123

Lieschke GJ et al (2002) Zebrafish SPI-1 (PU.1) marks a site of myeloid development independent of primitive erythropoiesis: implications for axial patterning. Dev Biol 246:274–295. https://doi.org/10.1006/dbio.2002.0657S0012160602906573

Lim KH et al (2016) Expression of CALR mutants causes mpl-dependent thrombocytosis in zebrafish. Blood Cancer J 6:e481. https://doi.org/10.1038/bcj.2016.83

Lin Q et al (2017) Establishment of a congenital amegakaryocytic thrombocytopenia model and a thrombocyte-specific reporter line in zebrafish. Leukemia 31:1206–1216. https://doi.org/10.1038/leu.2016.320

Liu C et al (2018) Lipoprotein lipase regulates hematopoietic stem progenitor cell maintenance through DHA supply. Nat Commun 9:1310. https://doi.org/10.1038/s41467-018-03775-y

Liu TX et al (2003) Knockdown of zebrafish Fancd2 causes developmental abnormalities via p53-dependent apoptosis. Dev Cell 5:903–914. https://doi.org/10.1016/s1534-5807(03)00339-3

Liu W et al (2017) c-myb hyperactivity leads to myeloid and lymphoid malignancies in zebrafish. Leukemia 31: 222–233. https://doi.org/10.1038/leu.2016.170

Liu Y et al (2007) The zebrafish udu gene encodes a novel nuclear factor and is essential for primitive erythroid cell development. Blood 110:99–106. https://doi.org/10.1182/blood-2006-11-059204

Liu Z et al (2019) Primary cilia regulate hematopoietic stem and progenitor cell specification through Notch signaling in zebrafish. Nat Commun 10:1839. https://doi.org/10.1038/s41467-019-09403-7

Lundin V et al (2020) YAP regulates hematopoietic stem cell formation in response to the biomechanical forces of blood flow. Dev Cell 52:446–460 e445. https://doi.org/10.1016/j.devcel.2020.01.006

Ma AC et al (2009) A novel zebrafish jak2a(V581F) model shared features of human JAK2(V617F) polycythemia vera. Exp Hematol 37:1379-1386 e1374. https://doi.org/10.1016/j.exphem.2009.08.008

Mahony CB, Fish RJ, Pasche C, Bertrand JY (2016) tfec controls the hematopoietic stem cell vascular niche during zebrafish embryogenesis. Blood 128:1336–1345. https://doi.org/10.1182/blood-2016-04-710137

Mandal L, Banerjee U, Hartenstein V (2004) Evidence for a fruit fly hemangioblast and similarities between lymph-gland hematopoiesis in fruit fly and mammal aorta-gonadal-mesonephros mesoderm. Nat Genet 36: 1019–1023. https://doi.org/10.1038/ng1404ng1404

Marconi C et al (2019) Loss-of-function mutations in PTPRJ cause a new form of inherited thrombocytopenia. Blood 133:1346–1357. https://doi.org/10.1182/blood-2018-07-859496

McGrath KE et al (2015) Distinct sources of hematopoietic progenitors emerge before HSCs and provide functional blood cells in the mammalian embryo. Cell Rep 11:1892–1904. https://doi.org/10.1016/j.celrep.2015.05.036S2211-1247(15)00572-0

Meng X, Noyes MB, Zhu LJ, Lawson ND, Wolfe SA (2008) Targeted gene inactivation in zebrafish using engineered zinc-finger nucleases. Nat Biotechnol 26: 695–701. https://doi.org/10.1038/nbt1398

Monteiro R et al (2016) Transforming growth factor beta drives hemogenic endothelium programming and the transition to hematopoietic stem cells. Dev Cell 38: 358–370. https://doi.org/10.1016/j.devcel.2016.06.024S1534-5807(16)30427-0

Moore JC et al (2016) T cell immune deficiency in zap70 mutant zebrafish. Mol Cell Biol 36:2868–2876. https://doi.org/10.1128/MCB.00281-16

Murayama E et al (2006) Tracing hematopoietic precursor migration to successive hematopoietic organs during zebrafish development. Immunity 25:963–975. https://doi.org/10.1016/j.immuni.2006.10.015

Murayama E et al (2015) NACA deficiency reveals the crucial role of somite-derived stromal cells in haematopoietic niche formation. Nat Commun 6: 8375. https://doi.org/10.1038/ncomms9375ncomms9375

Nasevicius A, Ekker SC (2000) Effective targeted gene 'knockdown' in zebrafish. Nat Genet 26:216–220. https://doi.org/10.1038/79951

Nguyen PD et al (2014) Haematopoietic stem cell induction by somite-derived endothelial cells controlled by meox1. Nature 512:314–318. https://doi.org/10.1038/nature13678nature13678

Nik S, Weinreb JT, Bowman TV (2017) Developmental HSC microenvironments: lessons from zebrafish. Adv Exp Med Biol 1041:33–53. https://doi.org/10.1007/978-3-319-69194-7_4

Nishikawa SI, Nishikawa S, Hirashima M, Matsuyoshi N, Kodama H (1998) Progressive lineage analysis by cell sorting and culture identifies FLK1+VE-cadherin+ cells at a diverging point of endothelial and hemopoietic lineages. Development 125:1747–1757

North TE et al (2007) Prostaglandin E2 regulates vertebrate haematopoietic stem cell homeostasis. Nature 447:1007–1011, . nature05883. https://doi.org/10.1038/nature05883

North TE et al (2009) Hematopoietic stem cell development is dependent on blood flow. Cell 137:736–748. https://doi.org/10.1016/j.cell.2009.04.023S0092-8674(09)00448-6

Oyarbide U, Topczewski J, Corey SJ (2019) Peering through zebrafish to understand inherited bone marrow failure syndromes. Haematologica 104:13–24. https://doi.org/10.3324/haematol.2018.196105

Oyarbide U et al (2020) Loss of Sbds in zebrafish leads to neutropenia and pancreas and liver atrophy. JCI Insight 5. https://doi.org/10.1172/jci.insight.134309

Padron-Barthe L et al (2014) Clonal analysis identifies hemogenic endothelium as the source of the blood-endothelial common lineage in the mouse embryo. Blood 124:2523–2532. https://doi.org/10.1182/blood-2013-12-545939blood-2013-12-545939

Pang MJ et al (2020) Light-sheet fluorescence imaging charts the gastrula origin of vascular endothelial cells in early zebrafish embryos. Cell Discov 6. https://doi.org/10.1038/s41421-020-00204-7

Parinov S, Kondrichin I, Korzh V, Emelyanov A (2004) Tol2 transposon-mediated enhancer trap to identify developmentally regulated zebrafish genes in vivo. Dev Dyn 231:449–459. https://doi.org/10.1002/dvdy.20157

Parvez S et al (2021) MIC-Drop: A platform for large-scale in vivo CRISPR screens. Science 373:1146–1151. https://doi.org/10.1126/science.abi8870

Pazhakh V, Clark S, Keightley MC, Lieschke GJ (2017) A GCSFR/CSF3R zebrafish mutant models the persistent basal neutrophil deficiency of severe congenital neutropenia. Sci Rep 7:44455. https://doi.org/10.1038/srep44455

Pereboom TC, van Weele LJ, Bondt A, MacInnes AW (2011) A zebrafish model of dyskeratosis congenita reveals hematopoietic stem cell formation failure resulting from ribosomal protein-mediated p53 stabilization. Blood 118:5458–5465. https://doi.org/10.1182/blood-2011-04-351460

Pouget C et al (2014) FGF signalling restricts haematopoietic stem cell specification via modulation of the BMP pathway. Nat Commun 5:5588. https://doi.org/10.1038/ncomms6588ncomms6588

Provost E et al (2012) Ribosomal biogenesis genes play an essential and p53-independent role in zebrafish pancreas development. Development 139:3232–3241. https://doi.org/10.1242/dev.077107

Qiu J et al (2016) Embryonic hematopoiesis in vertebrate somites gives rise to definitive hematopoietic stem cells. J Mol Cell Biol 8:288–301. https://doi.org/10.1093/jmcb/mjw024mjw024

Ramanagoudr-Bhojappa R et al (2018) Multiplexed CRISPR/Cas9-mediated knockout of 19 Fanconi anemia pathway genes in zebrafish revealed their roles in growth, sexual development and fertility. PLoS Genet 14:e1007821. https://doi.org/10.1371/journal.pgen.1007821

Ransom DG et al (1996) Characterization of zebrafish mutants with defects in embryonic hematopoiesis. Development 123:311–319

Reischauer S et al (2016) Cloche is a bHLH-PAS transcription factor that drives haemato-vascular specification. Nature 535:294–298. https://doi.org/10.1038/nature18614nature18614

Rhodes J et al (2005) Interplay of pu.1 and gata1 determines myelo-erythroid progenitor cell fate in zebrafish. Dev Cell 8:97–108. S1534-5807(04)00410-1. https://doi.org/10.1016/j.devcel.2004.11.014

Ridges S et al (2012) Zebrafish screen identifies novel compound with selective toxicity against leukemia. Blood 119:5621–5631. https://doi.org/10.1182/blood-2011-12-398818

Robu ME et al (2007) p53 activation by knockdown technologies. PLoS Genet 3:e78. https://doi.org/10.1371/journal.pgen.0030078

Rodriguez-Mari A et al (2011) Roles of brca2 (fancd1) in oocyte nuclear architecture, gametogenesis, gonad tumors, and genome stability in zebrafish. PLoS Genet 7:e1001357. https://doi.org/10.1371/journal.pgen.1001357

Rossi A et al (2015) Genetic compensation induced by deleterious mutations but not gene knockdowns. Nature 524:230–233. https://doi.org/10.1038/nature14580

Sabaawy HE et al (2006) TEL-AML1 transgenic zebrafish model of precursor B cell acute lymphoblastic leukemia. Proc Natl Acad Sci U S A 103:15166–15171. https://doi.org/10.1073/pnas.0603349103

Sander JD et al (2011) Targeted gene disruption in somatic zebrafish cells using engineered TALENs. Nat Biotechnol 29:697–698. https://doi.org/10.1038/nbt.1934

Sawamiphak S, Kontarakis Z, Stainier DY (2014) Interferon gamma signaling positively regulates hematopoietic stem cell emergence. Dev Cell 31:640–653. https://doi.org/10.1016/j.devcel.2014.11.007S1534-5807(14)00724-2

Schulz C et al (2012) A lineage of myeloid cells independent of Myb and hematopoietic stem cells. Science 336:86–90. https://doi.org/10.1126/science.1219179

Shah AN, Davey CF, Whitebirch AC, Miller AC, Moens CB (2015) Rapid reverse genetic screening using CRISPR in zebrafish. Nat Methods 12:535-+. https://doi.org/10.1038/nmeth.3360

Sheng J, Ruedl C, Karjalainen K (2015) Most tissue-resident macrophages except microglia are derived from fetal hematopoietic stem cells. Immunity 43:382–393. https://doi.org/10.1016/j.immuni.2015.07.016S1074-7613(15)00305-2

Shi X et al (2015) Functions of idh1 and its mutation in the regulation of developmental hematopoiesis in zebrafish. Blood 125:2974–2984. https://doi.org/10.1182/blood-2014-09-601187

Shin U et al (2021) Large-scale generation and phenotypic characterization of zebrafish CRISPR mutants of DNA repair genes. DNA Repair (Amst) 107:103173. https://doi.org/10.1016/j.dnarep.2021.103173

Shive HR et al (2010) brca2 in zebrafish ovarian development, spermatogenesis, and tumorigenesis. Proc Natl Acad Sci U S A 107:19350–19355. https://doi.org/10.1073/pnas.1011630107

Skokowa J, Dale DC, Touw IP, Zeidler C, Welte K (2017) Severe congenital neutropenias. Nat Rev Dis Primers 3:17032. https://doi.org/10.1038/nrdp.2017.32

Solman M et al (2022) Inflammatory response in hematopoietic stem and progenitor cells triggered by activating SHP2 mutations evokes blood defects. Elife 11. https://doi.org/10.7554/eLife.73040

Stachura DL et al (2013) The zebrafish granulocyte colony-stimulating factors (Gcsfs): 2 paralogous cytokines and their roles in hematopoietic development and maintenance. Blood 122:3918–3928. https://doi.org/10.1182/blood-2012-12-475392blood-2012-12-475392

Stainier DY, Weinstein BM, Detrich HW 3rd, Zon LI, Fishman MC (1995) Cloche, an early acting zebrafish gene, is required by both the endothelial and hematopoietic lineages. Development 121:3141–3150

Sun J et al (2013) Suppression of Pu.1 function results in expanded myelopoiesis in zebrafish. Leukemia 27:1913–1917. https://doi.org/10.1038/leu.2013.67

Taylor AM et al (2020) Calmodulin inhibitors improve erythropoiesis in Diamond-Blackfan anemia. Sci Transl Med 12. https://doi.org/10.1126/scitranslmed.abb5831

Tian Y et al (2017) The first wave of T lymphopoiesis in zebrafish arises from aorta endothelium independent of hematopoietic stem cells. J Exp Med 214:3347–3360. https://doi.org/10.1084/jem.20170488jem.20170488

Torihara H et al (2011) Erythropoiesis failure due to RPS19 deficiency is independent of an activated

Tp53 response in a zebrafish model of Diamond-Blackfan anaemia. Brit J Haematol 152:648–654. https://doi.org/10.1111/j.1365-2141.2010.08535.x
Travnickova J et al (2015) Primitive macrophages control HSPC mobilization and definitive haematopoiesis. Nat Commun 6:6227. https://doi.org/10.1038/ncomms7227ncomms7227
Trede NS et al (2008) Zebrafish mutants with disrupted early T-cell and thymus development identified in early pressure screen. Dev Dyn 237:2575–2584. https://doi.org/10.1002/dvdy.21683
Ueno H, Weissman IL (2006) Clonal analysis of mouse development reveals a polyclonal origin for yolk sac blood islands. Dev Cell 11:519–533. https://doi.org/10.1016/j.devcel.2006.08.001
Varshney GK et al (2015) High-throughput gene targeting and phenotyping in zebrafish using CRISPR/Cas9. Genome Research 25:1030–1042. https://doi.org/10.1101/gr.186379.114
Venkatasubramani N, Mayer AN (2008) A zebrafish model for the Shwachman-Diamond syndrome (SDS). Pediatr Res 63:348–352. https://doi.org/10.1203/PDR.0b013e3181659736
Vilboux T et al (2013) A congenital neutrophil defect syndrome associated with mutations in VPS45. N Engl J Med 369:54–65. https://doi.org/10.1056/NEJMoa1301296
Vogeli KM, Jin SW, Martin GR, Stainier DY (2006) A common progenitor for haematopoietic and endothelial lineages in the zebrafish gastrula. Nature 443:337–339. https://doi.org/10.1038/nature05045
Wang H, Long Q, Marty SD, Sassa S, Lin S (1998) A zebrafish model for hepatoerythropoietic porphyria. Nat Genet 20:239–243. https://doi.org/10.1038/3041
Wang K et al (2012) Large-scale forward genetic screening analysis of development of hematopoiesis in zebrafish. J Genet Genomics 39:473–480. https://doi.org/10.1016/j.jgg.2012.07.008
Wang L et al (2011) A blood flow-dependent klf2a-NO signaling cascade is required for stabilization of hematopoietic stem cell programming in zebrafish embryos. Blood 118:4102–4110. https://doi.org/10.1182/blood-2011-05-353235blood-2011-05-353235
Warga RM, Kane DA, Ho RK (2009) Fate mapping embryonic blood in zebrafish: multi- and unipotential lineages are segregated at gastrulation. Dev Cell 16: 744–755. https://doi.org/10.1016/j.devcel.2009.04.007S1534-5807(09)00169-5
Wattrus SJ, Zon LI (2018) Stem cell safe harbor: the hematopoietic stem cell niche in zebrafish. Blood Adv 2:3063–3069. https://doi.org/10.1182/bloodadvances.2018021725bloodadvances.2018021725
Wei W et al (2008) Gfi1.1 regulates hematopoietic lineage differentiation during zebrafish embryogenesis. Cell Res 18:677–685. https://doi.org/10.1038/cr.2008.60cr200860
Weinstein BM et al (1996) Hematopoietic mutations in the zebrafish. Development 123:303–309
Weng AP et al (2004) Activating mutations of NOTCH1 in human T cell acute lymphoblastic leukemia. Science 306:269–271. https://doi.org/10.1126/science.1102160
White RM et al (2008) Transparent adult zebrafish as a tool for in vivo transplantation analysis. Cell Stem Cell 2:183–189. https://doi.org/10.1016/j.stem.2007.11.002S1934-5909(07)00275-5
Wienholds E, Schulte-Merker S, Walderich B, Plasterk RH (2002) Target-selected inactivation of the zebrafish rag1 gene. Science 297:99–102. https://doi.org/10.1126/science.1071762
Woloszynek JR et al (2004) Mutations of the SBDS gene are present in most patients with Shwachman-Diamond syndrome. Blood 104:3588–3590. https://doi.org/10.1182/blood-2004-04-1516
Wu RS et al (2018) A rapid method for directed gene knockout for screening in G0 zebrafish. Dev Cell 46: 112-125 e114. https://doi.org/10.1016/j.devcel.2018.06.003
Xu J et al (2015) Temporal-spatial resolution fate mapping reveals distinct origins for embryonic and adult microglia in zebrafish. Dev Cell 34:632–641. https://doi.org/10.1016/j.devcel.2015.08.018S1534-5807(15)00552-3
Xue Y et al (2017) The vascular niche regulates hematopoietic stem and progenitor cell lodgment and expansion via klf6a-ccl25b. Dev Cell 42:349-362 e344. https://doi.org/10.1016/j.devcel.2017.07.012
Yang L et al (2022) Generation of a thrombopoietin-deficient thrombocytopenia model in zebrafish. J Thromb Haemost 20:1900–1909. https://doi.org/10.1111/jth.15772
Yeh JR et al (2008) AML1-ETO reprograms hematopoietic cell fate by downregulating scl expression. Development 135:401–410. https://doi.org/10.1242/dev.008904
Yeh JR et al (2009) Discovering chemical modifiers of oncogene-regulated hematopoietic differentiation. Nat Chem Biol 5:236–243. https://doi.org/10.1038/nchembio.147
Yoshimoto M, Yoder MC (2009) Developmental biology: Birth of the blood cell. Nature 457:801–803. https://doi.org/10.1038/457801a457801a
Yu SS et al (2019) BCAS2 is essential for hematopoietic stem and progenitor cell maintenance during zebrafish embryogenesis. Blood 133:805–815. https://doi.org/10.1182/blood-2018-09-876599
Yuan H et al (2011) Sumoylation of CCAAT/enhancer-binding protein alpha promotes the biased primitive hematopoiesis of zebrafish. Blood 117:7014–7020. https://doi.org/10.1182/blood-2010-12-325712blood-2010-12-325712
Yuan H et al (2019) Primitive macrophages are dispensable for HSPC mobilization and definitive hematopoiesis. Blood 134:782–784. https://doi.org/10.1182/blood.2018893974blood.2018893974
Zhan Y et al (2018) The caudal dorsal artery generates hematopoietic stem and progenitor cells via the

endothelial-to-hematopoietic transition in zebrafish. J Genet Genomics. https://doi.org/10.1016/j.jgg.2018.02.010

Zhang P et al (2015) G protein-coupled receptor 183 facilitates endothelial-to-hematopoietic transition via Notch1 inhibition. Cell Res 25:1093–1107. https://doi.org/10.1038/cr.2015.109cr2015109

Zhang Y, Jin H, Li L, Qin FX, Wen Z (2011) cMyb regulates hematopoietic stem/progenitor cell mobilization during zebrafish hematopoiesis. Blood 118:4093–4101. https://doi.org/10.1182/blood-2011-03-342501

Zhang Y, Morimoto K, Danilova N, Zhang B, Lin S (2012) Zebrafish models for dyskeratosis congenita reveal critical roles of p53 activation contributing to hematopoietic defects through RNA processing. PLoS One 7:e30188. https://doi.org/10.1371/journal.pone.0030188

Zhang Y et al (2014) Defects of protein production in erythroid cells revealed in a zebrafish Diamond-Blackfan anemia model for mutation in RPS19. Cell Death Dis 5:e1352. https://doi.org/10.1038/cddis.2014.318

Zhang Z et al (2018) Retinoblastoma 1 protects T cell maturation from premature apoptosis by inhibiting E2F1. Development 145. https://doi.org/10.1242/dev.158139

Zhao F et al (2018) Irf8 regulates the progression of myeloproliferative neoplasm-like syndrome via Mertk signaling in zebrafish. Leukemia 32:149–158. https://doi.org/10.1038/leu.2017.189

Zhao Y et al (2021) The spliceosome factor sart3 regulates hematopoietic stem/progenitor cell development in zebrafish through the p53 pathway. Cell Death Dis 12, . ARTN: 906. https://doi.org/10.1038/s41419-021-04215-4

Zhuravleva J et al (2008) MOZ/TIF2-induced acute myeloid leukaemia in transgenic fish. Br J Haematol 143: 378–382. https://doi.org/10.1111/j.1365-2141.2008.07362.x

Zon LI, Peterson R (2010) The new age of chemical screening in zebrafish. Zebrafish 7:1. https://doi.org/10.1089/zeb.2010.9996

Multi-lineage Differentiation from Hematopoietic Stem Cells

10

Xiaoshuang Wang, Siqi Liu, and Jia Yu

Abstract

The hematopoietic stem cells (HSCs) have the ability to differentiate and give rise to all mature blood cells. Commitment to differentiation progressively limits the self-renewal potential of the original HSCs by regulating the level of lineage-specific gene expression. In this review, we will summarize the current understanding of the molecular mechanisms underlying HSC differentiation toward erythroid, myeloid, and lymphocyte lineages. Moreover, we will decipher how the single-cell technologies advance the lineage-biased HSC subpopulations and their differentiation potential.

Keywords

Hematopoietic stem cell · Lineage differentiation · Single-cell technology · Transcription factor networks · Epigenetic regulation · Non-coding RNA

X. Wang (✉)
The State Key Laboratory for Complex, Severe, and Rare Diseases, Institute of Basic Medical Sciences, Chinese Academy of Medical Sciences / Peking Union Medical College, Beijing, China

The Institute of Blood Transfusion, Chinese Academy of Medical Sciences / Peking Union Medical College, Chengdu, China
e-mail: wangxs@ibms.pumc.edu.cn

S. Liu
State Key Laboratory of Common Mechanism Research for Major Diseases, Institute of Basic Medical Sciences, Chinese Academy of Medical Sciences / Peking Union Medical College, Beijing, China

J. Yu (✉)
State Key Laboratory of Common Mechanism Research for Major Diseases, Institute of Basic Medical Sciences, Chinese Academy of Medical Sciences / Peking Union Medical College, Beijing, China

The Institute of Blood Transfusion, Chinese Academy of Medical Sciences / Peking Union Medical College, Chengdu, China
e-mail: j-yu@ibms.pumc.edu.cn

10.1 Introduction

Hematopoietic stem cells (HSCs) differentiate into multipotent progenitors (MPPs), which further give rise to increasingly lineage-restricted progenitor cells and then undergo terminal differentiation. Mature blood and immune cells, from myeloid lineage or lymphoid lineage, are specialized to perform distinct physiological functions throughout the life span (Haas et al. 2018). Erythrocytes are the most abundant cells in the human body, accounting for around 70% of the total cell number, and 200 million erythrocytes are produced daily (Muckenthaler et al. 2017; Kato and Igarashi 2019). The production rate of white blood cells is also high, with the estimated production of 50–100 million neutrophils per day (Summers et al. 2010), which may be correlated with the short circulating half-life of these cells. Thus, the ability for HSCs

M. Zhao, P. Qian (eds.), *Hematopoietic Stem Cells*, Advances in Experimental Medicine and Biology 1442, https://doi.org/10.1007/978-981-99-7471-9_10

to differentiate into progenitor cells, while also maintaining an adequate pool of HSCs via balancing self-renewal and differentiation, is essential for maintaining the short life span of blood cells (Dharampuriya et al. 2017).

Traditionally, hematopoiesis is recognized as a hierarchical process with the relatively homogeneous HSCs at the apex, leading to the formation of hematopoietic tree model or branching model (Fig. 10.1). Compared with HSC (Yamamoto et al. 2013), MPPs lose the extended self-renewal capacity but can still create all adult hematopoietic lineages. Three multipotent subpopulations with decreasing repopulation potentials were classified according to the outcome of transplantation assays – LT-HSCs, ST-HSCs (also known as MPP1), and MPPs – all three populations allowing only short-term (up to 8 weeks) multi-lineage repopulation (Watcham et al. 2019; Pietras et al. 2015). On the contrary, committed progenitors are defined by the absence of extended self-renewal and a restricted lineage differentiation capacity (normally bi- or uni-lineage) (Doulatov et al. 2010). While shutting down self-renewal and turning on lineage programs, the steady cellular state of HSCs also alters in early progenitors. It is commonly accepted that HSCs are maintained in a quiescent, autophagy-dependent, and glycolytic state marked by low mitochondrial activity and tightly controlled levels of protein synthesis. In contrast, progenitors display higher proliferative and metabolic activity dependent on oxidative metabolism and mitochondrial function. For instance, cell cycle characteristics, molecular features, and functional biases are distinct among different MPP subpopulations (Signer et al. 2014; Ho et al. 2017; Cabezas-Wallscheid et al. 2017; Simsek et al. 2010).

Postnatally, HSCs migrate to the bone marrow, which is the primary site of HSC maintenance and hematopoiesis throughout the life (Cullen et al. 2014). The initial steps of HSC differentiation take place in the bone marrow, which is determined by cytokine signals that are provided by niche cells and that promote the progressive activation of the lineage-specifying transcriptional program. Afterward, terminal differentiation processes leading to the generation of mature blood cells occur in the blood or peripheral tissues and depend on the exposure of precursor cells to cytokines, antigens, and other factors. However, when in response to hematopoietic stress, the HSC niche can transfer to extramedullary sites, such as the liver, spleen, lymph nodes, heart, respiratory system, gastrointestinal tract, kidney, adrenal gland, breast, prostate gland, uterus, mediastinum, and arterial atherosclerotic plaques (Morrison and Scadden 2014; Yang et al. 2020).

Differentiation from HSCs and progenitor cells to fully differentiated blood cells is exquisitely controlled by the timely regulation of gene expression, the basis of which depends on the interplay of a variety of elements, including transcription factors, epigenetic regulation, and post-transcriptional control mechanisms. In this review, we summarize the findings concerning the mechanisms underlying adult hematopoiesis in mice and humans, focusing on classic gene regulatory networks composed of genes encoding key transcription factors and the epigenetic events, which compete for lineage identities. In addition, we review the roles of regulatory RNAs in hematopoiesis. Finally, we describe recently emerging data that point to a relative transitional differentiation model with fewer branch points of heterogeneous HSC and progenitors to the final mature cells, which may reshape our knowledge of hematopoietic differentiation landscape.

10.2 Regulatory Networks of Lineage Commitment

In the hierarchy model, the first branch point segregates lymphoid potential from all other lineages (myeloid, erythroid, and megakaryocytic), followed by several further branching steps on either side of the tree progressing from multi- to bi- and finally to uni-potent progenitor cells (Yamamoto et al. 2013; Akashi et al. 2000; Kondo et al. 1997). Fluorescence-activated cell sorting (FACS)-based cell surface marker detection enables the identification of functionally discrete populations in the hematopoietic system. HSCs and MPP1 are mainly characterized by a $CD150^{+}CD34^{-/low}Flt3^{-}lineage^{-}Sca\text{-}1^{+}c\text{-}Kit^{+}$ ($CD150^{+}CD34^{-}Flt3^{-}LSK$) phenotype and $Flt3^{-}CD34^{+}LSK$ phenotype (Christensen and Weissman 2001; Morrison and Weissman 1994; Kiel et al. 2005; Osawa et al. 1996), respectively.

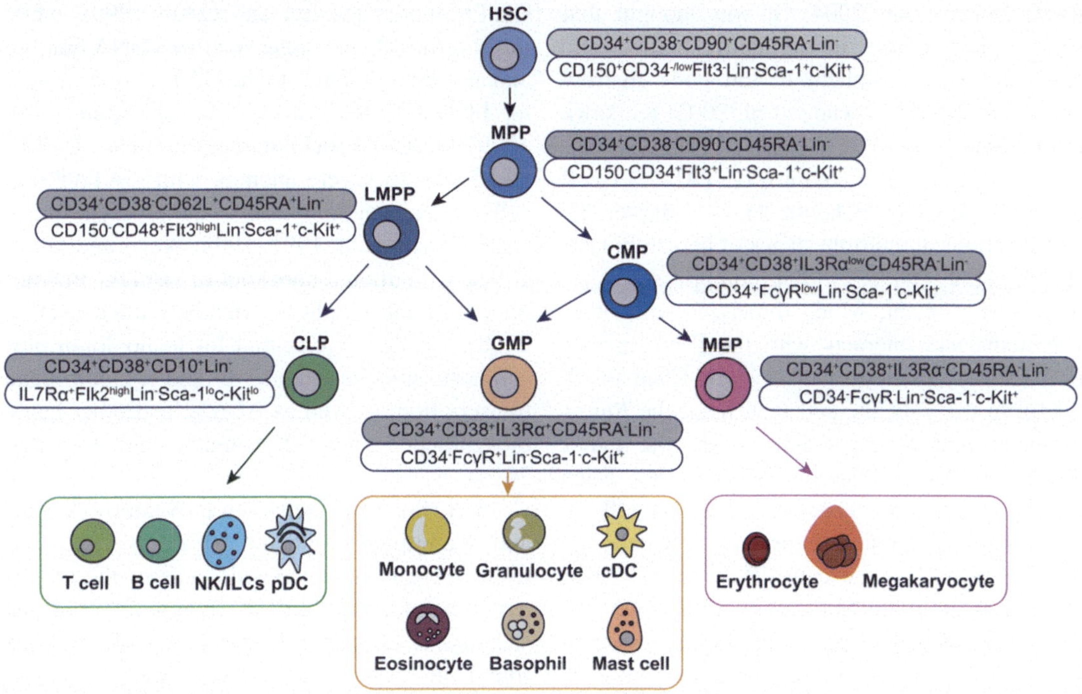

Fig. 10.1 The classic model of hematopoietic stem cell differentiation. The hematopoiesis hierarchy is organized in disconnected and progressive states that contain cells of identical potential, with hematopoietic stem cells sitting at the apex. *HSC* hematopoietic stem cell, *MMP* multipotent progenitor, *LMPP* lymphoid-primed multipotent progenitor, *CLP* common lymphoid progenitor, *CMP* common myeloid progenitor, *MEP* megakaryocyte/erythroid progenitor, *GMP* granulocyte/macrophage progenitor

MPP1 produces $Flt3^{+}CD34^{+}LSK$ lymphoid-primed multipotent progenitors (LMPPs) (Adolfsson et al. 2005); alternatively, differential expression of CD150, CD48, and Flk2 defines erythroid/megakaryocytic-biased MPP2 and myeloid-biased MPP3 and lymphoid-primed MPP4 (Pietras et al. 2015; Wilson et al. 2008; Cabezas-Wallscheid et al. 2014). These cells subsequently give rise to lineage-committed progenitors such as common myeloid progenitors (CMPs) (Akashi et al. 2000), common lymphoid progenitors (CLPs) (Kondo et al. 1997), or megakaryocyte-erythroid progenitors (MEPs), with loss of self-renewal activity.

10.2.1 Transcription Factors

This step-wise hematopoiesis process is driven by the interaction of external stimulus and internal regulation (Kaushansky 2006; Mikkola and Orkin 2006). Central to the internal regulation are transcription factors (TFs), which are expressed in a cell type- and developmental stage-dependent way (Orkin 2000). More than 50 TFs have been found to be functional during hematopoietic stem cell development (Gottgens 2015). Extensive studies on the roles of different TFs in this process have greatly promoted our understanding of cell fate determination. Here we focus on a variety of classical TFs whose roles have been established in HSCs and lineage commitment.

Runx1 is a member of the Runx family (Runx1, Runx2, and Runx3), and it was originally designated as CBFα (core-binding factor α) or PEBP2 (polyoma enhancer-binding protein 2) (van Wijnen et al. 2004). Both primitive and definitive hematopoiesis require Runx1 for their proper differentiation (de Bruijn and Dzierzak 2017). During embryonic development, Runx1 expression is dynamic and slightly preceding the appearance of blood cells (North et al. 1999), while in adult, all blood cells except for erythrocytes express Runx1 (Lorsbach et al.

2004; North et al. 2004). It was thought that Runx1 plays a role in regulating homeostatic HSC and HPC numbers in the bone marrow (Cai et al. 2011; Growney et al. 2005; Ichikawa et al. 2004).

SCL (stem cell leukemia), or TAL-1 (T-cell acute lymphocytic leukemia-1), is a master TF that functions at multiple stages of hematopoiesis (Begley et al. 1989a, b). SCL comprises a helix-loop-helix domain, which dimerizes with other E-proteins and interacts with the E-box motifs (CANNTG) (Begley et al. 1989b; Chen et al. 1990). In the yolk sac and fetal liver, the *Runx1* promoter and *Runx3* enhancer are the major targets of SCL (Kallianpur et al. 1994). Mice lacking SCL cannot generate red blood cells in the yolk sac, and $Scl^{-/-}$ embryonic stem cells fail to arise hematopoietic cells (Robb et al. 1995; Porcher et al. 1996; Shivdasani et al. 1995). In adult hematopoiesis, Scl is mainly expressed in HSCs, myeloid progenitors, and mature cells (erythroid cells and megakaryocytic cells) but absent in differentiating T and B lymphocytes (Green et al. 1992; Mouthon et al. 1993). SCL's involvement in lineage progenitors and their differentiation into erythroid cells, megakaryocytes, mast cells, and monocytes/macrophages have been clearly demonstrated (Porcher et al. 2017). SCL is also a key target of caspases in developing erythroblasts, and its cleavage leads to GATA1 degradation and impaired erythropoiesis (Zeuner et al. 2003).

GATA1 is considered as the master TF for erythropoiesis, which is involved in the control of differentiation, survival, and cell cycle during the process (Kerenyi and Orkin 2010; Ferreira et al. 2005). GATA1 is a member of the GATA family TFs including six members, GATA1 to GATA6. All members play important roles in the regulation of several biological processes, such as cell growth, cell differentiation, embryonic development, and disease (Tremblay et al. 2018). GATA family recognizes and binds to a specific DNA motif, (A/T)GATA(A/G) (WGATAR), through one of their two zinc fingers (Martin and Orkin 1990). GATA1 has three main functional domains: an N-terminal transactivation domain (N-TAD) and two zinc fingers. The C-terminal zinc finger recognizes GATA motifs (Martin and Orkin 1990), while the N-terminal zinc finger stabilizes DNA binding and also directly binds to the GATA1 main cofactor FOG-1 (Friend of GATA1) (Crispino et al. 1999). GATA1 knockout mice died at day E10.5–E11.5 due to severe anemia, with GATA1 null cells undergoing apoptosis at the proerythroblast stage (Pevny et al. 1991, 1995; Weiss and Orkin 1995). Conditional knockout of GATA1 in adult mice resulted in aplastic anemia (Gutierrez et al. 2008). GATA1 is essential for hemoglobin biosynthesis since it activates β- and α-globin gene transcription, as well as ALAS2 and other heme biosynthetic genes (Katsumura and Bresnick 2017).

PU.1 (Spi-1), a member of ETS family (Verger and Duterque-Coquillaud 2002), is expressed exclusively in hematopoietic cells of the monocytic/granulocytic and B lymphoid lineages (Klemsz et al. 1990). Lower levels of PU.1 are found in mature erythrocytes, and it is not expressed in mature T cells (Chen et al. 1995). PU.1 contains a variety of different functional domains, namely, ETS, which shows a winged helix-turn-helix recognizing DNA sequence GGAA motif (Kodandapani et al. 1996). PU.1 knockout mice died in a relatively late embryonic lethality (Scott et al. 1994). $PU.1^{-/-}$ mice generate reduced pool of HSCs and progenitors, which leads to the loss of mature macrophages, neutrophils, B cells, and T cells and to the mild defect of erythropoiesis (Kim et al. 2004; Iwasaki et al. 2005; Dakic et al. 2005). In adult bone marrow progenitors, conditional deletion of the PU.1 showed the reduction of monocyte and lymphoid progeny whereas the increasement of granulocytic like cells (Iwasaki et al. 2005; Dakic et al. 2007).

Studies on individual TFs also led to the recognition of transcriptional networks in hematopoiesis. For example, SCL can exist in a multiprotein complex consisting of LMO2, E47, Ldb1, and GATA1 in addition to heterodimerizing with E-protein (Wadman et al. 1997), and this complex is found in cells during different stages of erythropoiesis (Lahlil et al. 2004). PU.1 transcription networks have been shown to regulate T-cell differentiation through interacting with Myb, T-cell factor 7, and

GATA3 in response to Notch signaling (Rothenberg et al. 2016; Del Real and Rothenberg 2013). Genome-wide analysis of 10 key TFs in hematopoiesis progenitor cells revealed that there are previously unknown interactions between 7 of them (SCL, Lyl1, LMO2, GATA2, Runx1, Erg, and Fli-1), generating 221 combinational patterns in HSCs (Wilson et al. 2010). Single-cell profiling has revealed the network between GATA2, Gfi1, and Gfi1b, as well as that between Bcl11A, Sox4, and TEAD1 in B-cell differentiation (Moignard et al. 2013; Laurenti et al. 2013).

Hematopoietic TFs also interact with other chromatin-associated factors to regulate lineages commitment. The GATA1 (or GATA2)/FOG1 complex is associated with the NuRD chromatin remodeling complex in the erythroid and megakaryocytic lineages (Kim and Bresnick 2007). The erythroid lineage TF EKLF interacts with Brg1, a component of the Swi/Snf ATP-dependent chromatin remodeling complex (Brown et al. 2002). Another erythroid TF NF-E2 associates with the MLL2 complex, which is responsible for H3K4 histone methylation (Demers et al. 2007). Further studies on the way of hematopoietic TF-interacting proteins is an ongoing challenge.

10.2.2 Epigenetic Programming

Hematopoietic progenitor cells can become primed by opening up regulatory elements associated with genes that drive differentiation down a specific mature lineage (Table 10.1). DNA methylation profile analyses reveal the remarkable dynamic plasticity of methylation during early hematopoietic lineage specification (Ji et al. 2010; Bock et al. 2012; Farlik et al. 2016) and terminal neutrophil differentiation from CMP (Rönnerblad et al. 2014). First of all, during adult HSC differentiation, gene regulatory regions related to other lineages and to ESCs became gradually methylated. Second, gene regulatory elements of the preferred lineage were just moderately demethylated in HSCs and displayed significantly reduced DNA methylation levels in both committed progenitors and terminally differentiated cells. Third, a strong proliferative gene expression signature was identified in committed progenitor cells but not in stem cells and in terminally differentiated cells (Bock et al. 2012). Moreover, several TFs, such as MEIS1, were methylated and silenced during differentiation (Ji et al. 2010). DNA methylation can also epigenetically protect lymphoid progenitors from the default program of myeloid differentiation during myeloid-lymphoid lineage choice (Farlik et al. 2016). This is confirmed by the observation that HSCs from mice with reduced DNMT1 activity cannot suppress key myelo-erythroid regulators and thus can differentiate into myelo-erythroid, but not lymphoid, progeny (Bröske et al. 2009). The roles of de novo DNA methyltransferases DNMT3A and DNMT3B are distinct but complementary in governing HSC fate decisions. Dnmt3a but not Dnmt3b knockout alone contributes to persistent self-renewal and a differentiation block of HSCs; additionally, their simultaneous deletion further enhances the self-renewal versus differentiation of HSCs (Challen et al. 2011, 2014). TET2, the enzyme that reverses DNA methylation and methylation-driven gene silencing (Hu and Shilatifard 2016), is required to maintain hematopoietic homeostasis. Deletion of Tet2 in LSKs is linked with an increased hematopoietic repopulating capacity and an altered cell differentiation skewing toward monocytic/granulocytic lineages, which might facilitate the development of myeloid malignancies (Li et al. 2011). Similarly, knockdown of TET2 in human CD34^{+} progenitors skews their differentiation toward the granulomonocytic lineage, at the expense of the lymphoid and erythroid lineages (Pronier et al. 2011).

In addition to DNA methylation, histone post-translational modifications are also dynamically regulated to sustain HSC function and ensure proper lineage commitment. HDACs play a direct role in myeloid and erythro-megakaryocytic fate decision. Both HDAC1 and HDAC2 are component of the NuRD and SIN3 complex, and the total levels of Hdac1 and Hdac2 in the hematopoietic system are critical for erythrocyte-megakaryocyte differentiation (Wilting et al. 2010). Transcription of the

Table 10.1 Epigenetic regulation during HSC differentiation

Epilayers (Shema et al. 2019)	Effectors	Regulators	Stages
Chemical modification to genomic DNA	DNA methylation	DNMT1, DNMT3A, DNMT3B, TET2	Early decision of myeloid or lymphoid fate (Bröske et al. 2009), HSC proliferation and differentiation (Challen et al. 2011, 2014), monocytic/granulocytic lineage output (Li et al. 2011; Pronier et al. 2011)
Posttranslational modifications of histones	Histone methylation, ubiquitination, and deacetylase	PRC1 (BMI1), PRC2 (EZH2 and SUZ12), HDAC1, HDAC2	Myeloid and erythro-megakaryocytic fate decision (Wada et al. 2009), lymphoid differentiation (Oguro et al. 2010; Caganova et al. 2013; Tumes et al. 2013; Béguelin et al. 2013)
Chromatin accessibility	Enhancer activity	Transcription factors and multiple chromatin-associated proteins	Overall hematopoietic differentiation process in mouse and human (Lara-Astiaso et al. 2014; Corces et al. 2016; Buenrostro et al. 2018), establishment of early megakaryopoietic profiles (Heuston et al. 2018)
3D genome organization	Compartments, TADs, chromatin loops	*ThymoD*	Myeloid differentiation (Zhang et al. 2020) and lymphoid differentiation in mouse (Hu et al. 2018; Johanson et al. 2018; Isoda et al. 2017)

HDAC1 gene was repressed by C/EBPs during myeloid differentiation and activated by GATA1 during erythro-megakaryocytic differentiation (Wada et al. 2009). Members of PRC1 and PRC2 complex, implementing H2AK119 monoubiquitination and H3K27 methylation, respectively, are essential for HSC maintenance in a developmental stage-specific and dosage-dependent manner (Hu and Shilatifard 2016; Xie et al. 2014; Liu et al. 2009; Hidalgo et al. 2012). PRC1 and PRC2 are also indispensable in regulating lymphoid differentiation. BMI1 plays a crucial role to maintain lineage specification of HSCs, given that loss of Bmi1 resulted in a resolution of bivalent domains at the Ebf1 and Pax5 loci, leading to their premature expression in HSC/MPPs accompanied by accelerated lymphoid specification and a marked reduction in HSC/MPPs (Oguro et al. 2010). EZH2 is crucial in early B-cell development, germinal center formation, as well as differentiation of CD4+ T cells as revealed by lymphoid lineage-specific conditional deletion models (Caganova et al. 2013; Tumes et al. 2013; Béguelin et al. 2013). The essential role of SUZ12 in early B and T lymphoid development has also been highlighted (Lee et al. 2015).

Genome-wide chromatin profiling studies further illustrated the massive dynamic chromatin accessibility landscape underlying cell fate commitment in hematopoiesis. It is found that many poised enhancers are established in lineage progenitors before their activation and precede RNA expression in subsequent lineage differentiation (Lara-Astiaso et al. 2014; Corces et al. 2016). Moreover, the activity of TF motifs, such as the GATA motif in HSCs, may represent indicators of lineage priming pulling cells toward different developmentally committed states (Buenrostro et al. 2018).

Recently, increasing evidences revealed the important impact of the 3D organization of the genome, including compartments, topologically associating domains (TADs), and chromatin loops, in gene regulation during hematopoiesis. Along HSC differentiation, diverse hematopoietic cell types display significant differences in 3D architecture. MEP and granulocytes had more condensed chromatin structures and stronger tendency to adopt the Rabl configuration, possibly correlating with their physiological characteristics such as granulocyte migration and erythrocyte enucleation process (Zhang et al. 2020). Moreover, during the last stage of T lineage commitment, abrupt genome-wide changes at all three levels of chromatin organization occur to establish a chromatin barrier locking cell fate into the T lineages (Hu et al. 2018). A non-coding RNA named ThymoD can orchestrate chromatin folding and compartmentalization by promoting demethylation at CTCF-bound sites and activating cohesin-dependent looping to direct with high precision enhancer-promoter communication of Bcl11b and T-cell fate (Isoda et al. 2017). Studies on lymphoid cell differentiation also found that lineage-defining TFs such as Bcl11b and Pax5 are particularly required for the specific 3D architecture change during hematopoiesis (Hu et al. 2018; Johanson et al. 2018). Thus, the chromatin reorganization during hematopoiesis lags behind transcriptional responses, suggesting that 3D chromatin architectural changes are not drivers of gene expression changes, but rather that they ensure lineage commitment, preventing cells from reverting back to immature states or from being redirected to alternative lineages (Kloetgen et al. 2019).

10.2.3 Other Layers of Gene Regulation

With the discovery of a growing family of regulatory RNA species, increasing evidences have accumulated for the involvement of non-coding RNAs (ncRNAs) and post-transcriptional regulation in hematopoiesis. NcRNAs can be transcribed from different regions throughout the genome and display time- or tissue-specific expression pattern. MicroRNAs (miRNAs) are a set of small RNAs (19–24 nt long) that downregulate gene expression by regulating the translation and degradation of target mRNAs, and the involvement of miRNAs in the regulatory network of hematopoiesis has long been established (Wang and Zhang 2008). Three miRNAs, miR-181a, miR-142, and miR-223, were first reported to be specifically and dynamically expressed in hematopoietic tissue, and ectopic expression of these miRNAs in murine HSCs dramatically altered the proportion of differentiated cells (Chen et al. 2004). GATA1 can occupy the promotor of miR-144/451 and miR-23a~27a~24 miRNA loci to activate their transcription, which is essential for erythropoiesis (Dore et al. 2008; Zhu et al. 2013). Another famous myeloid miRNA required for granulocyte differentiation, miR-223, is under the coordinate control of various TFs including PU.1, NFI-A, and C/EBPα (Johnnidis et al. 2008; Fukao et al. 2007).

Long non-coding RNAs (lncRNAs) are a branch of >200 nt transcripts that act as scaffolds or as decoys to recruit or sequester effector proteins from their DNA, RNA, or protein targets, thus influencing gene expression at transcriptional or post-transcriptional levels (Alvarez-Dominguez and Lodish 2017). In the earliest fate choices of HSC, several lncRNAs, including *Spehd*, are found to be functional (Delás et al. 2019). Recently annotated LncHSC-1 and LncHSC-2 are involved in this early process by differentially regulating myeloid differentiation or HSC self-renewal and T-cell differentiation. LncHSC-2 functions as a recruiter and mediates TF E2A binding on certain target sites (Luo et al. 2015).

The high lineage specificity of lncRNAs also supports their biological activities in lineage specification. Multiple lncRNAs whose expression is dynamically regulated along erythropoiesis are essential for erythrocyte maturation and are targeted by key erythroid TFs GATA1, TAL1, or KLF1. Antisense lncRNA-EC7 is specifically needed for activation of the neighboring gene encoding BAND 3, which is a major anion transporter on erythrocyte membranes, while enhancer-derived lncRNA-EC3 functions as cis-regulator to promote KIF2A expression during erythropoiesis (Alvarez-Dominguez et al. 2014; Paralkar et al. 2014). Intergenic lncRNA (lincRNA)-EPS promotes terminal erythroid differentiation and inhibits apoptosis of mature erythrocytes by repressing many apoptotic genes, such as Pycard. In addition, lincRNA-EPS-deficient mice show enhanced inflammation in that lincRNA-EPS acts as a transcriptional brake to repress immune response gene expression by interacting with hnRNPL (Atianand et al. 2016; Hu et al. 2011). During myeloid differentiation, several lncRNAs promote lineage commitment by acting as decoys to sequester miRNAs or signal transducer proteins. Lnc-MC facilitates the differentiation of monocytes by enhancing the effect of PU.1 and sequestering miR-199-5p, resulting in the increased expression of ACVR1B complexes (Chen et al. 2015). LncRNA HOTAIRM1 is located between HOXA1 and HOXA2 and implicated in the regulation of granulocytic differentiation (Zhang et al. 2009). Mechanistic studies in acute promyelocytic leukemia demonstrated that HOTAIRM1 regulates the degradation of PML-RARA by binding to endogenous miRNAs that repress the autophagy pathway (Chen et al. 2017). Lnc-DC is exclusively expressed in classical antigen-presenting DCs (cDCs), and it is essential for human monocyte-derived DC (Mo-DC) differentiation. Lnc-DC can bind to STAT3 and isolates it from inhibitory phosphatases, favoring nuclear import and activation of numerous DC function-related genes (Wang et al. 2014). Moreover, a lncRNA highly expressed in mature eosinophils, EGO, is involved in eosinophil development. Diminished EGO leads to decreased basic protein expression and eosinophil-derived neurotoxins, which are essential components of eosinophil development (Wagner et al. 2007).

As a specific type of lncRNAs, circular RNAs (circRNAs) are covalently closed RNA molecules that are generated by a process called back-splicing (Bonizzato et al. 2016). A comprehensive analysis indicated that circRNA expression is widespread in hematopoietic cells. Their expression is cell type specific and altered during differentiation, with differentiated cells containing substantially higher levels of circRNAs (Nicolet et al. 2018). It is in line with reports that enucleated blood cells, i.e., platelets and erythrocytes, express the highest numbers of circRNAs to maintain their function, to respond to environmental factors, or to transmit signals to other cells via vesicles (Preußer et al. 2018; Panda and Gorospe 2018). As the close functional association of dysregulated circRNA expression and hematological malignancies, such as AML, MDS, and B-ALL, becomes clear (Bonizzato et al. 2016), further functional characterization of circRNAs in normal hematopoiesis is expected.

10.3 Evolving Models of Hematopoietic Hierarchy

Nearly 20 years ago, clonal analysis identified lineage-biased (Lin-bi) HSCs that can generate cells of all hematopoietic lineages, but with skewed ratios of lymphoid to myeloid cells. The progeny of myeloid-biased (My-bi) HSCs and lymphoid-biased (Ly-bi) HSCs are different in number, composition, and function (Müller-Sieburg et al. 2002, 2004), which was further confirmed by transplantation assays (Dykstra et al. 2007). The subsequent identification of platelet-biased HSCs that predominantly differentiate toward megakaryocytes and platelets added another layer of complex to the differentiation behaviors of HSC and MPPs (Yamamoto et al. 2013; Sanjuan-Pla et al. 2013). Collectively, these studies indicate that oligo-, bi-, and uni-potent cells co-exist in HSC populations. On the other hand, it has been shown that many cells within progenitor gates are already lineage restricted (Adolfsson et al. 2005), indicating most cell fate decisions are taken earlier than expected from the classical hematopoietic tree model (Watcham et al. 2019).

10.3.1 Heterogeneous in Hematopoietic Cell Population

As methods for single-cell transcriptomic analysis became available (Macosko et al. 2015; Picelli et al. 2014; Ziegenhain et al. 2017; Tang et al. 2010), hematopoietic cell heterogeneity and their hierarchical organization can be investigated by comparing the similarity in gene expression, which enables the discovery of several re-annotated cell types in both lymphoid and myeloid lineages. Dendritic cells (DCs) are heterogeneous and consist of multiple subtypes with unique functions, which could be seen among mouse splenic plasmacytoid DCs by scRNA-seq

(Jaitin et al. 2014). A recent study refined the classification of human DCs and monocytes in peripheral blood, with a new DC population responsible for T-cell activation previously misclassified as plasmacytoid DCs and a new conventional DC progenitor population ($CD100^{Hi}$, $CD34^{Int}$), functionally distinct from the $CD34^{Hi}$ HSCs (Villani et al. 2017). Heterogeneity of innate lymphoid cell (ILC) subpopulations in human was also identified, with two new small intestine ILC subpopulations marked by high levels of expression of NKp46, retinoic acid receptor-related orphan receptor-γ t (RORγt) and interferon-γ (IFNγ) or IL-2 and CCL22 (Björklund et al. 2016).

scRNA-seq investigations of HSCs and MPPs have described previously uncharacterized transitional developmental cell states, consistent with the foregoing functional results (Dykstra et al. 2007; Sanjuan-Pla et al. 2013). The transition of HSCs from dormancy toward cell cycle entry is a continuous developmental path associated with upregulation of biosynthetic processes rather than a stepwise progression (Cabezas-Wallscheid et al. 2017). In addition, studies in human revealed that early hematopoiesis is represented by a cellular Continuum of LOw-primed UnDifferentiated ("CLOUD")-HSPCs, which contains phenotypic MPPs and MLPs at transitory states. Distinct lineages emerge directly from the fluid cloud state HSPCs, without passing through a series of discrete, stable progenitor (Velten et al. 2017). Such findings also explained previous observations by clonal assays and xenograft transplantations of oligopotent progenitors (Notta et al. 2016). Importantly, the MPP subpopulations shift toward more myeloid bias during aging or after external stress mainly because of a reduction in self-renewal activity (Pietras et al. 2015; Kowalczyk et al. 2015). Aging can also induce increased platelet priming and functional platelet bias of HSCs, including a significant increase in a class of HSCs that exclusively produce platelets. Moreover, this phenotype can be rescued by deletion of the platelet TF FOG1 (Grover et al. 2016).

Similar to HSC, several studies have identified distinct CMP, MEP, and myeloid subpopulations with varying differentiation potential. Nineteen cell clusters with either distinct or transitional cellular states were found in bone marrow $lineage^{-}Sca\text{-}1^{-}c\text{-}Kit^{+}$ cell population, with the aiding of TFs CEBPα and CEBPε to determine granulocyte-monocyte and neutrophil specification, respectively (Paul et al. 2015). When focusing on MEP, three distinct subpopulations were identified, and differentiation assays revealed only a minority give rise to mixed lineage colonies, while the majority of MEP cells are transcriptionally primed to generate exclusively single-lineage output (Psaila et al. 2016). Dynamic changes in gene expression among single cells also help in reconstructing a map showing the differentiation trajectories of HSCs and progenitor cells, in which upregulation of cell cycle and oxidative phosphorylation transcriptional programs are common between erythroid and granulocyte-macrophage differentiation trajectories (Nestorowa et al. 2016).

Since clonal tracking has shown that HSC function is pre-determined by epigenetic configuration (Yu et al. 2016), conjugate measurements of transcriptional and epigenomic states at single cell can present a better understanding of the heterogeneity that exists in the regulatory mechanisms during hematopoiesis (Liggett and Sankaran 2020; Buenrostro et al. 2015; Kelsey et al. 2017). These methods not only uncovered key cis- and trans-effectors governing hematopoietic differentiation but also identified the evolution of regulatory elements during disease progression in acute myeloid leukemia (Buenrostro et al. 2018; Granja et al. 2019). Collaborative analysis by scATAC-seq, multi-omics single-cell methylation, and scRNA-seq demonstrated that mutations in *TET2* and *DNMT3A* genes cause opposite shifts in the frequencies of erythroid versus myelomonocytic progenitors possibly mediated by methylation of Myc and Myb (Izzo et al. 2020).

10.3.2 New Sights into Lineage Commitment

Nowadays, capturing cells in an unbiased way across multiple developmental stages and then reconstructing their developmental progressions have provided a powerful means to study cellular decision-making and differentiation (Papalexi and Satija 2018). Such studies led to a deeper understanding of the regulators governing early myeloid (Drissen et al. 2016), lymphoid (Tsang et al. 2015), and megakaryocytic differentiation (Psaila et al. 2016) and the correlation among multi-lineage differentiation (Dahlin et al. 2018; Tusi et al. 2018). In general, certain cell fate potentials are interconnected, supporting a hierarchical view of hematopoiesis, with MPPs diverging either toward myeloid and lymphoid fates or toward the erythroid, megakaryocyte, and Ba/Mast cell fates (Tusi et al. 2018). Indeed, the separation of Gata1-expressing lineages (megakaryocytes, erythrocytes, eosinophils, and mast cells) and Gata1-non-expressing lineages (lymphocytes, neutrophils, monocytes) may represent an early lineage bifurcation of the multipotent HSCs (Drissen et al. 2016). In addition, basophils are considered to originate from either the neutrophil/monocyte branch or from the megakaryocyte/erythroid branch (Wolf et al. 2019).

However, the above studies failed to dynamically track adult hematopoiesis as it occurs in native hosts, only providing the snapshot of lineage output to infer the hierarchy of cell states. The development of in vivo single-cell lineage tracing solved such problem. In model organisms, most approaches rely on an engineered genetic label to tag individual cells with heritable marks (Kester and van Oudenaarden 2018; Carrelha et al. 2018; Rodriguez-Fraticelli et al. 2018), and the combination of such tracing methods with scRNA-seq is adopted to interrogate both lineage correlations and cell states (Pei et al. 2020; Weinreb et al. 2020). These studies indicate the characteristic of a "continuum" between progressive lineage divergences and full precommitment in the process of lineage commitment (Fig. 10.2), in which some HSCs and progenitors may be biased toward certain cell fates at the beginning while others may be actually multipotent and able to undergo stepwise cell fate choices at each differentiation stage (Pucella et al. 2020).

Lineage tracing also confirmed that a differentiated cell type can be produced via multiple pathways. Two distinct pathways of monocyte differentiation were observed, namely, the DC-like and neutrophil-like trajectories, which likely represent either monocyte-dendritic progenitors (MDPs) or granulocyte-monocyte progenitors (GMPs) origin of monocyte differentiation (Weinreb et al. 2020). Likewise, megakaryocytes can be generated from early HSPCs along the classic path of differentiation through progenitor states on the way to producing mature cells or alternatively directly from HSCs that is irrelevant with classic differentiation path (Carrelha et al. 2018). As much as half of the megakaryocyte lineage arises largely independently of other hematopoietic fates under steady state, a phenomenon not observed in any other hematopoietic lineage (Rodriguez-Fraticelli et al. 2018). The close relationship of megakaryocyte lineage and the native fate of long-term HSCs may also explain different kinetics among adult HSC differentiation in vivo, with particularly fast and slow contributions to platelets and lymphocytes, respectively (Upadhaya et al. 2018). Therefore, lymphoid and myeloid potentials are largely determined at the level of HSCs, which is subjected to modulation in conditions such as inflammation and aging (Pucella et al. 2020).

In humans, lineage tracing studies have relied on the detection of naturally occurring somatic mutations, including single-nucleotide variants (SNVs), copy number variants (CNVs), variation in short tandem repeat sequences (microsatellites or STRs), and even somatic mutations in mitochondrial DNA (mtDNA), which are stably propagated to daughter cells but are absent in distantly related cells (Ju et al. 2017; Lodato et al. 2015; Ludwig et al. 2019). Despite the scarce application to human native hematopoiesis, existing studies have demonstrated the

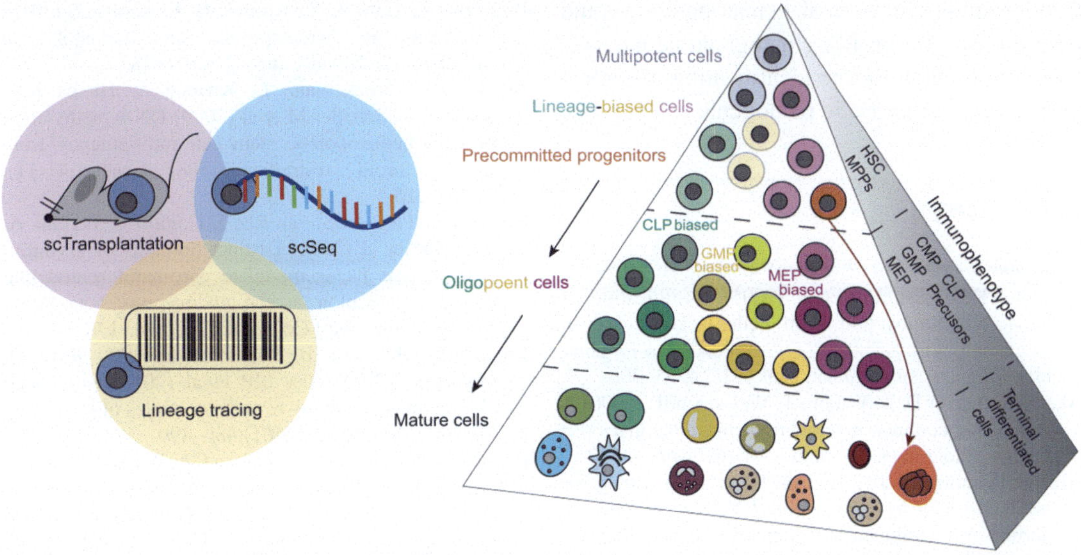

Fig. 10.2 Evolved hematopoiesis model renewed by single-cell technologies. Advances in single-cell technologies (scRNA-seq, scTransplatation, and lineage tracing) and their combined application profoundly revised the classic hematopoietic hierarchy. In the evolved model, hematopoietic stem cells and progenitors are considered to be heterogeneous, different in lineage-biased or multipotent potential. Myeloid, lymphoid, or erythroid, megakaryocyte, and Ba/Mast cell fates diverge early at MPP stages, and megakaryocyte lineage can arise independent of other hematopoietic fates

existence of multi-lineage output from adult human stem cell clones and progenitors in vivo, and the time between symmetric stem cell divisions to fall in the range is estimated to be 2–20 months (Lee-Six et al. 2018; Biasco et al. 2016).

Since clonal tracking has shown that HSC function is pre-determined by epigenetic configuration (Yu et al. 2016), conjugate measurements of transcriptional and epigenomic states at single cell can present a better understanding of the heterogeneity that exists in the regulatory mechanisms during hematopoiesis (Liggett and Sankaran 2020; Buenrostro et al. 2015; Kelsey et al. 2017). These methods not only uncovered key cis- and trans-effectors governing hematopoietic differentiation but also identified the evolution of regulatory elements during disease progression in acute myeloid leukemia (Buenrostro et al. 2018; Granja et al. 2019). Collaborative analysis by scATAC-seq, multi-omics single-cell methylation, and scRNA-seq demonstrated that mutations in *TET2* and *DNMT3A* genes cause opposite shifts in the frequencies of erythroid versus myelomonocytic progenitors possibly mediated by methylation of Myc and Myb (Izzo et al. 2020).

10.4 Perspective

Nowadays, the revised hematopoiesis models demonstrate fewer branch points and display varied differentiation paths of heterogeneous HSC to the characteristic hematopoietic lineages. It is critical to further clarify the time the earliest lineage choices appear, the cellular intermediates, and the resulting lineage trees that emerge from them (Rodriguez-Fraticelli et al. 2018). Another question is whether functional discrepancy exists among mature cells (i.e., basophils and megakaryocytes) that originated from different progenitors. More importantly, modulators underlying the cell heterogeneity and regulating maturation stages at both transcriptional and post-transcriptional levels are not fully understood.

Current studies in normal hematopoiesis would pave the way for profiling malignant hematopoiesis, which allow for the identification of molecular drivers of disease in pathogenic cell subsets.

References

Adolfsson J, Månsson R, Buza-Vidas N, Hultquist A, Liuba K, Jensen CT et al (2005) Identification of Flt3+ lympho-myeloid stem cells lacking erythro-megakaryocytic potential a revised road map for adult blood lineage commitment. Cell 121(2):295–306

Akashi K, Traver D, Miyamoto T, Weissman IL (2000) A clonogenic common myeloid progenitor that gives rise to all myeloid lineages. Nature 404(6774):193–197

Alvarez-Dominguez JR, Lodish HF (2017) Emerging mechanisms of long noncoding RNA function during normal and malignant hematopoiesis. Blood 130(18): 1965–1975

Alvarez-Dominguez JR, Hu W, Yuan B, Shi J, Park SS, Gromatzky AA et al (2014) Global discovery of erythroid long noncoding RNAs reveals novel regulators of red cell maturation. Blood 123(4):570–581

Atianand MK, Hu W, Satpathy AT, Shen Y, Ricci EP, Alvarez-Dominguez JR et al (2016) A Long noncoding RNA lincRNA-EPS acts as a transcriptional brake to restrain inflammation. Cell 165(7):1672–1685

Begley CG, Aplan PD, Davey MP, Nakahara K, Tchorz K, Kurtzberg J et al (1989a) Chromosomal translocation in a human leukemic stem-cell line disrupts the T-cell antigen receptor delta-chain diversity region and results in a previously unreported fusion transcript. Proc Natl Acad Sci U S A 86(6):2031–2035

Begley CG, Aplan PD, Denning SM, Haynes BF, Waldmann TA, Kirsch IR (1989b) The gene SCL is expressed during early hematopoiesis and encodes a differentiation-related DNA-binding motif. Proc Natl Acad Sci U S A 86(24):10128–10132

Béguelin W, Popovic R, Teater M, Jiang Y, Bunting KL, Rosen M et al (2013) EZH2 is required for germinal center formation and somatic EZH2 mutations promote lymphoid transformation. Cancer Cell 23(5):677–692

Biasco L, Pellin D, Scala S, Dionisio F, Basso-Ricci L, Leonardelli L et al (2016) In vivo tracking of human hematopoiesis reveals patterns of clonal dynamics during early and steady-state reconstitution phases. Cell Stem Cell 19(1):107–119

Björklund ÅK, Forkel M, Picelli S, Konya V, Theorell J, Friberg D et al (2016) The heterogeneity of human CD127(+) innate lymphoid cells revealed by single-cell RNA sequencing. Nat Immunol 17(4):451–460

Bock C, Beerman I, Lien WH, Smith ZD, Gu H, Boyle P et al (2012) DNA methylation dynamics during in vivo differentiation of blood and skin stem cells. Mol Cell 47(4):633–647

Bonizzato A, Gaffo E, Te Kronnie G, Bortoluzzi S (2016) CircRNAs in hematopoiesis and hematological malignancies. Blood Cancer J 6(10):e483

Bröske AM, Vockentanz L, Kharazi S, Huska MR, Mancini E, Scheller M et al (2009) DNA methylation protects hematopoietic stem cell multipotency from myeloerythroid restriction. Nat Genet 41(11): 1207–1215

Brown RC, Pattison S, van Ree J, Coghill E, Perkins A, Jane SM et al (2002) Distinct domains of erythroid Kruppel-like factor modulate chromatin remodeling and transactivation at the endogenous beta-globin gene promoter. Mol Cell Biol 22(1):161–170

Buenrostro JD, Wu B, Litzenburger UM, Ruff D, Gonzales ML, Snyder MP et al (2015) Single-cell chromatin accessibility reveals principles of regulatory variation. Nature 523(7561):486–490

Buenrostro JD, Corces MR, Lareau CA, Wu B, Schep AN, Aryee MJ et al (2018) Integrated single-cell analysis maps the continuous regulatory landscape of human hematopoietic differentiation. Cell 173(6):1535–48.e16

Cabezas-Wallscheid N, Klimmeck D, Hansson J, Lipka DB, Reyes A, Wang Q et al (2014) Identification of regulatory networks in HSCs and their immediate progeny via integrated proteome, transcriptome, and DNA methylome analysis. Cell Stem Cell 15(4): 507–522

Cabezas-Wallscheid N, Buettner F, Sommerkamp P, Klimmeck D, Ladel L, Thalheimer FB et al (2017) Vitamin A-retinoic acid signaling regulates hematopoietic stem cell dormancy. Cell 169(5): 807–23.e19

Caganova M, Carrisi C, Varano G, Mainoldi F, Zanardi F, Germain PL et al (2013) Germinal center dysregulation by histone methyltransferase EZH2 promotes lymphomagenesis. J Clin Invest 123(12):5009–5022

Cai X, Gaudet JJ, Mangan JK, Chen MJ, De Obaldia ME, Oo Z et al (2011) Runx1 loss minimally impacts long-term hematopoietic stem cells. PLoS One 6(12): e28430

Carrelha J, Meng Y, Kettyle LM, Luis TC, Norfo R, Alcolea V et al (2018) Hierarchically related lineage-restricted fates of multipotent haematopoietic stem cells. Nature 554(7690):106–111

Challen GA, Sun D, Jeong M, Luo M, Jelinek J, Berg JS et al (2011) Dnmt3a is essential for hematopoietic stem cell differentiation. Nat Genet 44(1):23–31

Challen GA, Sun D, Mayle A, Jeong M, Luo M, Rodriguez B et al (2014) Dnmt3a and Dnmt3b have overlapping and distinct functions in hematopoietic stem cells. Cell Stem Cell 15(3):350–364

Chen Q, Yang CY, Tsan JT, Xia Y, Ragab AH, Peiper SC et al (1990) Coding sequences of the tal-1 gene are disrupted by chromosome translocation in human T cell leukemia. J Exp Med 172(5):1403–1408

Chen H, Ray-Gallet D, Zhang P, Hetherington CJ, Gonzalez DA, Zhang DE et al (1995) PU.1 (Spi-1)

autoregulates its expression in myeloid cells. Oncogene 11(8):1549–1560
Chen CZ, Li L, Lodish HF, Bartel DP (2004) MicroRNAs modulate hematopoietic lineage differentiation. Science 303(5654):83–86
Chen MT, Lin HS, Shen C, Ma YN, Wang F, Zhao HL et al (2015) PU.1-regulated long noncoding RNA lnc-MC controls human monocyte/macrophage differentiation through interaction with MicroRNA 199a-5p. Mol Cell Biol 35(18):3212–3224
Chen ZH, Wang WT, Huang W, Fang K, Sun YM, Liu SR et al (2017) The lncRNA HOTAIRM1 regulates the degradation of PML-RARA oncoprotein and myeloid cell differentiation by enhancing the autophagy pathway. Cell Death Differ 24(2):212–224
Christensen JL, Weissman IL (2001) Flk-2 is a marker in hematopoietic stem cell differentiation: a simple method to isolate long-term stem cells. Proc Natl Acad Sci U S A 98(25):14541–14546
Corces MR, Buenrostro JD, Wu B, Greenside PG, Chan SM, Koenig JL et al (2016) Lineage-specific and single-cell chromatin accessibility charts human hematopoiesis and leukemia evolution. Nat Genet 48(10): 1193–1203
Crispino JD, Lodish MB, MacKay JP, Orkin SH (1999) Use of altered specificity mutants to probe a specific protein-protein interaction in differentiation: the GATA-1:FOG complex. Mol Cell 3(2):219–228
Cullen SM, Mayle A, Rossi L, Goodell MA (2014) Hematopoietic stem cell development: an epigenetic journey. Curr Top Dev Biol 107:39–75
Dahlin JS, Hamey FK, Pijuan-Sala B, Shepherd M, Lau WWY, Nestorowa S et al (2018) A single-cell hematopoietic landscape resolves 8 lineage trajectories and defects in kit mutant mice. Blood 131(21):e1–e11
Dakic A, Metcalf D, Di Rago L, Mifsud S, Wu L, Nutt SL (2005) PU.1 regulates the commitment of adult hematopoietic progenitors and restricts granulopoiesis. J Exp Med 201(9):1487–1502
Dakic A, Wu L, Nutt SL (2007) Is PU.1 a dosage-sensitive regulator of haemopoietic lineage commitment and leukaemogenesis? Trends Immunol 28(3):108–114
de Bruijn M, Dzierzak E (2017) Runx transcription factors in the development and function of the definitive hematopoietic system. Blood 129(15):2061–2069
Del Real MM, Rothenberg EV (2013) Architecture of a lymphomyeloid developmental switch controlled by PU.1, Notch and Gata3. Development 140(6): 1207–1219
Delás MJ, Jackson BT, Kovacevic T, Vangelisti S, Munera Maravilla E, Wild SA et al (2019) lncRNA Spehd regulates hematopoietic stem and progenitor cells and is required for multilineage differentiation. Cell Rep 27(3):719–29.e6
Demers C, Chaturvedi CP, Ranish JA, Juban G, Lai P, Morle F et al (2007) Activator-mediated recruitment of the MLL2 methyltransferase complex to the beta-globin locus. Mol Cell 27(4):573–584
Dharampuriya PR, Scapin G, Wong C, John Wagner K, Cillis JL, Shah DI (2017) Tracking the origin, development, and differentiation of hematopoietic stem cells. Curr Opin Cell Biol 49:108–115
Dore LC, Amigo JD, Dos Santos CO, Zhang Z, Gai X, Tobias JW et al (2008) A GATA-1-regulated microRNA locus essential for erythropoiesis. Proc Natl Acad Sci U S A 105(9):3333–3338
Doulatov S, Notta F, Eppert K, Nguyen LT, Ohashi PS, Dick JE (2010) Revised map of the human progenitor hierarchy shows the origin of macrophages and dendritic cells in early lymphoid development. Nat Immunol 11(7):585–593
Drissen R, Buza-Vidas N, Woll P, Thongjuea S, Gambardella A, Giustacchini A et al (2016) Distinct myeloid progenitor-differentiation pathways identified through single-cell RNA sequencing. Nat Immunol 17(6):666–676
Dykstra B, Kent D, Bowie M, McCaffrey L, Hamilton M, Lyons K et al (2007) Long-term propagation of distinct hematopoietic differentiation programs in vivo. Cell Stem Cell 1(2):218–229
Farlik M, Halbritter F, Müller F, Choudry FA, Ebert P, Klughammer J et al (2016) DNA methylation dynamics of human hematopoietic stem cell differentiation. Cell Stem Cell 19(6):808–822
Ferreira R, Ohneda K, Yamamoto M, Philipsen S (2005) GATA1 function, a paradigm for transcription factors in hematopoiesis. Mol Cell Biol 25(4):1215–1227
Fukao T, Fukuda Y, Kiga K, Sharif J, Hino K, Enomoto Y et al (2007) An evolutionarily conserved mechanism for microRNA-223 expression revealed by microRNA gene profiling. Cell 129(3):617–631
Gottgens B (2015) Regulatory network control of blood stem cells. Blood 125(17):2614–2620
Granja JM, Klemm S, McGinnis LM, Kathiria AS, Mezger A, Corces MR et al (2019) Single-cell multiomic analysis identifies regulatory programs in mixed-phenotype acute leukemia. Nat Biotechnol 37(12):1458–1465
Green AR, Lints T, Visvader J, Harvey R, Begley CG (1992) SCL is coexpressed with GATA-1 in hemopoietic cells but is also expressed in developing brain. Oncogene 7(4):653–660
Grover A, Sanjuan-Pla A, Thongjuea S, Carrelha J, Giustacchini A, Gambardella A et al (2016) Single-cell RNA sequencing reveals molecular and functional platelet bias of aged haematopoietic stem cells. Nat Commun 7:11075
Growney JD, Shigematsu H, Li Z, Lee BH, Adelsperger J, Rowan R et al (2005) Loss of Runx1 perturbs adult hematopoiesis and is associated with a myeloproliferative phenotype. Blood 106(2):494–504
Gutierrez L, Tsukamoto S, Suzuki M, Yamamoto-Mukai-H, Yamamoto M, Philipsen S et al (2008) Ablation of Gata1 in adult mice results in aplastic crisis, revealing its essential role in steady-state and stress erythropoiesis. Blood 111(8):4375–4385

Haas S, Trumpp A, Milsom MD (2018) Causes and consequences of hematopoietic stem cell heterogeneity. Cell Stem Cell 22(5):627–638

Heuston EF, Keller CA, Lichtenberg J, Giardine B, Anderson SM, Hardison RC et al (2018) Establishment of regulatory elements during erythro-megakaryopoiesis identifies hematopoietic lineage-commitment points. Epigenetics Chromatin 11(1):22

Hidalgo I, Herrera-Merchan A, Ligos JM, Carramolino L, Nuñez J, Martinez F et al (2012) Ezh1 is required for hematopoietic stem cell maintenance and prevents senescence-like cell cycle arrest. Cell Stem Cell 11(5):649–662

Ho TT, Warr MR, Adelman ER, Lansinger OM, Flach J, Verovskaya EV et al (2017) Autophagy maintains the metabolism and function of young and old stem cells. Nature 543(7644):205–210

Hu D, Shilatifard A (2016) Epigenetics of hematopoiesis and hematological malignancies. Genes Dev 30(18): 2021–2041

Hu W, Yuan B, Flygare J, Lodish HF (2011) Long non-coding RNA-mediated anti-apoptotic activity in murine erythroid terminal differentiation. Genes Dev 25(24):2573–2578

Hu G, Cui K, Fang D, Hirose S, Wang X, Wangsa D et al (2018) Transformation of accessible chromatin and 3D Nucleome underlies lineage commitment of early T cells. Immunity 48(2):227–42.e8

Ichikawa M, Asai T, Saito T, Seo S, Yamazaki I, Yamagata T et al (2004) AML-1 is required for megakaryocytic maturation and lymphocytic differentiation, but not for maintenance of hematopoietic stem cells in adult hematopoiesis. Nat Med 10(3):299–304

Isoda T, Moore AJ, He Z, Chandra V, Aida M, Denholtz M et al (2017) Non-coding transcription instructs chromatin folding and compartmentalization to dictate enhancer-promoter communication and T cell fate. Cell 171(1):103–19.e18

Iwasaki H, Somoza C, Shigematsu H, Duprez EA, Iwasaki-Arai J, Mizuno S et al (2005) Distinctive and indispensable roles of PU.1 in maintenance of hematopoietic stem cells and their differentiation. Blood 106(5):1590–1600

Izzo F, Lee SC, Poran A, Chaligne R, Gaiti F, Gross B et al (2020) DNA methylation disruption reshapes the hematopoietic differentiation landscape. Nat Genet 52(4):378–387

Jaitin DA, Kenigsberg E, Keren-Shaul H, Elefant N, Paul F, Zaretsky I et al (2014) Massively parallel single-cell RNA-seq for marker-free decomposition of tissues into cell types. Science 343(6172):776–779

Ji H, Ehrlich LI, Seita J, Murakami P, Doi A, Lindau P et al (2010) Comprehensive methylome map of lineage commitment from haematopoietic progenitors. Nature 467(7313):338–342

Johanson TM, Lun ATL, Coughlan HD, Tan T, Smyth GK, Nutt SL et al (2018) Transcription-factor-mediated supervision of global genome architecture maintains B cell identity. Nat Immunol 19(11): 1257–1264

Johnnidis JB, Harris MH, Wheeler RT, Stehling-Sun S, Lam MH, Kirak O et al (2008) Regulation of progenitor cell proliferation and granulocyte function by microRNA-223. Nature 451(7182):1125–1129

Ju YS, Martincorena I, Gerstung M, Petljak M, Alexandrov LB, Rahbari R et al (2017) Somatic mutations reveal asymmetric cellular dynamics in the early human embryo. Nature 543(7647):714–718

Kallianpur AR, Jordan JE, Brandt SJ (1994) The SCL/TAL-1 gene is expressed in progenitors of both the hematopoietic and vascular systems during embryogenesis. Blood 83(5):1200–1208

Kato H, Igarashi K (2019) To be red or white: lineage commitment and maintenance of the hematopoietic system by the "inner myeloid". Haematologica 104(10):1919–1927

Katsumura KR, Bresnick EH (2017) The GATA factor revolution in hematology. Blood 129(15):2092–2102

Kaushansky K (2006) Lineage-specific hematopoietic growth factors. N Engl J Med 354(19):2034–2045

Kelsey G, Stegle O, Reik W (2017) Single-cell epigenomics: recording the past and predicting the future. Science 358(6359):69–75

Kerenyi MA, Orkin SH (2010) Networking erythropoiesis. J Exp Med 207(12):2537–2541

Kester L, van Oudenaarden A (2018) Single-cell transcriptomics meets lineage tracing. Cell Stem Cell 23(2):166–179

Kiel MJ, Yilmaz OH, Iwashita T, Yilmaz OH, Terhorst C, Morrison SJ (2005) SLAM family receptors distinguish hematopoietic stem and progenitor cells and reveal endothelial niches for stem cells. Cell 121(7): 1109–1121

Kim SI, Bresnick EH (2007) Transcriptional control of erythropoiesis: emerging mechanisms and principles. Oncogene 26(47):6777–6794

Kim HG, de Guzman CG, Swindle CS, Cotta CV, Gartland L, Scott EW et al (2004) The ETS family transcription factor PU.1 is necessary for the maintenance of fetal liver hematopoietic stem cells. Blood 104(13):3894–3900

Klemsz MJ, McKercher SR, Celada A, Van Beveren C, Maki RA (1990) The macrophage and B cell-specific transcription factor PU.1 is related to the ets oncogene. Cell 61(1):113–124

Kloetgen A, Thandapani P, Tsirigos A, Aifantis I (2019) 3D chromosomal landscapes in hematopoiesis and immunity. Trends Immunol 40(9):809–824

Kodandapani R, Pio F, Ni CZ, Piccialli G, Klemsz M, McKercher S et al (1996) A new pattern for helix-turn-helix recognition revealed by the PU.1 ETS-domain-DNA complex. Nature 380(6573):456–460

Kondo M, Weissman IL, Akashi K (1997) Identification of clonogenic common lymphoid progenitors in mouse bone marrow. Cell 91(5):661–672

Kowalczyk MS, Tirosh I, Heckl D, Rao TN, Dixit A, Haas BJ et al (2015) Single-cell RNA-seq reveals changes in cell cycle and differentiation programs upon aging of hematopoietic stem cells. Genome Res 25(12): 1860–1872

Lahlil R, Lecuyer E, Herblot S, Hoang T (2004) SCL assembles a multifactorial complex that determines glycophorin A expression. Mol Cell Biol 24(4): 1439–1452

Lara-Astiaso D, Weiner A, Lorenzo-Vivas E, Zaretsky I, Jaitin DA, David E et al (2014) Immunogenetics. Chromatin state dynamics during blood formation. Science 345(6199):943–949

Laurenti E, Doulatov S, Zandi S, Plumb I, Chen J, April C et al (2013) The transcriptional architecture of early human hematopoiesis identifies multilevel control of lymphoid commitment. Nat Immunol 14(7):756–763

Lee SC, Miller S, Hyland C, Kauppi M, Lebois M, Di Rago L et al (2015) Polycomb repressive complex 2 component Suz12 is required for hematopoietic stem cell function and lymphopoiesis. Blood 126(2): 167–175

Lee-Six H, Øbro NF, Shepherd MS, Grossmann S, Dawson K, Belmonte M et al (2018) Population dynamics of normal human blood inferred from somatic mutations. Nature 561(7724):473–478

Li Z, Cai X, Cai CL, Wang J, Zhang W, Petersen BE et al (2011) Deletion of Tet2 in mice leads to dysregulated hematopoietic stem cells and subsequent development of myeloid malignancies. Blood 118(17):4509–4518

Liggett LA, Sankaran VG (2020) Unraveling hematopoiesis through the lens of genomics. Cell 182(6): 1384–1400

Liu J, Cao L, Chen J, Song S, Lee IH, Quijano C et al (2009) Bmi1 regulates mitochondrial function and the DNA damage response pathway. Nature 459(7245): 387–392

Lodato MA, Woodworth MB, Lee S, Evrony GD, Mehta BK, Karger A et al (2015) Somatic mutation in single human neurons tracks developmental and transcriptional history. Science 350(6256):94–98

Lorsbach RB, Moore J, Ang SO, Sun W, Lenny N, Downing JR (2004) Role of RUNX1 in adult hematopoiesis: analysis of RUNX1-IRES-GFP knock-in mice reveals differential lineage expression. Blood 103(7): 2522–2529

Ludwig LS, Lareau CA, Ulirsch JC, Christian E, Muus C, Li LH et al (2019) Lineage tracing in humans enabled by mitochondrial mutations and single-cell genomics. Cell 176(6):1325–39.e22

Luo M, Jeong M, Sun D, Park HJ, Rodriguez BA, Xia Z et al (2015) Long non-coding RNAs control hematopoietic stem cell function. Cell Stem Cell 16(4):426–438

Macosko EZ, Basu A, Satija R, Nemesh J, Shekhar K, Goldman M et al (2015) Highly parallel genome-wide expression profiling of individual cells using nanoliter droplets. Cell 161(5):1202–1214

Martin DI, Orkin SH (1990) Transcriptional activation and DNA binding by the erythroid factor GF-1/NF-E1/Eryf 1. Genes Dev 4(11):1886–1898

Mikkola HK, Orkin SH (2006) The journey of developing hematopoietic stem cells. Development 133(19): 3733–3744

Moignard V, Macaulay IC, Swiers G, Buettner F, Schutte J, Calero-Nieto FJ et al (2013) Characterization of transcriptional networks in blood stem and progenitor cells using high-throughput single-cell gene expression analysis. Nat Cell Biol 15(4):363–372

Morrison SJ, Scadden DT (2014) The bone marrow niche for haematopoietic stem cells. Nature 505(7483): 327–334

Morrison SJ, Weissman IL (1994) The long-term repopulating subset of hematopoietic stem cells is deterministic and isolatable by phenotype. Immunity 1(8):661–673

Mouthon MA, Bernard O, Mitjavila MT, Romeo PH, Vainchenker W, Mathieu-Mahul D (1993) Expression of tal-1 and GATA-binding proteins during human hematopoiesis. Blood 81(3):647–655

Muckenthaler MU, Rivella S, Hentze MW, Galy B (2017) A red carpet for iron metabolism. Cell 168(3):344–361

Müller-Sieburg CE, Cho RH, Thoman M, Adkins B, Sieburg HB (2002) Deterministic regulation of hematopoietic stem cell self-renewal and differentiation. Blood 100(4):1302–1309

Muller-Sieburg CE, Cho RH, Karlsson L, Huang JF, Sieburg HB (2004) Myeloid-biased hematopoietic stem cells have extensive self-renewal capacity but generate diminished lymphoid progeny with impaired IL-7 responsiveness. Blood 103(11):4111–4118

Nestorowa S, Hamey FK, Pijuan Sala B, Diamanti E, Shepherd M, Laurenti E et al (2016) A single-cell resolution map of mouse hematopoietic stem and progenitor cell differentiation. Blood 128(8):e20–e31

Nicolet BP, Engels S, Aglialoro F, van den Akker E, von Lindern M, Wolkers MC (2018) Circular RNA expression in human hematopoietic cells is widespread and cell-type specific. Nucleic Acids Res 46(16): 8168–8180

North T, Gu TL, Stacy T, Wang Q, Howard L, Binder M et al (1999) Cbfa2 is required for the formation of intra-aortic hematopoietic clusters. Development 126(11): 2563–2575

North TE, Stacy T, Matheny CJ, Speck NA, de Bruijn MF (2004) Runx1 is expressed in adult mouse hematopoietic stem cells and differentiating myeloid and lymphoid cells, but not in maturing erythroid cells. Stem Cells 22(2):158–168

Notta F, Zandi S, Takayama N, Dobson S, Gan OI, Wilson G et al (2016) Distinct routes of lineage development reshape the human blood hierarchy across ontogeny. Science 351(6269):aab2116

Oguro H, Yuan J, Ichikawa H, Ikawa T, Yamazaki S, Kawamoto H et al (2010) Poised lineage specification in multipotential hematopoietic stem and progenitor cells by the polycomb protein Bmi1. Cell Stem Cell 6(3):279–286

Orkin SH (2000) Diversification of haematopoietic stem cells to specific lineages. Nat Rev Genet 1(1):57–64

Osawa M, Hanada K, Hamada H, Nakauchi H (1996) Long-term lymphohematopoietic reconstitution by a single CD34-low/negative hematopoietic stem cell. Science 273(5272):242–245

Panda AC, Gorospe M (2018) Detection and analysis of circular RNAs by RT-PCR. Bio Protoc 8(6):e2775

Papalexi E, Satija R (2018) Single-cell RNA sequencing to explore immune cell heterogeneity. Nat Rev Immunol 18(1):35–45

Paralkar VR, Mishra T, Luan J, Yao Y, Kossenkov AV, Anderson SM et al (2014) Lineage and species-specific long noncoding RNAs during erythro-megakaryocytic development. Blood 123(12):1927–1937

Paul F, Arkin Y, Giladi A, Jaitin DA, Kenigsberg E, Keren-Shaul H et al (2015) Transcriptional heterogeneity and lineage commitment in myeloid progenitors. Cell 163(7):1663–1677

Pei W, Shang F, Wang X, Fanti AK, Greco A, Busch K et al (2020) Resolving fates and single-cell transcriptomes of hematopoietic stem cell clones by PolyloxExpress barcoding. Cell Stem Cell 27(3): 383–95.e8

Pevny L, Simon MC, Robertson E, Klein WH, Tsai SF, D'Agati V et al (1991) Erythroid differentiation in chimaeric mice blocked by a targeted mutation in the gene for transcription factor GATA-1. Nature 349(6306):257–260

Pevny L, Lin CS, D'Agati V, Simon MC, Orkin SH, Costantini F (1995) Development of hematopoietic cells lacking transcription factor GATA-1. Development 121(1):163–172

Picelli S, Faridani OR, Björklund AK, Winberg G, Sagasser S, Sandberg R (2014) Full-length RNA-seq from single cells using Smart-seq2. Nat Protoc 9(1): 171–181

Pietras EM, Reynaud D, Kang YA, Carlin D, Calero-Nieto FJ, Leavitt AD et al (2015) Functionally distinct subsets of lineage-biased multipotent progenitors control blood production in normal and regenerative conditions. Cell Stem Cell 17(1):35–46

Porcher C, Swat W, Rockwell K, Fujiwara Y, Alt FW, Orkin SH (1996) The T cell leukemia oncoprotein SCL/tal-1 is essential for development of all hematopoietic lineages. Cell 86(1):47–57

Porcher C, Chagraoui H, Kristiansen MS (2017) SCL/TAL1: a multifaceted regulator from blood development to disease. Blood 129(15):2051–2060

Preußer C, Hung LH, Schneider T, Schreiner S, Hardt M, Moebus A et al (2018) Selective release of circRNAs in platelet-derived extracellular vesicles. J Extracell Vesicles 7(1):1424473

Pronier E, Almire C, Mokrani H, Vasanthakumar A, Simon A, da Costa Reis Monte Mor B et al (2011) Inhibition of TET2-mediated conversion of 5-methylcytosine to 5-hydroxymethylcytosine disturbs erythroid and granulomonocytic differentiation of human hematopoietic progenitors. Blood 118(9): 2551–2555

Psaila B, Barkas N, Iskander D, Roy A, Anderson S, Ashley N et al (2016) Single-cell profiling of human megakaryocyte-erythroid progenitors identifies distinct megakaryocyte and erythroid differentiation pathways. Genome Biol 17:83

Pucella JN, Upadhaya S, Reizis B (2020) The source and dynamics of adult hematopoiesis: insights from lineage tracing. Annu Rev Cell Dev Biol 36:529–550

Robb L, Lyons I, Li R, Hartley L, Kontgen F, Harvey RP et al (1995) Absence of yolk sac hematopoiesis from mice with a targeted disruption of the scl gene. Proc Natl Acad Sci U S A 92(15):7075–7079

Rodriguez-Fraticelli AE, Wolock SL, Weinreb CS, Panero R, Patel SH, Jankovic M et al (2018) Clonal analysis of lineage fate in native haematopoiesis. Nature 553(7687):212–216

Rönnerblad M, Andersson R, Olofsson T, Douagi I, Karimi M, Lehmann S et al (2014) Analysis of the DNA methylome and transcriptome in granulopoiesis reveals timed changes and dynamic enhancer methylation. Blood 123(17):e79–e89

Rothenberg EV, Ungerback J, Champhekar A (2016) Forging T-lymphocyte identity: intersecting networks of transcriptional control. Adv Immunol 129:109–174

Sanjuan-Pla A, Macaulay IC, Jensen CT, Woll PS, Luis TC, Mead A et al (2013) Platelet-biased stem cells reside at the apex of the haematopoietic stem-cell hierarchy. Nature 502(7470):232–236

Scott EW, Simon MC, Anastasi J, Singh H (1994) Requirement of transcription factor PU.1 in the development of multiple hematopoietic lineages. Science 265(5178):1573–1577

Shema E, Bernstein BE, Buenrostro JD (2019) Single-cell and single-molecule epigenomics to uncover genome regulation at unprecedented resolution. Nat Genet 51(1):19–25

Shivdasani RA, Mayer EL, Orkin SH (1995) Absence of blood formation in mice lacking the T-cell leukaemia oncoprotein tal-1/SCL. Nature 373(6513):432–434

Signer RA, Magee JA, Salic A, Morrison SJ (2014) Haematopoietic stem cells require a highly regulated protein synthesis rate. Nature 509(7498):49–54

Simsek T, Kocabas F, Zheng J, Deberardinis RJ, Mahmoud AI, Olson EN et al (2010) The distinct metabolic profile of hematopoietic stem cells reflects their location in a hypoxic niche. Cell Stem Cell 7(3): 380–390

Summers C, Rankin SM, Condliffe AM, Singh N, Peters AM, Chilvers ER (2010) Neutrophil kinetics in health and disease. Trends Immunol 31(8):318–324

Tang F, Barbacioru C, Nordman E, Li B, Xu N, Bashkirov VI et al (2010) RNA-Seq analysis to capture the transcriptome landscape of a single cell. Nat Protoc 5(3):516–535

Tremblay M, Sanchez-Ferras O, Bouchard M (2018) GATA transcription factors in development and disease. Development 145(20):dev164384

Tsang JC, Yu Y, Burke S, Buettner F, Wang C, Kolodziejczyk AA et al (2015) Single-cell transcriptomic reconstruction reveals cell cycle and multi-lineage differentiation defects in Bcl11a-deficient hematopoietic stem cells. Genome Biol 16: 178

Tumes DJ, Onodera A, Suzuki A, Shinoda K, Endo Y, Iwamura C et al (2013) The polycomb protein Ezh2 regulates differentiation and plasticity of CD4(+) T helper type 1 and type 2 cells. Immunity 39(5): 819–832

Tusi BK, Wolock SL, Weinreb C, Hwang Y, Hidalgo D, Zilionis R et al (2018) Population snapshots predict early haematopoietic and erythroid hierarchies. Nature 555(7694):54–60

Upadhaya S, Sawai CM, Papalexi E, Rashidfarrokhi A, Jang G, Chattopadhyay P et al (2018) Kinetics of adult hematopoietic stem cell differentiation in vivo. J Exp Med 215(11):2815–2832

van Wijnen AJ, Stein GS, Gergen JP, Groner Y, Hiebert SW, Ito Y et al (2004) Nomenclature for Runt-related (RUNX) proteins. Oncogene 23(24):4209–4210

Velten L, Haas SF, Raffel S, Blaszkiewicz S, Islam S, Hennig BP et al (2017) Human haematopoietic stem cell lineage commitment is a continuous process. Nat Cell Biol 19(4):271–281

Verger A, Duterque-Coquillaud M (2002) When Ets transcription factors meet their partners. BioEssays 24(4): 362–370

Villani AC, Satija R, Reynolds G, Sarkizova S, Shekhar K, Fletcher J et al (2017) Single-cell RNA-seq reveals new types of human blood dendritic cells, monocytes, and progenitors. Science 356(6335):eaah4573

Wada T, Kikuchi J, Nishimura N, Shimizu R, Kitamura T, Furukawa Y (2009) Expression levels of histone deacetylases determine the cell fate of hematopoietic progenitors. J Biol Chem 284(44):30673–30683

Wadman IA, Osada H, Grutz GG, Agulnick AD, Westphal H, Forster A et al (1997) The LIM-only protein Lmo2 is a bridging molecule assembling an erythroid, DNA-binding complex which includes the TAL1, E47, GATA-1 and Ldb1/NLI proteins. EMBO J 16(11):3145–3157

Wagner LA, Christensen CJ, Dunn DM, Spangrude GJ, Georgelas A, Kelley L et al (2007) EGO, a novel, noncoding RNA gene, regulates eosinophil granule protein transcript expression. Blood 109(12): 5191–5198

Wang XS, Zhang JW (2008) The microRNAs involved in human myeloid differentiation and myelogenous/myeloblastic leukemia. J Cell Mol Med 12(5a):1445–1455

Wang P, Xue Y, Han Y, Lin L, Wu C, Xu S et al (2014) The STAT3-binding long noncoding RNA lnc-DC controls human dendritic cell differentiation. Science 344(6181):310–313

Watcham S, Kucinski I, Gottgens B (2019) New insights into hematopoietic differentiation landscapes from single-cell RNA sequencing. Blood 133(13): 1415–1426

Weinreb C, Rodriguez-Fraticelli A, Camargo FD, Klein AM (2020) Lineage tracing on transcriptional landscapes links state to fate during differentiation. Science 367(6479):eaaw3381

Weiss MJ, Orkin SH (1995) Transcription factor GATA-1 permits survival and maturation of erythroid precursors by preventing apoptosis. Proc Natl Acad Sci U S A 92(21):9623–9627

Wilson A, Laurenti E, Oser G, van der Wath RC, Blanco-Bose W, Jaworski M et al (2008) Hematopoietic stem cells reversibly switch from dormancy to self-renewal during homeostasis and repair. Cell 135(6):1118–1129

Wilson NK, Foster SD, Wang X, Knezevic K, Schutte J, Kaimakis P et al (2010) Combinatorial transcriptional control in blood stem/progenitor cells: genome-wide analysis of ten major transcriptional regulators. Cell Stem Cell 7(4):532–544

Wilting RH, Yanover E, Heideman MR, Jacobs H, Horner J, van der Torre J et al (2010) Overlapping functions of Hdac1 and Hdac2 in cell cycle regulation and haematopoiesis. EMBO J 29(15):2586–2597

Wolf FA, Hamey FK, Plass M, Solana J, Dahlin JS, Göttgens B et al (2019) PAGA: graph abstraction reconciles clustering with trajectory inference through a topology preserving map of single cells. Genome Biol 20(1):59

Xie H, Xu J, Hsu JH, Nguyen M, Fujiwara Y, Peng C et al (2014) Polycomb repressive complex 2 regulates normal hematopoietic stem cell function in a developmental-stage-specific manner. Cell Stem Cell 14(1):68–80

Yamamoto R, Morita Y, Ooehara J, Hamanaka S, Onodera M, Rudolph KL et al (2013) Clonal analysis unveils self-renewing lineage-restricted progenitors generated directly from hematopoietic stem cells. Cell 154(5):1112–1126

Yang X, Chen D, Long H, Zhu B (2020) The mechanisms of pathological extramedullary hematopoiesis in diseases. Cell Mol Life Sci 77(14):2723–2738

Yu VWC, Yusuf RZ, Oki T, Wu J, Saez B, Wang X et al (2016) Epigenetic memory underlies cell-autonomous heterogeneous behavior of hematopoietic stem cells. Cell 167(5):1310–22.e17

Zeuner A, Eramo A, Testa U, Felli N, Pelosi E, Mariani G et al (2003) Control of erythroid cell production via caspase-mediated cleavage of transcription factor SCL/Tal-1. Cell Death Differ 10(8):905–913

Zhang X, Lian Z, Padden C, Gerstein MB, Rozowsky J, Snyder M et al (2009) A myelopoiesis-associated regulatory intergenic noncoding RNA transcript within the human HOXA cluster. Blood 113(11):2526–2534

Zhang C, Xu Z, Yang S, Sun G, Jia L, Zheng Z et al (2020) tagHi-C reveals 3D chromatin architecture dynamics during mouse hematopoiesis. Cell Rep 32(13):108206

Zhu Y, Wang D, Wang F, Li T, Dong L, Liu H et al (2013) A comprehensive analysis of GATA-1-regulated miRNAs reveals miR-23a to be a positive modulator of erythropoiesis. Nucleic Acids Res 41(7):4129–4143

Ziegenhain C, Vieth B, Parekh S, Reinius B, Guillaumet-Adkins A, Smets M et al (2017) Comparative analysis of single-cell RNA sequencing methods. Mol Cell 65(4):631–43.e4

Gene Editing in Hematopoietic Stem Cells

11

Jiaoyang Liao and Yuxuan Wu

Abstract

Hematopoietic stem cells (HSCs) can be isolated and collected from the body, genetically modified, and expanded ex vivo. The invention of innovative and powerful gene editing tools has provided researchers with great convenience in genetically modifying a wide range of cells, including hematopoietic stem and progenitor cells (HSPCs). In addition to being used to modify genes to study the functional role that specific genes play in the hematopoietic system, the application of gene editing platforms in HSCs is largely focused on the development of cell-based gene editing therapies to treat diseases such as immune deficiency disorders and inherited blood disorders. Here, we review the application of gene editing tools in HSPCs. In particular, we provide a broad overview of the development of gene editing tools, multiple strategies for the application of gene editing tools in HSPCs, and exciting clinical advances in HSPC gene editing therapies. We also outline the various challenges integral to clinical translation of HSPC gene editing and provide the possible corresponding solutions.

Keywords

HSC · Gene editing · Clinical trials · Challenges · Perspectives

11.1 Hematopoietic Stem Cells: The Application Frontier of Gene Editing Therapy

Many tissues in the adult body retain a small number of cells responsible for maintaining and replenishing the type and number of tissue cells, and they are called adult stem cells (Post and Clevers 2019). Blood cells perform vital functions such as the transport of substances and play a pivotal role in immune response; a huge number of cells are lost owing to wear and tear, damage, or disease resistance (Bryder et al. 2006). Hematopoietic stem cells (HSCs) in the adult are a class of adult stem cells present in the bone marrow and the peripheral blood that can repopulate all types of blood cells. HSCs possess the abilities of self renewal and multi-potency. In addition to being able to multiply through a process known as self-renewal to maintain the HSC pool size properly, HSCs can also differentiate into all types of functional blood cells through a process called multi-potency to

J. Liao · Y. Wu (✉)
Shanghai Frontiers Science Center of Genome Editing and Cell Therapy, Shanghai Key Laboratory of Regulatory Biology, Institute of Biomedical Sciences and School of Life Sciences, East China Normal University, Shanghai, China
e-mail: yxwu@bio.ecnu.edu.cn

M. Zhao, P. Qian (eds.), *Hematopoietic Stem Cells*, Advances in Experimental Medicine and Biology 1442, https://doi.org/10.1007/978-981-99-7471-9_11

maintain hematopoietic homeostasis (Seita and Weissman 2010).

HSC research achieved initial success in the 1950s when scientists discovered that intravenously injected bone marrow cells could rescue irradiated mice from lethality by re-establishing blood cell production (Jacobson et al. 1951; Ford et al. 1956). The researchers then conducted experiments to try to identify the specific cell types responsible for reconstituting blood cells. They detected donor-derived multi-lineage colonies on the surface of the spleen 1 to 2 weeks after transplantation, and these colonies could be engrafted sequentially (Till and McCulloch 1961; Becker et al. 1963). Although the identification of pluripotent cell subsets in the bone marrow still takes considerable effort, a series of pioneering experiments such as bone marrow transplantation have provided us with proof of concept for self-renewal and multipotency (Wu et al. 1967, 1968; Siminovitch et al. 1963; Fowler et al. 1967; Till et al. 1964). Advances in analytical techniques and the development of multiple experimental systems have collectively helped us to accomplish the discrimination and functional assessment of cell subpopulations at the clonal level, including cell lineage tracing (Lawson et al. 1986), fluorescence-activated cell sorting (Muller-Sieburg et al. 1986), single-cell omics (Kumar et al. 2017), monoclonal antibody labeling (Spangrude et al. 1988), and the establishment of in vivo and in vitro systems for assessing self-renewal and lineage potential analysis (Osawa et al. 1996; Smith et al. 1991). These breakthroughs enable researchers to understand how to harness the body's regenerative abilities to heal or treat tissues damaged by disease or trauma (Mao and Mooney 2015). HSCs were the first tissue-specific adult stem cells to be isolated (Spangrude et al. 1988). HSC transplantation is widely used in the treatment of blood disorders such as autoimmune diseases and leukemia and even many types of cancers (Duarte et al. 2019). Moreover, HSC is the only type of stem cell used in conventional clinical treatment (Barriga et al. 2012). Additionally, extensive research has shed light on the characterization of the basic molecular and cellular characteristics of HSCs. This has further provided novel insights into the field of regenerative medicine that is closely related to our human health (Muller-Sieburg et al. 2012).

In summary, hematopoietic stem cells maintain the self-renewal and long-term multi-lineage repopulation capacities, thus providing avenues for cell replacement-based therapeutic modalities. Gene editing tools provide researchers with genetic modification strategies that enable unraveling of the causative factors of many inherited blood disorders and immune dysfunction diseases. The combination of the aforementioned properties has propelled hematopoietic stem cell-based gene editing therapies at the forefront of cellular and genetic drug development. The ultimate goal is developing innovative strategies for the treatment of genetic, malignant, and degenerative diseases using gene-edited hematopoietic stem and progenitor cells (HSPCs) as a cellular platform to address currently unmet clinical needs.

11.2 Gene Editing in HSC

Allogeneic hematopoietic stem cell transplantation (HSCT) is a curative approach using healthy allogeneic HSCs as donors to repopulate and replace damaged blood cells in patients to treat related diseases. However, allogeneic HSCT does not represent an optimal treatment option especially when considering its dependence on the availability of a well-matching sibling or related donor with highly compatible human leukocyte antigen (HLA), the chances of which are below 30%. In addition, unexpected complications or even death due to immune rejection is associated with allogeneic transplantation (Copelan 2006; Angelucci et al. 2014). Hence, genetic modification of autologous HSCs to eliminate the underlying factors of disease represents a new alternative curative option. This includes myeloablative autologous HSCT after ex vivo genetic manipulation and in vivo gene therapy of HSCs.

Due to the remarkable progress made in research in the application of HSCs, gene therapy based on it has been carried out for a long time, and tremendous strides have been made in this field (Cartier et al. 2009; Aiuti et al. 2013; Biffi et al. 2013; Eichler et al. 2017; Gentner et al. 2021; Marktel et al. 2019). However, clinical researchers have also uncovered the limitations of gene addition therapy based on viral vectors (Morgan et al. 2017), such as the possible risks posed by random insertions, gradual silencing, and insertion copy number variation (Morgan et al. 2021). In 2019, Zynteglo, a LentiGlobin-based lentiviral gene therapy developed by bluebird bio, was approved for marketing in the European Union for the treatment of transfusion-dependent β-thalassemia (TDT) patients aged 12 years and older with non-β^0/β^0 genotype. This is the world's first gene therapy approved for the treatment of thalassemia. The therapy first extracts HSCs from the patient's bone marrow and then adds a functional copy of a modified form of the β-globin gene (the βA-T87Q-globin gene) to the patient's own hematopoietic stem cells via lentivirus, thereby restoring hemoglobin production (Thompson et al. 2018). On February 17, 2021, bluebird bio announced the suspension of two of its clinical trials for the treatment of sickle cell disease after a patient who underwent clinical trials of its LentiGlobin gene therapy 5 years earlier was diagnosed with acute myeloid cell leukemia (AML) (Goyal et al. 2022) and one patient developed myelodysplastic syndrome (MDS) (Hsieh et al. 2020). After nearly a month of investigation, bluebird bio resumed clinical trials after announcing that its LentiGlobin gene therapy is "very unlikely" to cause AML in treated sickle cell disease patients, but cannot eliminate people's concerns about the safety of viral vector-mediated gene therapy.

Compared with gene addition therapy, gene editing can provide precise modification of specific sites such as nucleotide(s) deletion, insertion, and substitution (Anzalone et al. 2020), allowing for individualized gene modification for the treatment of various disorders. Based on the platform of collection and in vitro manipulation of HSPCs, genetic modification with powerful gene editing tools encourages the development of promising gene editing therapies to target a variety of immune and inherited blood disorders, such as Wiskott-Aldrich syndrome (WAS), severe combined immunodeficiencies (ADA-SCID, X-SCID), β-hemoglobinopathies (β thalassemia and sickle cell disease (SCD)), etc. (Koniali et al. 2021).

11.2.1 Development of Gene Editing Tools

Genes essentially determine the physiological state of cells. Repairing damaged genes to restore cells' normal biological functions is a coveted goal of biologists and medical researchers. Gene editing is the process of rewriting the genetic code of a species at a specific site in a desired way. There are a variety of tools available for gene editing, most of which possess catalytic nuclease activity, that can act on double-stranded DNA. Gene editing tools share a similar feature in that they can bring the effector domain with catalytic nuclease activity to specific sites in the genome via specific DNA-binding motifs or guide sequences.

Although gene editing may seem like a simple and convenient experimental tool for nascent biologists, scientists have been searching for ways to conveniently alter genetic information since the discovery of the double helix structure of DNA. The discovery and application of homing endonucleases was the initial stage of the development of gene editing tools (Belfort and Roberts 1997). Homing endonucleases can cut DNA apart at specific sites by recognizing specific nucleotide sequence patterns, providing an opportunity to artificially insert foreign DNA fragments into the genome (Belfort and Roberts 1997; Chames et al. 2005). Homing endonucleases have limitations in their application in the field of gene editing due to their fixed specific recognition patterns. Thus, there is still a need to develop a tool that can perform intended site-specific modification of the genome.

After prolonged and sustained efforts, a programmable zinc finger nuclease (ZFN) was developed in 1996 to achieve the goal of site-specific gene editing (Kim et al. 1996). ZFNs are composed of a zinc finger domain containing a DNA recognition module and a cleavage domain of the homing endonuclease FokI and function as a dimer (Carroll 2011). The DNA recognition module of ZFNs usually includes multiple zinc finger domains, with each zinc finger domain recognizing a different DNA base triplet, and multiple zinc finger domains are linked together by connecting peptides to generate polyfingered zinc finger proteins, thus expanding the recognition range of targetable DNA sequences (Carroll 2011). Although targeted gene editing can be achieved at some loci using ZFNs, their design and assembly are time-consuming and labor-intensive, which limits the application of ZFNs in non-specialized laboratories (Gupta and Musunuru 2014; Ramirez et al. 2008). Moreover, ZFNs are prone to off-target cleavage by creation of dimers at non-specific sites due to the lack of a stringent recognition code. This has limited the usage of ZFN-based method for genome engineering in clinical trials (Gupta and Musunuru 2014; Ramirez et al. 2008).

TALENs are another gene editing tool structurally similar to ZFNs. But unlike ZFNs, TALENs use transcriptional activation-like effectors (TALEs) to recognize specific sequences of DNA (Bogdanove and Voytas 2011). Each of these TALE repeats recognizes a single nucleotide, thereby expanding targeting specificity and accessibility (Bogdanove and Voytas 2011; Gaj et al. 2013). Despite the higher resolution of DNA sequence identification, it still takes a great deal of time and money to complete a successful targeted editing using TALENs (Wright et al. 2014). Moreover, TALEN is much larger than ZFN, which makes it difficult for it to be efficiently cloned and delivered by viral vectors (Gupta and Musunuru 2014).

Subsequently, the clustered regularly interspaced short palindromic repeats (CRISPR)/CRISPR-associated (Cas) adaptive immune systems that mediate adaptive immunity in bacteria and archaea have gradually attracted widespread interest, because they can be re-purposed into RNA-guided gene editing in diverse organisms including humans. The Cas9 protein from *Streptococcus pyogenes* (SpCas9) belongs to the type II family of CRISPR/Cas systems and is currently the most widely exploited platform for CRISPR-based targeted genome editing (Chylinski et al. 2014). Mechanistically, short fragments of exogenous DNA are integrated into the CRISPR locus and transcribed into CRISPR-related RNA (crRNA), which is then annealed to transactivating crRNA (tracrRNA) forming crRNA: tracrRNA complexes that direct the Cas9 protein for sequence-specific degradation of target DNA (Horvath and Barrangou 2010). In 2012, the pioneering work of Jennifer Doudna and Emmanuelle Charpentier et al. revealed the biochemical properties of CRISPR systems in bacteria that help defend against phage invasion. They found that target recognition by the Cas9 protein requires only a seed sequence within the crRNA and a conserved protospacer adjacent motif (PAM) immediately following the target DNA sequence, and engineered RNA molecules can be programmed to guide the Cas9 endonuclease to cut specific DNA sequences, resulting in double-stranded DNA breaks (Jinek et al. 2012). Following this observation, Feng Zhang and colleagues immediately demonstrated how the CRISPR-Cas9 system could be utilized for gene editing in eukaryotic cells (Cong et al. 2013). These findings astounded the scientific world and revolutionarily transformed the area of gene editing. Henceforth, the CRISPR/Cas system achieved unparalleled popularity as a gene editing technique.

The engineered CRISPR/Cas9 system is composed of the Cas9 endonuclease, which is the enzyme required to mediate target DNA cleavage, and a single guide RNA (sgRNA) to direct the Cas9 endonuclease to the target site. Compared with DNA-mediated targeting by protein recognition, DNA recognition by sgRNA is more concise and elegant. The engineered CRISPR/Cas9 system is flexible, as it is easy to design and synthesize a sgRNA by combining tracrRNA and crRNA, which will undoubtedly reduce the application threshold of gene editing tools to a large extent. It will also provide a wide range of innovative applications, such as editing tool optimization, high-throughput editing, and the combination of molecular tags and next-generation sequencing (Hsu et al. 2014). CRISPR-Cas9 has rapidly changed the research landscape of gene editing due to its simplicity and robustness. It is becoming the most widely used engineered nuclease for editing mammalian genomes. So far, the CRISPR-Cas9 system has enabled convenient and efficient gene editing in a wide range of species and cell types, including hematopoietic stem cells (Gundry et al. 2016; Wu et al. 2019) (Fig. 11.1).

11.2.2 Double-Strand Break (DSB)-Dependent Gene Editing

Most gene editing tools work by cutting at specific sites in the genome under the guidance of DNA recognition motifs (ZFNs and TALENs) or

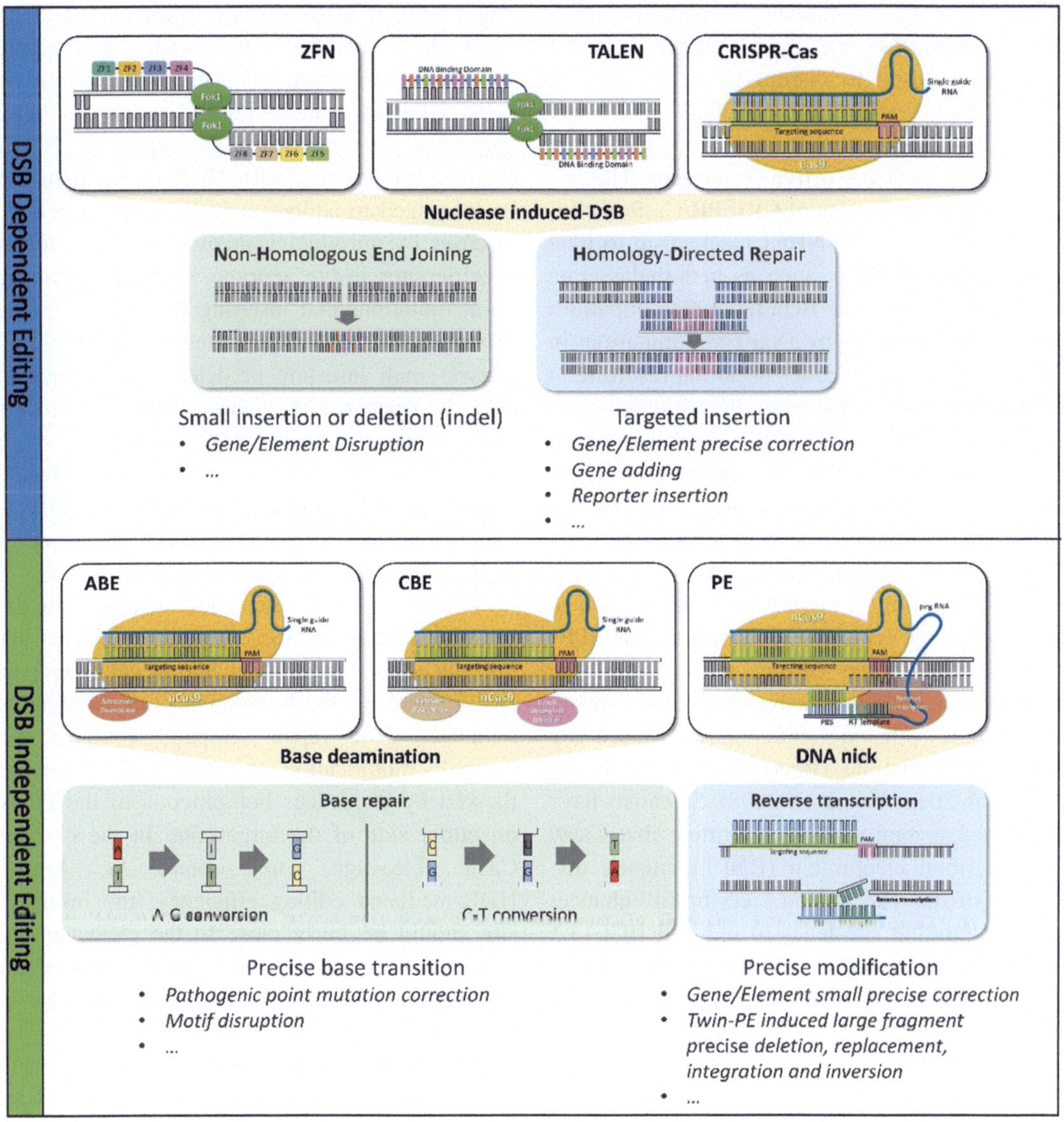

Fig. 11.1 Schematic of the working mechanisms of current gene editing tools

RNA (CRISPR-Cas systems) to generate DSBs. Endogenous cellular repair mechanisms address broken DNA ends, which provides the opportunity to rewrite genomic sequence information of the organism. Typically, cells repair DSBs by two main pathways, namely, non-homologous end joining (NHEJ) and homology-directed repair (HDR) (Iliakis et al. 2004). NHEJ occurs at a relatively high frequency in cells and is an error-prone rapid repair mechanism that results in nucleotide insertions or deletions, often leading to frameshift mutations that alter the reading frame of genes and completely or partially inhibit gene function (Ran et al. 2013). Therefore, NHEJ is primarily used to disrupt target genes. The NHEJ pathway is active in all phases of the cell cycle, including in non-dividing cells, and is the major repair pathway active in quiescent HSCs.

A typical application of CRISPR/Cas9 in clinical trials is using the NHEJ mechanism to treat β-hemoglobinopathies, such as beta-thalassemia and sickle cell anemia. Beta-hemoglobinopathies are genetically caused by a variety of mutations in the *HBB* gene that encodes β-globin, resulting in reduced or absent erythrocyte β-globin synthesis (Piel 2016; Taher et al. 2018). Scientists previously found that functionally compensating for the impaired beta-globin by inducing the expression of gamma-globin alleviates the symptoms of beta-thalassemia (Bauer and Orkin 2011; Platt 2008). Disruption of the GATA1-binding site of the +58 *BCL11A* erythroid enhancer by CRISPR-Cas9-mediated gene editing induces γ-globin expression and is a promising therapeutic strategy to alleviate β-hemoglobinopathies caused by HBB gene mutations (Bauer et al. 2013; Bauer and Orkin 2015; Uda et al. 2008). Scientists have conducted several studies confirming that Cas9/sgRNA ribonucleoprotein (RNP)-mediated disruption of the +58 *BCL11A* erythroid enhancer GATA1-binding site leads to reduced BCL11A expression and induction of fetal γ-globin; this is hence a viable therapeutic strategy for the treatment of β-thalassemia and sickle cell disease (Wu et al. 2019; Brendel et al. 2016, 2020). Several clinical studies based on this strategy have been conducted, such as CTX001 of CRISPR Therapeutics (Frangoul et al. 2021) and BRL-101 of BRL Medicine (Fu et al. 2022). This gene editing therapy has shown promising results in a wide range of β-globin genotypes and in a wide range of patient ages.

HDR is another separable pathway by which cells repair DNA DSBs. It relies on homologous sequences in the repair template to complete homologous recombination, which mainly occurs in the S and G2 phases of the cell cycle (Salsman and Dellaire 2017). HDR can use intracellular sister chromatids as templates, or it can use externally imported DNA sequences as templates to repair broken DNA double strands (Hsu et al. 2014). HDR-mediated repair occurs significantly less frequently than NHEJ-mediated repair in cells. HDR is a more faithful repair modality compared to the error-prone repair of NHEJ (Sander and Joung 2014). HDR mechanism can be leveraged to address a wide range of genetic diseases by introducing precise modifications at specific sites in the genome, such as correcting point mutations and inserting desired fragments (Zhang et al. 2014). Compared to NHEJ, which causes small insertion or deletion mutations to alter the open reading frame of genes and thus cause gene disruption, HDR allows for more diverse and more precise gene editing. Hence, enhancing the frequency of HDR in cells is widely pursued in current research (Khan et al. 2016).

When Cas9 and other endonucleases cut the genome, a large number of repair templates are introduced into cells, which allows cells to repair DSBs referring to the sequence of exogenous templates. The repair template contains the desired modification or insertion fragment flanked by sequences homologous to the DNA on either side of the target site. In the case of Cas9 cleavage, to obtain a higher HDR-mediated editing efficiency, the insertion site should be fairly close to the cleavage site and ideally less than 10 bp if possible (Ran et al. 2013; Mali et al. 2013). It is important to bear in mind that target DNA cleavage can still be mediated by CRISPR enzymes after the resulting DSB has been repaired, which may hinder obtaining accurate editing results as long as the protospacer and PAM sequences remain intact.

Two common ways, mutating the seed sequence of the sgRNA and the PAM sequence, can prevent further editing (Paquet et al. 2016).

When designing the repair template, it is necessary to design homologous arms of appropriate length according to the target modifications to obtain more efficient HDR repair. Both double-stranded DNA and single-stranded DNA can be introduced into cells as HDR repair templates (Hashimoto and Takemoto 2015; Renaud et al. 2016). Typically, single-stranded oligodeoxynucleotides (ssODNs) have at least three advantages over double-stranded DNA, viz., lower cytotoxicity, generally higher editing efficiency, and no risk of genomic integration (Song and Stieger 2017; Chen et al. 2011). The appropriate form of template should be chosen depending on the desired gene modification, and ssODNs are generally used as a template for small modifications (e.g., point mutation correction). A homology arm as short as 30–50 bases can mediate small insertions and edits, but this also varies with the site to be inserted and the experimental system, with a slightly longer homology arm contributing to the frequency of HDR occurrence (Okamoto et al. 2019; Remy et al. 2017). Despite the many advantages of ssODNs, longer homology arms may be required to mediate efficient HDR repair when inserting larger fragments, such as fluorescent protein reporter genes or expression cassettes of selected genes and well-structured functional genes. However, the direct synthesis of ssODNs exceeding 200 nt is still not commercialized due to technical difficulty and cost considerations (Hao et al. 2020). Researchers are also developing convenient and cost-effective ways to prepare long ssDNA, such as Easi-CRISPR (Quadros et al. 2017), a technique that transcribes a template in vitro and then reverse-transcribes it to obtain a single strand of complementary DNA. It is more common to use plasmids carrying insertion elements with long homology arms or PCR products directly as templates, although cytotoxicity or genomic integration risks can compromise editing benefits to some extent.

Using gene editing techniques to mark individual genes or genetic regions is a fundamental strategy for understanding gene function. To explore the functions of hematopoietic genes, researchers commonly insert molecular tags or reporter genes at specified genomic loci by HDR-mediated gene editing, which may provide deep insights that help elucidate the genesis, differentiation, and other cytological aspects of HSC. Furthermore, hematopoietic disease models can also be constructed by inserting specific fragments by HDR-based gene editing (Christen et al. 2022; Rissone and Burgess 2018). In addition to fundamental research, the clinical application of gene editing in HSC represents enormous potential in the treatment of associated diseases. Given that HDR can result in precise gene editing, this technique has the potential to address a wide range of disorders. The HDR-mediated genome editing can repair numerous point mutations that cause β-hemoglobinopathies in the *HBB* gene (Pattabhi et al. 2019; Antony et al. 2018) as well as pathogenic mutations in many other genes (Scharenberg et al. 2020; Pavel-Dinu et al. 2019; Rai et al. 2020). Alternatively, functional substitution of damaged genes can be accomplished using HDR by inserting functional intact gene expression cassettes into the genomic safe harbor (Gomez-Ospina et al. 2019).

The HDR pathway allows for specific gene modifications as desired, which allows researchers to fully exploit the DSBs robustly caused by gene editing tools. In its natural state, the cell represses the HDR pathway in favor of NHEJ repair (Liu et al. 2019), which would logically reduce the editing products of precisely modified genes. Researchers have developed various strategies to circumvent this issue, including but not limited to using small molecules or overexpressing related factors to inhibit the NHEJ pathway (Yang et al. 2020; Bischoff et al. 2020) or directly enhancing the HDR pathway (Zhang et al. 2020a). In addition, the cell cycle also significantly affects the frequency of HDR, and especially long-term hematopoietic stem cells are still in quiescent phase unfavorable to the occurrence of HDR repair. Thus, synchronizing the cell cycle of HSPC has proven to be a feasible strategy (Salisbury-Ruf and Larochelle 2021).

Moreover, loosening the chromatin state and increasing donor availability at target sites can also increase the frequency of HDR editing (Liu et al. 2020; Fu et al. 2021; Park et al. 2020). A detailed description of the challenges faced in HDR-mediated gene editing and the appropriate solutions explored can be found in an elegant review by Azhagiri et al (2021). Although various strategies have been successfully developed to improve the editing efficiency of HDR, to what extent the in vivo implantation and long-term repopulation capacity of HSPC will be affected by these pretreatments still needs further exploration, and the balance between enhancing editing efficiency and reducing cytotoxicity needs to be considered for clinical applications.

11.2.3 DSB-Independent Base Editing and Prime Editing

Base editing technology is a new target gene modification technology based on the CRISPR/Cas system. Base editors can introduce single base substitutions without introducing DNA double-strand breaks. Base editors are mainly composed of two parts, Cas proteins and DNA-modifying enzymes. So far, the cytosine base editor (CBE) (Komor et al. 2016) and adenine base editor (ABE) (Gaudelli et al. 2017) have been developed by David R. Liu's group in 2016 and 2017, respectively. These two types of base editing systems use cytosine deaminase or artificially evolved adenine deaminase to perform precise base editing on the target site and finally achieve cytosine base editing, replacing C-G with T-A, and adenine base editing, replacing A-T with G-C.

The efficient third-generation CBE, BE3 (the APOBEC1-Cas9 nickase-UGI fusion), generated during the initial CBE development process, achieved an average of 37% C → T conversion at the target test site in 293T cells (Komor et al. 2016). To further increase product purity with no apparent loss of activity, Liu and coworkers designed fourth-generation base editor (BE4) by appending an additional copy of UGI to the C-terminus of BE3 (Komor et al. 2017). BE4 yielded C-to-T editing percentages comparable to those of BE3 with non-T editing products decreased by an average of 2.2 ± 0.8-fold relative to BE3, across the four loci tested in 293T cells (Komor et al. 2017). ABE7.10, a base editor with high efficiency obtained during the initial ABE development, can efficiently convert A-T to C-G, with efficiencies as high as 50% at certain sites in human 293T cells (Gaudelli et al. 2017). The vast majority of known pathogenic human genetic variants are point mutations, and four point mutation types, T → C (complementary strand A → G) and G → A (complementary strand C → T), account for 61% of the human pathogenic single nucleotide polymorphisms (SNPs) in the ClinVar database (Landrum et al. 2014, 2016). The invention of the base editor allows direct correction of these four types of point mutations, which makes it an important potential for clinical applications. However, the editing efficiency of BEs is vulnerable to the cell line and the DNA sequence of the editing site, and the low editing efficiency in a specific cell line or at a specific site is the biggest obstacle to the application of them. The cytidine (BE4) and adenine (ABE7.10) base editors were optimized by changing nuclear localization signals, codon use, and ancestral rebuilding of the deaminase component (Koblan et al. 2018). The resulting BE4max, AncBE4max, and ABEmax editors considerably improved efficiency and corrected harmful SNPs in a number of mammalian cell types. Jason M. Gehrke et al. replaced the rat-derived cytosine deaminase (rat APOBEC1, rAPO1) of BE3 with a modified human-derived cytosine deaminase (engineered human APOBEC3A, eA3A) to construct a more specific base editor, eA3A-BE3 (Gehrke et al. 2018). The eA3A-BE3 was applied to repair −28 site A > G mutation (i.e., complementary strand is T > C mutation) at HBB promoter region by electroporation delivery of eA3A-BE3 protein with sgRNA into hHSPCs, obtaining higher repair efficiency and precision compared to BE3. Yuxuan Wu et al. demonstrated that the method of electroporation of RNP to deliver eA3A-BE3 into $CD34^+$ hHSPCs can achieve high-efficiency base editing in the enhancer region of BCL11A, and in vitro differentiation experiments detected high levels of γ-inducible expression (Zeng et al. 2020). After two consecutive transplantations in immunodeficient mice, it was found that long-term HSCs could be successfully edited without affecting the multi-lineage differentiation potential (Zeng et al. 2020).

Thereafter, to further improve the activity and compatibility of ABE with multiple Cas analogs, in March 2020, David R. Liu's group reported the evolution of the eighth-generation ABE mutant ABE8e based on ABE7.10max using the phage-assisted sequential evolution system (Richter et al. 2020). Compared to the deaminase of ABE7.10max, the deaminase of ABE8e (TadA8e) contains 8 additional mutations that increase the deaminase activity by 590-fold. Investigators combined the TadA8e with various Cas homologs, including SpCas9, to generate new generation of ABEs with greatly improved editing efficiency. Nicole M. Gaudelli et al. obtained ABE8s mutant with significantly higher editing efficiency based on ABE7.10 by TadA saturation mutation screening (Gaudelli et al. 2020). Delivering ABE8s mRNA and sgRNA to hHSPCs by electroporation can achieve up to 60% editing efficiency at the promoter-198 target site of *HBG1/2* gene, which is 2–3 times higher than that of ABE7.10. The expression of γ-globin in hHSPCs edited by ABE8s was 3.5 times higher than that of the ABE7.10-treated group. In order to further relieve the PAM limitation of the ABE system and expand its application range, we constructed a nearly PAM-free variant ABE8e-SpRY and successfully applied it to efficiently correct (up to 60%) two common β-thalassemia mutations, HbE mutation (the 26th codon G to A mutation in the exon region of the *HBB* gene) and the IVS2-654 mutation (the 654th C to T mutation in the second intron region of the *HBB* gene; the complementary chain is a G to A mutation), and normal β-globin expression was restored to a great extent.

Currently, there are two primary approaches for gene repair for human pathogenic gene mutations, namely, homology-mediated recombination and base editing. These two methods address the repair of gene abnormalities caused by mutations, deletions, and insertions, but they also have certain application limitations. For example, homology-mediated recombination is limited by the delivery and length of the donor DNA template and is also limited by the cell cycle, as well as high unintended indels due to the need for DNA double-strand breaks in the repair process. However, base editors can only achieve certain kinds of base conversion editing, and there is a problem of limited editing scope. Therefore, these application limitations suggest the need for the development of a gene editing tool that can achieve base transversions and long-range, high-precision editing without inducing DSBs. Andrew Anzalone et al. developed an editing system called prime editing (PE) (Anzalone et al. 2019). The researchers fused the murine reverse transcriptase MLV at the C-terminus of Cas9 H840A to form Cas9-MLV while extending the 3′ end of the sgRNA so that it contains two parts. One part targets and binds a specific DNA region responsible for binding the target DNA and acting as a primer for reverse transcription of the reverse transcriptase, called PBS; the other part carries the desired sequence of the desired editing result. This information is reverse-transcribed by reverse transcriptase into the newly generated DNA strand. This extended sgRNA is named pegRNA, which together with Cas9-reverse transcriptase implements the reverse transcription process to edit the gene. The editing efficiency of the first-generation PE (PE1) at the test site in 293T cells depends on the length of the PBS, and the editing efficiency is relatively low, up to 5.5%. By screening reverse transcriptase mutants, optimizing the PBS length and template length of pegRNA, and adding a conventional sgRNA to nick the non-editing strand, the researchers obtained the third-generation powerful priming editing tool PE3, which can achieve multi-function, long-distance, high-efficiency editing with low indel rate. The PE system can successfully edit HEK293, K562, U2OS, Hela, and other cell lines as well as mouse primary cerebral cortex cells. However, a wide variety of cell attempts have not been carried out. For example, the application of PE system in human hematopoietic stem and progenitor cells has not yet been reported. We purified the Cas9-MLV protein of PE3 and delivered it together with chemically synthesized pegRNA by electroporation, and no editing efficiency was detected in CD34^{+} human hematopoietic stem progenitor cells. It is hypothesized that the PE system may be differentially inhibited in reverse transcription activity in different cell types, resulting in variable editing efficiency. PE-based manipulation strategies, especially with HSCs, need rigorous improvization for future clinical translation.

11.2.4 In Vivo and In Vitro Gene Editing Element Delivery

Hematopoietic stem cells mainly reside in the bone marrow, which is the major site of hematopoiesis in humans. HSCs are responsible for the life-long production of all types of mature blood cells. In the realm of clinical applications, genome editing platforms manifest in two distinct modalities: the first entails the acquisition of donor or patient cells for ex vivo genetic manipulation, followed by their reinfusion into the patient's physiological milieu(ex-vivo), while the second method encompasses the direct in vivo delivery of editing components into the patient's system(in-vivo). Recent clinical trials with genome editors in HSCs have focused on in vitro genetic manipulation of the donor's allogenic or patient's autologous CD34^{+} cells, followed by reinfusion into the patient. Allogeneic hematopoietic stem cell transplantation involves transplanting hematopoietic stem cells from healthy donors to replace the defective hematopoietic stem cells of patients. Allogeneic HSCT is a curative approach, which results in lifelong phenotype alleviation of various hematological diseases, such as sickle cell anemia and β-thalassemia, after one-time transplant (Strocchio and Locatelli 2018; Locatelli et al. 2016; Angelucci and Pilo 2016). However, the success of allogenic HSCT depends on the availability of highly compatible HLA-matched donors, who are accessible to less than 30%; thus, the majority of patients lack a well-matched donor (Strocchio and Locatelli 2018; Locatelli et al. 2016; Angelucci and Pilo 2016). With the rapidly expanding technological advances in the gene therapy field, autologous HSCT represents an alternative curative option as it is based on the gene correction of the patient's own HSCs, eliminating the requirement for highly compatible donors and effectively preventing adverse immune effects. With the development of innovative and specific tools and gene editing approaches, the existing technology can achieve precise DNA surgery.

Another issue that must be taken into consideration is how to deliver these editing components to the genome in the nucleus with the ultimate goals of maximizing editing efficiency and minimizing treatment toxicity. Gene editing machinery can be delivered as DNA, mRNA, or RNP. In general, the existing delivery methods can be classified into viral approaches (such as adenoviruses (Ads), adeno-associated viruses (AAVs), and lentiviruses (LVs)), physical approaches (such as electroporation and microinjection), and chemical approaches (such as lipid nanoparticles (LNPs)) (Yip 2020). In addition, a delivery vehicle termed virus-like particles (VLPs) has been developed in the past several years (Yip 2020).

AAVs as a delivery means have very mild immunogenicity, and their broad range of serotype specificity can guide them effectively to infect certain tissue cells in human body. However, there is limitation of their payload (approximately up to 4.7 kb) (Yang et al. 2016; Ran et al. 2015; Bak and Porteus 2017). The limited cloning capacity of AAVs can be circumvented by utilizing smaller Cas proteins or dual AAV delivery systems (splitting recombinant large gene editors into two separate AAV vectors) (Yang et al. 2016; Ran et al. 2015; Bak and Porteus 2017).

LVs are another viral vector for gene editor delivery. LVs are derived from the enveloped viruses of HIV-1 from which replication and integrase-related genes are deleted to ensure their safety and are mainly used for delivery in vitro (Kotterman et al. 2015; Popescu et al. 1990; Rothe et al. 2013; Check 2002, 2005; Hacein-Bey-Abina et al. 2003, 2008; Liu et al. 2014). Conversely, LVs welcome larger payloads (approximately up to 10 kb), suitable for not only Cas protein but also multiple sgRNAs to achieve multi-gene editing. However, LVs carry the risk of viral genome integration, which leads to prolonged expression of gene editing tools after integration, increasing the chance of off-target effects (Kotterman et al. 2015; Popescu et al. 1990; Rothe et al. 2013; Check 2002, 2005; Hacein-Bey-Abina et al. 2003, 2008; Liu et al. 2014).

Ads have a generous loading capacity of nearly 36 kb, are genetically stable, and have high transduction efficiency (Lee et al. 2017). Various serotypes are used to target different human tissue cells. However, the major challenge that limits the application of using Ads for delivery is that they trigger a high level of innate immune response, which may mediate cytotoxicity (Muruve 2004).

Electroporation and microinjection are common physical strategies of delivery (Dever et al. 2016; de Melo and Blackshaw 2018). The former increases the permeability of cells by applying an electric field and transiently introduces DNA, RNA, or RNP into the cell. Existing HSPC gene editing clinical trials generally use electroporation to introduce gene editing tools into isolated HSPCs to achieve gene surgery. Direct delivery of RNPs has numerous advantages, including rapid action, high gene editing efficiency, reduced off-target effects, limited toxicity, and low level of immune responses. The microinjection method is a means of directly injecting gene editing tools into cells or nuclei under the microscope, but its application is limited by the operating throughput and is mainly used in early embryonic cells such as fertilized eggs (Horii et al. 2014).

LNPs are commonly used for in vivo delivery (Pensado et al. 2014). Liposomes are positively charged spherical lipid bilayer structures in which negatively charged nucleic acids are encased and mainly composed of cationic lipids, polyethylene glycol, cholesterol, and helper lipids. By changing the ratio of these ingredients or replacing lipid types, targeting at different organ types can be achieved. LNPs are the most popular choice for in vivo therapeutic delivery of RNA-related vaccines and drugs. LNP-mediated delivery of gene editing system in the form of DNA, RNA, or RNP results in reduced off-targets due to transient expression properties. It is less immunogenic than viral vectors and can also be administered repeatedly. Currently, its mainly targeted organ is the liver, and exploring new targets of this strategy is a hot research topic. In particular, targeting HSCs in the bone marrow is being explored, which can avoid the process of cell collection, in vitro culturing and gene manipulation, and transplantation (Yip 2020).

VLPs are also potentially promising delivery vehicles for gene editing tools (Mohsen et al. 2017). They are assembled from non-infectious parts of viral proteins and can be loaded with mRNA, protein, or RNP (Yip 2020). VLP-based delivery method represents a compromise between viral and nonviral delivery approaches and possesses the strengths of both approaches. The advantage lies in the lower chances of off-target properties. Delivery to different organs can be achieved by altering their different envelope glycoproteins.

Overall, these platforms offer a broad spectrum of cargo capacities and may be suitable for different editing strategies. Indeed, all these delivery platforms have intrinsic advantages and disadvantages, which would allow investigators to assess whether the proposed vehicles meet the gene therapeutic expectations with a favorable risk/benefit ratio before clinical transition (Fig. 11.2).

11.3 Advances in Clinical Trials

The rapid progression of gene editing tools has led to an escalation in registered HSPC gene editing clinical trials. Here we review the state of the art of ex vivo gene editing with programmable nucleases in human hematopoietic stem and progenitor cells.

ZFNs were the first nucleases to be used for gene editing in clinical trials. In the first clinical study for genome editing of HSCs using ZFNs, patients with chronic aviremic HIV infection were infused with autologous $CD4^{+}$ T cells in which the *CCR5* gene was rendered permanently dysfunctional by a ZFN (NCT00842634) (Tebas et al. 2014). HIV RNA became undetectable in

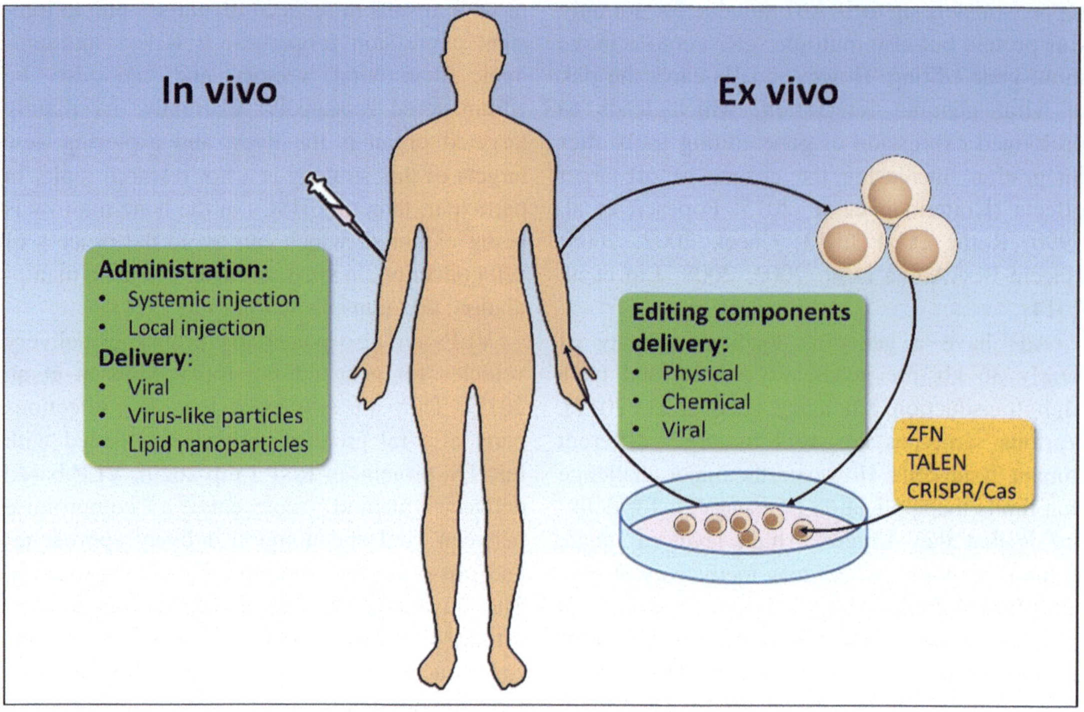

Fig. 11.2 Delivery strategies for gene editing therapy

one of the four patients who could be evaluated, and the blood level of HIV DNA decreased in most patients. In another clinical study, participants with SCD ended up "without recurrence of previous SCD symptoms" following treatment with a one-time infusion of autologous $CD34^+$ HSPCs transfected ex vivo with ZFN mRNAs targeting the B-cell lymphoma/leukemia 11A (*BCL11A*) enhancer region (NCT03653247).

In recent years, clinical trials for genome editing of HSCs have mainly used CRISPR/Cas9. In 2019, Hongkui Deng's team reported a successful allogeneic transplantation and long-term engraftment of CRISPR-Cas9-edited, *CCR5*-ablated HSPCs in a patient with HIV-1 infection and acute lymphoblastic leukemia (NCT03164135) (Xu et al. 2019). In this case, the patient had been followed for 19 months at the time of report, and the acute lymphoblastic leukemia was in complete remission. Off-target effects of the gene editing, meanwhile, were not noted. However, the percentage of *CCR5* disruption in lymphocytes was less than 8%, and the percentage of $CD4^+$ cells with CCR5 ablation increased only by a small degree during a period of antiretroviral therapy interruption, which indicates the need for further research to address this issue. In 2020, preliminary clinical results reported that the first two adult patients—one with TDT and the other with SCD—had been treated with CTX001 (developed jointly by CRISPR Therapeutics and Vertex), the autologous $CD34^+$ cells edited with CRISPR-Cas9 targeting the same *BCL11A* enhancer (NCT03655678 for TDT and NCT03745287 for SCD) (Frangoul et al. 2021). The obtained follow-up data for more than 1 year have shown the intended CRISPR-Cas9 editing of *BCL11A* in long-term HSCs, with both patients having high efficiency of allelic editing in the bone marrow and blood, high levels of fetal hemoglobin expression, elimination of vaso-occlusive episodes (in the patient with SCD), and transfusion independence. A most recent clinical trial reported clinical transition of CRISPR-Cas9-mediated, autologous HSPC gene editing for

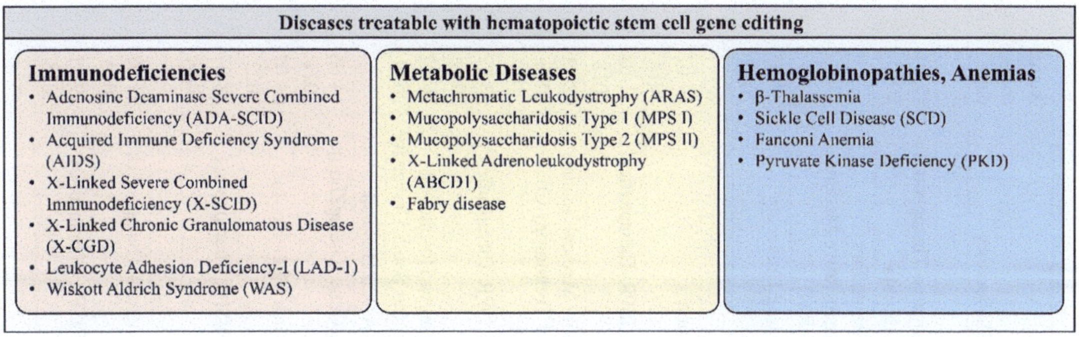

Fig. 11.3 Diseases treatable with hematopoietic stem cell gene editing

pediatric patients (Fu et al. 2022). Two children with TDT, one carrying the β0/β0 genotype, which was classified as the most severe type of TDT, received *BCL11A* enhancer-edited autologous HSPCs (BRL-101, developed by BRL Medicine) and had been followed for 18 months at the report time (NCT04211480) (Fu et al. 2022). In this clinical study, successful transplantation and long-term engraftment of CRISPR-Cas9-edited autologous HSPCs were achieved, and sustained elevations in HbF levels were enough for transfusion independence. The aforementioned clinical treatment methods for TDT and SCD using Cas9-based *BCL11A* enhancer-editing strategy indicate that curative levels of γ-globin expression can be reached by this approach, regardless of genotype or age, demonstrating remarkable advances in clinical application of HSPC gene editing mediated by CRISPR-Cas9. However, further experimental testing and longer follow-up are still needed to more extensively characterize the long-term efficacy and safety profiles of these gene editing treatments.

There are other ongoing HSPC gene editing clinical trials. One study is evaluating the safety and efficacy of a CRISPR-Cas9-edited autologous HSPC product—OTQ923 (developed by Novartis)—to reduce the biologic activity of BCL11A, increasing fetal hemoglobin (HbF) and reducing complications of sickle cell disease (NCT04443907). Another clinical trial is studying the safety and efficacy of a single dose of autologous CRISPR-Cas9-modified CD34$^+$ HSPCs with disrupted *BCL11A* enhancer (ET-01, developed by EdiGene) in subjects with TDT (NCT04925206). Instead of using *BCL11A* disruption to elevate HbF level, CRISPR-Cas9-mediated genome editing can also be exploited for targeted correction of pathogenic mutations in β-globin coding gene (*HBB*), and some companies are conducting safety and efficacy investigations of genome-corrected autologous HSPCs in patients with β-hemoglobinopathies such as TDT (NCT03728322, Allife Medical Science and Technology) and SCD (NCT04774536, Mark Walters; NCT04819841, Graphite Bio). Recently, the clinical application of CRISPR/Cas12a, a Cas9 homolog, has been tested. And a study is being conducted to evaluate the safety, tolerability, and efficacy of a single dose of autologous CRISPR/Cas12a-edited CD34$^+$ HSPC with disrupted HBG1/2 promoters (EDIT-301, developed by Editas Medicine) to reactivate the HbF expression in participants with TDT (NCT05444894) and SCD (NCT04853576).

Preliminary results of these clinical trials support clinical application of gene editing approaches and hold the promise of gene therapy for treatment of genetic diseases. More and more novel gene editors and technologies are being evaluated and progressing into clinical trials. We believe that there will be optimal gene editing tools and therapeutic strategies for the treatment of patients with inherited genetic diseases in the near future. Eliminating the burden of a life-long therapy by receiving one-time treatment to get a therapeutic effect for life is the ultimate goal of hematopoietic stem-cell gene therapy (Fig. 11.3) (Table 11.1).

Table 11.1 Gene editing of HSC (clinical trials)

NCT number	Phase	Nuclease	Strategy	Target gene	Country	Delivery	Sponsor	Collaborators	Actual study start data
Human immunodeficiency virus 1 (HIV-1)									
NCT02500849	I	ZFN	Ex vivo	CCR5	USA	mRNA electroporation	City of Hope Medical Center	Sangamo Therapeutics	17-07-2015
								California	
								Institute for Regenerative Medicine (CIRM)	
NCT03164135	Not applicable	Cas9	Ex vivo	CCR5	China	Unknown	Affiliated Hospital to Academy of Military Medical Sciences	Peking University	23-05-2017
Transfusion-dependent beta-thalassemia (TDT)									
NCT03432364	I/II	ZFN	Ex vivo	BCL11A enhancer	USA	mRNA electroporation	Sangamo Therapeutics	Sanofi	14-02-2018
NCT03655678	I/II/III	Cas9	Ex vivo	BCL11A enhancer	USA, UK	RNP electroporation	Vertex Pharmaceuticals Incorporated	CRISPR Therapeutics	31-08-2018
NCT03728322	I	Cas9	Ex vivo	HBB	Unknown	Unknown	Allife Medical Science and Technology Co., Ltd.	Allife Medical Science and Technology Co., Ltd.	02-11-2018
NCT04208529	Follow-up	Cas9	Ex vivo	BCL11A enhancer	USA, UK	RNP electroporation	Vertex Pharmaceuticals Incorporated	CRISPR Therapeutics	23-12-2019
NCT04211480	Not applicable	Cas9	Ex vivo	BCL11A enhancer	China	RNP electroporation	BRL Medicine	Xiangya Hospital of Central South University	26-12-2019
								PLA 923 Hospital	
NCT04925206	I	Cas9	Ex vivo	BCL11A enhancer	China	RNP electroporation	EdiGene (GuangZhou) Inc.	EdiGene (GuangZhou) Inc.	14-06-2021
NCT05444894	I/II	Cas12a	Ex vivo	HBG promotor	USA	RNP electroporation	Editas Medicine, Inc.	Editas Medicine, Inc.	06-07-2022
NCT05477563	III	Cas9	Ex vivo	BCL11A enhancer	USA	RNP electroporation	Vertex Pharmaceuticals Incorporated	CRISPR Therapeutics	28-07-2022
Sickle cell disease (SCD)									
NCT03653247	I/II	ZFN	Ex vivo	BCL11A	USA	mRNA electroporation	Sangamo Therapeutics	Sangamo Therapeutics	31-08-2018
NCT03745287	I/II	Cas9	Ex vivo	BCL11A enhancer	USA	RNP electroporation	Vertex Pharmaceuticals Incorporated	CRISPR Therapeutics	19-11-2018

NCT04208529	Follow-up	Cas9	Ex vivo	BCL11A enhancer	USA, UK	RNP electroporation	Vertex Pharmaceuticals Incorporated	CRISPR Therapeutics	23-12-2019
NCT04443907	I/II	Cas9	Ex vivo	BCL11A enhancer	USA	RNP electroporation	Novartis Pharmaceuticals	Novartis Pharmaceuticals	23-06-2020
NCT04774536	I/II	Cas9	Ex vivo	HBB	USA	RNP electroporation	Mark Walters, MD	University of California, Los Angeles	01-03-2021
								University of California, Berkeley	
NCT04819841	I/II	Cas9	Ex vivo	HBB	USA	RNP electroporation	Graphite Bio, Inc.	Graphite Bio, Inc.	29-03-2021
NCT04853576	I/II	Cas12a	Ex vivo	HBG promotor	USA, Canada	RNP electroporation	Editas Medicine, Inc.	Editas Medicine, Inc.	21-04-2021
NCT05477563	III	Cas9	Ex vivo	BCL11A enhancer	USA	RNP electroporation	Vertex Pharmaceuticals Incorporated	CRISPR Therapeutics	28-07-2022

ZFN zinc finger nuclease, *BCL11A* disrupt B-cell lymphoma/leukemia 11A, *HBB* hemoglobin subunit beta, *HBG* hemoglobin subunit gamma

11.4 Challenges Toward Clinical Translation of Gene Editing

At the stage of laboratory, efficiency of gene editing was the primary concern. However, at the clinical stage, its safety is a critical point that must not be ignored. Risks, such as off-target effects, chromosomal rearrangement, activation of the p53 DNA damage response caused by DSBs, and immune stimulation, must be assessed before clinical translation of gene editing technologies.

Programmable nuclease platforms mediate targeted genome editing based on recognizing targetable DNA sequences. An ideal gene editor should in principle exhibit perfect specificity for the target sequence without editing at any other sites of the genome. However, none of the current gene editing platforms are free from off-target events. Off-target events primarily arise from the targeted sequence binding with the tolerance of sequence mismatches (1–5 bps have been shown to be tolerated) and cause unintended deletions, insertions, or mutations in other regions of the genome. Therefore, specificity of programmable nucleases is defined as the ratio between on-target and off-target activity. Many tools have been developed for identifying off-target events such as in silico prediction algorithms (e.g., Cas-OFFinder (Bae et al. 2014), MIT CRISPR, E-CRISP) based on "rules" about mismatch number and position and in vitro (e.g., DIGENOME-seq (Kim et al. 2015), CIRCLE-seq (Tsai et al. 2017)) and in cellulo assays (e.g., GUIDE-seq (Tsai et al. 2015)). Indeed, none of these methods are free from limitations; hence, the combination of more than one method is generally being recommended. Biological consequences of nuclease off-target activity are expected to differ depending on the overall editing strategy and disease setting and thus require case-by-case evaluation. Furthermore, engineered HSPCs should be long-lived cells possessing the capacity to support life-long hematopoiesis by undergoing several cycles of self-renewal and differentiation in the patient. Hence, careful assessment of nuclease specificity in engineered HSPCs is mandatory for the success of gene editing tools in clinical applications.

Many practicable gene editing approaches rely on the generation of DSBs, which can give rise to chromosome structural abnormalities. Studies have shown that CRISPR-Cas9 generates structural defects of the nucleus, micronuclei, and chromosome bridges, which initiate a mutational process called chromothripsis (Leibowitz et al. 2021). Chromothripsis is an extensive chromosome rearrangement restricted to one or a few chromosomes and could lead to human diseases. The results have some practical implications, one of which is prompting the further development of genome editing strategies that do not generate DSBs; this may minimize the potential for inducing chromothripsis.

p53 is arguably the most important tumor suppressor in human cells and acts as a master regulator in DNA damage responses (DDRs), which mediates DNA repair and ultimately defines cell fate. Indeed, induction of DNA DSBs by programmable nucleases such as CRISPR/Cas9 triggers the DDR and induces a robust p53-dependent DDRs in HSPCs, resulting in a remarkable impact on clonogenic capacity. Thus, this may represent a significant barrier to therapeutic application (Schiroli et al. 2019). Recent research has found approaches of transient p53 inhibition during the genetic editing process could preserve HSPC multi-lineage potential. However, recent research has also reported that Cas9 introduction alone can lead to the upregulation of p53. These findings may have implications for motivating the usage of strategies reducing editing system exposure of CRISPR components so as to escape from eliciting p53 activation (Enache et al. 2020).

Immune stimulation is also an essential concern that should be taken into consideration for therapeutic purposes in HSPCs. Considering that the mice used in animals' experiments in preclinical research are all severely immunodeficient strains (such as NCG), it is necessary to evaluate the safety of immune stimulation after HSPC transplantation. Components of the editing system, such as Cas protein/mRNA and sgRNA, may trigger immune stimulation. Besides, either viral (such as AAV and Ads) or nonviral delivery approaches (such as electroporation and LNP) stimulate the immune system to varying degrees (Mehta and Merkel 2020). Electroporating RNP complex consisting of Cas9 protein and in vitro transcribed sgRNA (IVT sgRNA) targeting into $CD34^+$ HSPCs triggers type I IFN production, reducing HSPC stemness in addition to causing cell death (Mu et al. 2019). Conversely, dampening of these responses has been obtained by switching IVT sgRNAs to chemically synthetized sgRNAs (Hendel et al. 2015; Scott et al. 2020). Some viral vectors are still affected by neutralizing antibodies in the body, and the immune system may eliminate the viral delivery vectors (Verdera et al. 2020). This drawback can also lead to the limitation of repeat administration. Expression of exogenous genes, such as nucleases, will activate immune responses, leading to inflammation of tissues and subsequent removal of engineered cells (Wagner et al. 2021). However, one of the current popular delivery methods, LNP, uses PEG lipid as a capsule to protect the external ionizable cationic lipid from contacting with immune cells (Suk et al. 2016); thus, it shows much lower immunogenicity compared with viral vectors and can be suitable for repeat drug administrations (Kenjo et al. 2021).

In addition to the aforementioned safety concerns, there are a number of other issues that must be addressed before clinical transition, including sourcing and culturing of the cells, delivery of the nuclease platform, and generating robust site-specific gene editing, particularly in the long-term repopulating HSC fraction. Depending on the disease setting, each of these aspects must be considered for comprehensive risk/benefit evaluation of the therapeutic strategy. Studies in the last few years have still focused on the exploration of strategies that maximize gene editing efficiency and specificity and minimize risks in long-term HSCs.

11.5 Future Perspectives

The latest precise gene editing system—primer editor (Anzalone et al. 2019)—has been developed to generate the fifth generation (PE5) (Chen et al. 2021). Compared with PE3, PE5 has added MLH1dn (MLH1 △754–756) to the PE system, reducing the DNA mismatch repair (MMR) during editing and thereby improving the efficiency. The study has shown that PE5 could achieve 40% and 60% repair efficiency in human iPSC and human primary T cells, respectively.

For correction of specific pathogenic mutations, researchers have developed dual-functional base editing tools A&C-BEmax (Zhang et al. 2020b) and Target-ACEmax (Sakata et al. 2020) with combined functions of both adenine base editors (ABEs) and cytosine base editors (CBEs). They could achieve efficient A to G and C to T conversions at the same target site simultaneously and showed similar level of on-target activities to those of existing single-function base editors.

For genome editing applications, however, the necessity of PAM recognition by DNA-targeting CRISPR-Cas enzymes constrains targeting and affects editing efficiency and flexibility. The prototypical SpCas9 naturally recognizes target sites with NGG PAMs. To overcome PAM-related limitation with CRISPR/Cas9, Cas9 homologs utilizing different PAMs (e.g., Cas12a/Cpf1) (Zetsche et al. 2015; Teng et al. 2019) and Cas9 variants recognizing relaxed NG PAMs (Cas-NG and xCas) (Nishimasu et al. 2018; Hu et al. 2018) or targeting non-canonical PAMs (SpCas9-VQR, VRQR, and VRER) (Kleinstiver et al. 2015, 2016) or targeting an expanded set of NGN PAMs (SpG) (Walton et al. 2020) or targeting almost all PAMs (SpRY) (Walton et al. 2020) have been developed. These Cas9 homologs and variants expand the repertoire of potential targets and provide access to editing on previously inaccessible genetic sites.

The ShCAST system that consists of Tn7-like transposase subunits and the type V-K CRISPR effector catalyzes RNA-guided DNA transposition by unidirectionally inserting segments of DNA 60 to 66 base pairs downstream of the protospacer in the *Escherichia coli* genome with frequencies of up to 80% (Strecker et al. 2019). INTEGRATE system consists of Tn7-like transposon and a type I-F CRISPR-Cas system for programmable RNA-guided transposition. The optimized insertion of transposable elements by INTEGRATE system achieves ~100% unidirectional efficiency in bacteria (Vo et al. 2021). These works have expanded our understanding of the functional diversity of CRISPR-Cas systems and established a paradigm for precision DNA insertion.

The rapid development of gene editing technology in HSPCs will enable investigators to clarify the development mechanism of inherited hematopoietic diseases and provide additional therapeutic options. Meanwhile, targeted genome editing in HSPCs is progressing into clinical trials and currently being tested with encouraging results. These tremendous advances in exploration and development of optimal gene editing platforms and delivery methods will enable the clinical transition of gene editing technology in HSPCs, which will bring renewed hope to many patients.

Acknowledgments This work was supported by the National Key R&D Program of China 2019YFA0110803 (Y.W.) and 2019YFA0109901 (Y.W.), grants from the Shanghai Municipal Commission for Science and Technology 19PJ1403500 (Y.W.), the National Science Foundation of China grants 32001061 (S.C.), and China Postdoctoral Science Foundation grants 2019M661430 (S.C.), 2019TQ0096 (S.C.), and 2020M681231 (J.L.).

References

Aiuti A et al (2013) Lentiviral hematopoietic stem cell gene therapy in patients with Wiskott-Aldrich syndrome. Science 341:1233151

Angelucci E, Pilo F (2016) Management of iron overload before, during, and after hematopoietic stem cell transplantation for thalassemia major. Ann N Y Acad Sci 1368:115–121. https://doi.org/10.1111/nyas.13027

Angelucci E et al (2014) Hematopoietic stem cell transplantation in thalassemia major and sickle cell disease: indications and management recommendations from an international expert panel. Haematologica 99:811–820. https://doi.org/10.3324/haematol.2013.099747

Antony JS et al (2018) Gene correction of HBB mutations in CD34+ hematopoietic stem cells using Cas9 mRNA and ssODN donors. Mol Cell Pediatr 5:1–7

Anzalone AV et al (2019) Search-and-replace genome editing without double-strand breaks or donor DNA. Nature 576:149–157. https://doi.org/10.1038/s41586-019-1711-4

Anzalone AV, Koblan LW, Liu DR (2020) Genome editing with CRISPR–Cas nucleases, base editors, transposases and prime editors. Nat Biotechnol 38:824–844

Azhagiri MKK, Babu P, Venkatesan V, Thangavel S (2021) Homology-directed gene-editing approaches for hematopoietic stem and progenitor cell gene therapy. Stem Cell Res Ther 12:1–12

Bae S, Park J, Kim JS (2014) Cas-OFFinder: a fast and versatile algorithm that searches for potential off-target sites of Cas9 RNA-guided endonucleases. Bioinformatics 30:1473–1475. https://doi.org/10.1093/bioinformatics/btu048

Bak RO, Porteus MH (2017) CRISPR-mediated integration of large gene cassettes using AAV donor vectors. Cell Rep 20:750–756. https://doi.org/10.1016/j.celrep.2017.06.064

Barriga F, Ramírez P, Wietstruck A, Rojas N (2012) Hematopoietic stem cell transplantation: clinical use and perspectives. Biol Res 45:307–316

Bauer DE, Orkin SH (2011) Update on fetal hemoglobin gene regulation in hemoglobinopathies. Curr Opin Pediatr 23:1

Bauer DE, Orkin SH (2015) Hemoglobin switching's surprise: the versatile transcription factor BCL11A is a master repressor of fetal hemoglobin. Curr Opin Genet Dev 33:62–70. https://doi.org/10.1016/j.gde.2015.08.001

Bauer DE et al (2013) An erythroid enhancer of BCL11A subject to genetic variation determines fetal hemoglobin level. Science 342:253–257. https://doi.org/10.1126/science.1242088

Becker AJ, McCulloch EA, Till JE (1963) Cytological demonstration of the clonal nature of spleen colonies derived from transplanted mouse marrow cells. Nature 197:452

Belfort M, Roberts RJ (1997) Homing endonucleases: keeping the house in order. Nucleic Acids Res 25:3379–3388

Biffi A et al (2013) Lentiviral hematopoietic stem cell gene therapy benefits metachromatic leukodystrophy. Science 341:1233158

Bischoff N, Wimberger S, Maresca M, Brakebusch C (2020) Improving precise CRISPR genome editing by small molecules: is there a magic potion? Cell 9:1318

Bogdanove AJ, Voytas DF (2011) TAL effectors: customizable proteins for DNA targeting. Science 333:1843–1846

Brendel C et al (2016) Lineage-specific BCL11A knockdown circumvents toxicities and reverses sickle phenotype. J Clin Invest 126:3868–3878. https://doi.org/10.1172/JCI87885

Brendel C et al (2020) Preclinical evaluation of a novel lentiviral vector driving lineage-specific BCL11A knockdown for sickle cell gene therapy. Mol Ther Methods Clin Dev 17:589–600. https://doi.org/10.1016/j.omtm.2020.03.015

Bryder D, Rossi DJ, Weissman IL (2006) Hematopoietic stem cells: the paradigmatic tissue-specific stem cell. Am J Pathol 169:338–346

Carroll D (2011) Genome engineering with zinc-finger nucleases. Genetics 188:773–782

Cartier N et al (2009) Hematopoietic stem cell gene therapy with a lentiviral vector in X-linked adrenoleukodystrophy. Science 326:818–823

Chames P et al (2005) In vivo selection of engineered homing endonucleases using double-strand break induced homologous recombination. Nucleic Acids Res 33:e178–e178

Check E (2002) Regulators split on gene therapy as patient shows signs of cancer. Nature 419:545–546. https://doi.org/10.1038/419545a

Check E (2005) Gene therapy put on hold as third child develops cancer. Nature 433:561. https://doi.org/10.1038/433561a

Chen F et al (2011) High-frequency genome editing using ssDNA oligonucleotides with zinc-finger nucleases. Nat Methods 8:753–755

Chen PJ et al (2021) Enhanced prime editing systems by manipulating cellular determinants of editing outcomes. Cell 184:5635–5652.e5629. https://doi.org/10.1016/j.cell.2021.09.018

Christen F et al (2022) Modeling clonal hematopoiesis in umbilical cord blood cells by CRISPR/Cas9. Leukemia 36:1102–1110

Chylinski K, Makarova KS, Charpentier E, Koonin EV (2014) Classification and evolution of type II CRISPR-Cas systems. Nucleic Acids Res 42:6091–6105

Cong L et al (2013) Multiplex genome engineering using CRISPR/Cas systems. Science 339:819–823

Copelan EA (2006) Hematopoietic stem-cell transplantation. N Engl J Med 354:1813–1826

de Melo J, Blackshaw S (2018) In vivo electroporation of developing mouse retina. Methods Mol Biol 1715: 101–111. https://doi.org/10.1007/978-1-4939-7522-8_8

Dever DP et al (2016) CRISPR/Cas9 beta-globin gene targeting in human haematopoietic stem cells. Nature 539:384–389. https://doi.org/10.1038/nature20134

Duarte RF et al (2019) Indications for haematopoietic stem cell transplantation for haematological diseases, solid tumours and immune disorders: current practice in Europe, 2019. Bone Marrow Transplant 54:1525–1552

Eichler F et al (2017) Hematopoietic stem-cell gene therapy for cerebral adrenoleukodystrophy. N Engl J Med 377:1630–1638

Enache OM et al (2020) Cas9 activates the p53 pathway and selects for p53-inactivating mutations. Nat Genet 52:662–668. https://doi.org/10.1038/s41588-020-0623-4

Ford C, Hamerton J, Barnes D, Loutit J (1956) Cytological identification of radiation-chimaeras. Nature 177:452–454

Fowler J, Wu A, Till J, McCulloch E, Siminovitch L (1967) The cellular composition of hematopoietic spleen colonies. J Cell Physiol 69:65–71

Frangoul H et al (2021) CRISPR-Cas9 gene editing for sickle cell disease and β-thalassemia. N Engl J Med 384:252–260

Fu Y-W et al (2021) Dynamics and competition of CRISPR–Cas9 ribonucleoproteins and AAV donor-mediated NHEJ, MMEJ and HDR editing. Nucleic Acids Res 49:969–985

Fu B et al (2022) CRISPR–Cas9-mediated gene editing of the BCL11A enhancer for pediatric β0/β0 transfusion-dependent β-thalassemia. Nat Med 1-8:1573

Gaj T, Gersbach CA, Barbas CF III (2013) ZFN, TALEN, and CRISPR/Cas-based methods for genome engineering. Trends Biotechnol 31:397–405

Gaudelli NM et al (2017) Programmable base editing of A*T to G*C in genomic DNA without DNA cleavage. Nature 551:464–471. https://doi.org/10.1038/nature24644

Gaudelli NM et al (2020) Directed evolution of adenine base editors with increased activity and therapeutic application. Nat Biotechnol 38:892–900. https://doi.org/10.1038/s41587-020-0491-6

Gehrke JM et al (2018) An APOBEC3A-Cas9 base editor with minimized bystander and off-target activities. Nat Biotechnol 36:977–982. https://doi.org/10.1038/nbt.4199

Gentner B et al (2021) Hematopoietic stem-and progenitor-cell gene therapy for Hurler syndrome. N Engl J Med 385:1929–1940

Gomez-Ospina N et al (2019) Human genome-edited hematopoietic stem cells phenotypically correct mucopolysaccharidosis type I. Nat Commun 10:1–14

Goyal S et al (2022) Acute myeloid leukemia case after gene therapy for sickle cell disease. N Engl J Med 386:138–147

Gundry MC et al (2016) Highly efficient genome editing of murine and human hematopoietic progenitor cells by CRISPR/Cas9. Cell Rep 17:1453–1461

Gupta RM, Musunuru K (2014) Expanding the genetic editing tool kit: ZFNs, TALENs, and CRISPR-Cas9. J Clin Invest 124:4154–4161

Hacein-Bey-Abina S et al (2003) LMO2-associated clonal T cell proliferation in two patients after gene therapy for SCID-X1. Science 302:415–419. https://doi.org/10.1126/science.1088547

Hacein-Bey-Abina S et al (2008) Insertional oncogenesis in 4 patients after retrovirus-mediated gene therapy of SCID-X1. J Clin Invest 118:3132–3142. https://doi.org/10.1172/JCI35700

Hao M, Qiao J, Qi H (2020) Current and emerging methods for the synthesis of single-stranded DNA. Genes 11:116

Hashimoto M, Takemoto T (2015) Electroporation enables the efficient mRNA delivery into the mouse zygotes and facilitates CRISPR/Cas9-based genome editing. Sci Rep 5:1–8

Hendel A et al (2015) Chemically modified guide RNAs enhance CRISPR-Cas genome editing in human primary cells. Nat Biotechnol 33:985–989. https://doi.org/10.1038/nbt.3290

Horii T et al (2014) Validation of microinjection methods for generating knockout mice by CRISPR/Cas-mediated genome engineering. Sci Rep 4:4513. https://doi.org/10.1038/srep04513
Horvath P, Barrangou R (2010) CRISPR/Cas, the immune system of bacteria and archaea. Science 327:167–170
Hsieh MM et al (2020) Myelodysplastic syndrome unrelated to lentiviral vector in a patient treated with gene therapy for sickle cell disease. Blood Adv 4:2058–2063. https://doi.org/10.1182/bloodadvances.2019001330
Hsu PD, Lander ES, Zhang F (2014) Development and applications of CRISPR-Cas9 for genome engineering. Cell 157:1262–1278
Hu JH et al (2018) Evolved Cas9 variants with broad PAM compatibility and high DNA specificity. Nature 556:57–63. https://doi.org/10.1038/nature26155
Iliakis G et al (2004) Mechanisms of DNA double strand break repair and chromosome aberration formation. Cytogenet Genome Res 104:14–20
Jacobson L, Simmons E, Marks E, Eldredge J (1951) Recovery from radiation injury. Science 113:510–511
Jinek M et al (2012) A programmable dual-RNA–guided DNA endonuclease in adaptive bacterial immunity. Science 337:816–821
Kenjo E et al (2021) Low immunogenicity of LNP allows repeated administrations of CRISPR-Cas9 mRNA into skeletal muscle in mice. Nat Commun 12:7101. https://doi.org/10.1038/s41467-021-26714-w
Khan FA et al (2016) CRISPR/Cas9 therapeutics: a cure for cancer and other genetic diseases. Oncotarget 7:52541
Kim Y-G, Cha J, Chandrasegaran S (1996) Hybrid restriction enzymes: zinc finger fusions to Fok I cleavage domain. Proc Natl Acad Sci 93:1156–1160
Kim D et al (2015) Digenome-seq: genome-wide profiling of CRISPR-Cas9 off-target effects in human cells. Nat Methods 12:237–243. https://doi.org/10.1038/nmeth.3284, 231 p following 243
Kleinstiver BP et al (2015) Engineered CRISPR-Cas9 nucleases with altered PAM specificities. Nature 523:481–485. https://doi.org/10.1038/nature14592
Kleinstiver BP et al (2016) High-fidelity CRISPR-Cas9 nucleases with no detectable genome-wide off-target effects. Nature 529:490–495. https://doi.org/10.1038/nature16526
Koblan LW et al (2018) Improving cytidine and adenine base editors by expression optimization and ancestral reconstruction. Nat Biotechnol 36:843–846. https://doi.org/10.1038/nbt.4172
Komor AC, Kim YB, Packer MS, Zuris JA, Liu DR (2016) Programmable editing of a target base in genomic DNA without double-stranded DNA cleavage. Nature 533:420–424. https://doi.org/10.1038/nature17946
Komor AC et al (2017) Improved base excision repair inhibition and bacteriophage Mu Gam protein yields C:G-to-T:A base editors with higher efficiency and product purity. Sci Adv 3:eaao4774. https://doi.org/10.1126/sciadv.aao4774
Koniali L, Lederer CW, Kleanthous M (2021) Therapy development by genome editing of hematopoietic stem cells. Cell 10:1492
Kotterman MA, Chalberg TW, Schaffer DV (2015) Viral vectors for gene therapy: translational and clinical outlook. Annu Rev Biomed Eng 17:63–89. https://doi.org/10.1146/annurev-bioeng-071813-104938
Kumar P, Tan Y, Cahan P (2017) Understanding development and stem cells using single cell-based analyses of gene expression. Development 144:17–32
Landrum MJ et al (2014) ClinVar: public archive of relationships among sequence variation and human phenotype. Nucleic Acids Res 42:D980–D985. https://doi.org/10.1093/nar/gkt1113
Landrum MJ et al (2016) ClinVar: public archive of interpretations of clinically relevant variants. Nucleic Acids Res 44:D862–D868. https://doi.org/10.1093/nar/gkv1222
Lawson KA, Meneses JJ, Pedersen RA (1986) Cell fate and cell lineage in the endoderm of the presomite mouse embryo, studied with an intracellular tracer. Dev Biol 115:325–339
Lee CS et al (2017) Adenovirus-mediated gene delivery: potential applications for gene and cell-based therapies in the new era of personalized medicine. Genes Dis 4:43–63. https://doi.org/10.1016/j.gendis.2017.04.001
Leibowitz ML et al (2021) Chromothripsis as an on-target consequence of CRISPR-Cas9 genome editing. Nat Genet 53:895–905. https://doi.org/10.1038/s41588-021-00838-7
Liu KC et al (2014) Integrase-deficient lentivirus: opportunities and challenges for human gene therapy. Curr Gene Ther 14:352–364. https://doi.org/10.2174/1566523214666140825124311
Liu M et al (2019) Methodologies for improving HDR efficiency. Front Genet 9:691
Liu B et al (2020) Inhibition of histone deacetylase 1 (HDAC1) and HDAC2 enhances CRISPR/Cas9 genome editing. Nucleic Acids Res 48:517–532
Locatelli F, Merli P, Strocchio L (2016) Transplantation for thalassemia major: alternative donors. Curr Opin Hematol 23:515–523. https://doi.org/10.1097/MOH.0000000000000280
Mali P et al (2013) RNA-guided human genome engineering via Cas9. Science 339:823–826
Mao AS, Mooney DJ (2015) Regenerative medicine: current therapies and future directions. Proc Natl Acad Sci 112:14452–14459
Marktel S et al (2019) Intrabone hematopoietic stem cell gene therapy for adult and pediatric patients affected by transfusion-dependent ß-thalassemia. Nat Med 25:234–241
Mehta A, Merkel OM (2020) Immunogenicity of Cas9 protein. J Pharm Sci 109:62–67. https://doi.org/10.1016/j.xphs.2019.10.003
Mohsen MO, Zha L, Cabral-Miranda G, Bachmann MF (2017) Major findings and recent advances in virus-like particle (VLP)-based vaccines. Semin Immunol

34:123–132. https://doi.org/10.1016/j.smim.2017.08.014

Morgan RA, Gray D, Lomova A, Kohn DB (2017) Hematopoietic stem cell gene therapy: progress and lessons learned. Cell Stem Cell 21:574–590

Morgan MA, Galla M, Grez M, Fehse B, Schambach A (2021) Retroviral gene therapy in Germany with a view on previous experience and future perspectives. Gene Ther 28:494–512

Mu W et al (2019) In vitro transcribed sgRNA causes cell death by inducing interferon release. Protein Cell 10: 461–465. https://doi.org/10.1007/s13238-018-0605-9

Muller-Sieburg CE, Whitlock CA, Weissman IL (1986) Isolation of two early B lymphocyte progenitors from mouse marrow: a committed pre-pre-B cell and a clonogenic Thy-1lo hematopoietic stem cell. Cell 44: 653–662

Muller-Sieburg CE, Sieburg HB, Bernitz JM, Cattarossi G (2012) Stem cell heterogeneity: implications for aging and regenerative medicine. Blood 119:3900–3907

Muruve DA (2004) The innate immune response to adenovirus vectors. Hum Gene Ther 15:1157–1166. https://doi.org/10.1089/hum.2004.15.1157

Nishimasu H et al (2018) Engineered CRISPR-Cas9 nuclease with expanded targeting space. Science 361: 1259–1262. https://doi.org/10.1126/science.aas9129

Okamoto S, Amaishi Y, Maki I, Enoki T, Mineno J (2019) Highly efficient genome editing for single-base substitutions using optimized ssODNs with Cas9-RNPs. Sci Rep 9:1–11

Osawa M, Hanada K-I, Hamada H, Nakauchi H (1996) Long-term lymphohematopoietic reconstitution by a single CD34-low/negative hematopoietic stem cell. Science 273:242–245

Paquet D et al (2016) Efficient introduction of specific homozygous and heterozygous mutations using CRISPR/Cas9. Nature 533:125–129

Park H, Shin J, Choi H, Cho B, Kim J (2020) Valproic acid significantly improves CRISPR/Cas9-mediated gene editing. Cell 9:1447

Pattabhi S et al (2019) In vivo outcome of homology-directed repair at the HBB gene in HSC using alternative donor template delivery methods. Mol Ther Nucleic Acids 17:277–288

Pavel-Dinu M et al (2019) Gene correction for SCID-X1 in long-term hematopoietic stem cells. Nat Commun 10: 1–15

Pensado A, Seijo B, Sanchez A (2014) Current strategies for DNA therapy based on lipid nanocarriers. Expert Opin Drug Deliv 11:1721–1731. https://doi.org/10.1517/17425247.2014.935337

Piel FB (2016) The present and future global burden of the inherited disorders of hemoglobin. Hematol Oncol Clin North Am 30:327–341

Platt OS (2008) Hydroxyurea for the treatment of sickle cell anemia. N Engl J Med 358:1362–1369

Popescu NC, Zimonjic D, DiPaolo JA (1990) Viral integration, fragile sites, and proto-oncogenes in human neoplasia. Hum Genet 84:383–386. https://doi.org/10.1007/BF00195804

Post Y, Clevers H (2019) Defining adult stem cell function at its simplest: the ability to replace lost cells through mitosis. Cell Stem Cell 25:174–183. https://doi.org/10.1016/j.stem.2019.07.002

Quadros RM et al (2017) Easi-CRISPR: a robust method for one-step generation of mice carrying conditional and insertion alleles using long ssDNA donors and CRISPR ribonucleoproteins. Genome Biol 18:1–15

Rai R et al (2020) Targeted gene correction of human hematopoietic stem cells for the treatment of Wiskott-Aldrich syndrome. Nat Commun 11:1–15

Ramirez CL et al (2008) Unexpected failure rates for modular assembly of engineered zinc fingers. Nat Methods 5:374–375

Ran F et al (2013) Genome engineering using the CRISPR-Cas9 system. Nat Protoc 8:2281–2308

Ran FA et al (2015) In vivo genome editing using Staphylococcus aureus Cas9. Nature 520:186–191. https://doi.org/10.1038/nature14299

Remy S et al (2017) Generation of gene-edited rats by delivery of CRISPR/Cas9 protein and donor DNA into intact zygotes using electroporation. Sci Rep 7:1–13

Renaud J-B et al (2016) Improved genome editing efficiency and flexibility using modified oligonucleotides with TALEN and CRISPR-Cas9 nucleases. Cell Rep 14:2263–2272

Richter MF et al (2020) Phage-assisted evolution of an adenine base editor with improved Cas domain compatibility and activity (vol 15, pg 891, 2020). Nat Biotechnol 38:901–901. https://doi.org/10.1038/s41587-020-0562-8

Rissone A, Burgess SM (2018) Rare genetic blood disease modeling in zebrafish. Front Genet 9:348

Rothe M, Modlich U, Schambach A (2013) Biosafety challenges for use of lentiviral vectors in gene therapy. Curr Gene Ther 13:453–468. https://doi.org/10.2174/15665232113136660006

Sakata RC et al (2020) Base editors for simultaneous introduction of C-to-T and A-to-G mutations. Nat Biotechnol 38:865–869. https://doi.org/10.1038/s41587-020-0509-0

Salisbury-Ruf CT, Larochelle A (2021) Advances and obstacles in homology-mediated gene editing of hematopoietic stem cells. J Clin Med 10:513

Salsman J, Dellaire G (2017) Precision genome editing in the CRISPR era. Biochem Cell Biol 95:187–201

Sander JD, Joung JK (2014) CRISPR-Cas systems for editing, regulating and targeting genomes. Nat Biotechnol 32:347–355

Scharenberg SG et al (2020) Engineering monocyte/macrophage-specific glucocerebrosidase expression in human hematopoietic stem cells using genome editing. Nat Commun 11:1–14

Schiroli G et al (2019) Precise gene editing preserves hematopoietic stem cell function following transient p53-mediated DNA damage response. Cell Stem Cell

24:551–565.e558. https://doi.org/10.1016/j.stem.2019.02.019

Scott T, Soemardy C, Morris KV (2020) Development of a facile approach for generating chemically modified CRISPR/Cas9 RNA. Mol Ther Nucleic Acids 19: 1176–1185. https://doi.org/10.1016/j.omtn.2020.01.004

Seita J, Weissman IL (2010) Hematopoietic stem cell: self-renewal versus differentiation. Wiley Interdiscip Rev Syst Biol Med 2:640–653

Siminovitch L, McCulloch EA, Till JE (1963) The distribution of colony-forming cells among spleen colonies. J Cell Comp Physiol 62:327

Smith LG, Weissman IL, Heimfeld S (1991) Clonal analysis of hematopoietic stem-cell differentiation in vivo. Proc Natl Acad Sci 88:2788–2792

Song F, Stieger K (2017) Optimizing the DNA donor template for homology-directed repair of double-strand breaks. Mol Ther Nucleic Acids 7:53–60

Spangrude GJ, Heimfeld S, Weissman IL (1988) Purification and characterization of mouse hematopoietic stem cells. Science 241:58–62

Strecker J et al (2019) RNA-guided DNA insertion with CRISPR-associated transposases. Science 365:48–53. https://doi.org/10.1126/science.aax9181

Strocchio L, Locatelli F (2018) Hematopoietic stem cell transplantation in thalassemia. Hematol Oncol Clin North Am 32:317–328. https://doi.org/10.1016/j.hoc.2017.11.011

Suk JS, Xu Q, Kim N, Hanes J, Ensign LM (2016) PEGylation as a strategy for improving nanoparticle-based drug and gene delivery. Adv Drug Deliv Rev 99: 28–51. https://doi.org/10.1016/j.addr.2015.09.012

Taher AT, Weatherall DJ, Cappellini MD (2018) Thalassaemia. Lancet 391:155–167

Tebas P et al (2014) Gene editing of CCR5 in autologous CD4 T cells of persons infected with HIV. N Engl J Med 370:901–910. https://doi.org/10.1056/NEJMoa1300662

Teng F et al (2019) Enhanced mammalian genome editing by new Cas12a orthologs with optimized crRNA scaffolds. Genome Biol 20:15. https://doi.org/10.1186/s13059-019-1620-8

Thompson AA et al (2018) Gene therapy in patients with transfusion-dependent β-thalassemia. N Engl J Med 378:1479–1493

Till JE, McCulloch EA (1961) A direct measurement of the radiation sensitivity of normal mouse bone marrow cells. Radiat Res 14:213–222

Till JE, McCulloch EA, Siminovitch L (1964) A stochastic model of stem cell proliferation, based on the growth of spleen colony-forming cells. Proc Natl Acad Sci 51: 29–36

Tsai SQ et al (2015) GUIDE-seq enables genome-wide profiling of off-target cleavage by CRISPR-Cas nucleases. Nat Biotechnol 33:187–197. https://doi.org/10.1038/nbt.3117

Tsai SQ et al (2017) CIRCLE-seq: a highly sensitive in vitro screen for genome-wide CRISPR-Cas9 nuclease off-targets. Nat Methods 14:607–614. https://doi.org/10.1038/nmeth.4278

Uda M et al (2008) Genome-wide association study shows BCL11A associated with persistent fetal hemoglobin and amelioration of the phenotype of beta-thalassemia. Proc Natl Acad Sci U S A 105:1620–1625. https://doi.org/10.1073/pnas.0711566105

Verdera HC, Kuranda K, Mingozzi F (2020) AAV vector immunogenicity in humans: a long journey to successful gene transfer. Mol Ther 28:723–746. https://doi.org/10.1016/j.ymthe.2019.12.010

Vo PLH et al (2021) CRISPR RNA-guided integrases for high-efficiency, multiplexed bacterial genome engineering. Nat Biotechnol 39:480–489. https://doi.org/10.1038/s41587-020-00745-y

Wagner DL, Peter L, Schmueck-Henneresse M (2021) Cas9-directed immune tolerance in humans-a model to evaluate regulatory T cells in gene therapy? Gene Ther 28:549–559. https://doi.org/10.1038/s41434-021-00232-2

Walton RT, Christie KA, Whittaker MN, Kleinstiver BP (2020) Unconstrained genome targeting with near-PAMless engineered CRISPR-Cas9 variants. Science 368:290–296. https://doi.org/10.1126/science.aba8853

Wright DA, Li T, Yang B, Spalding MH (2014) TALEN-mediated genome editing: prospects and perspectives. Biochem J 462:15–24

Wu A, Till J, Siminovitch L, McCulloch E (1967) A cytological study of the capacity for differentiation of normal hemopoietic colony-forming cells. J Cell Physiol 69:177–184

Wu AM, Till JE, Siminovitch L, McCulloch EA (1968) Cytological evidence for a relationship between normal hematopoietic colony-forming cells and cells of the lymphoid system. J Exp Med 127:455–464

Wu Y et al (2019) Highly efficient therapeutic gene editing of human hematopoietic stem cells. Nat Med 25:776–783. https://doi.org/10.1038/s41591-019-0401-y

Xu L et al (2019) CRISPR-edited stem cells in a patient with HIV and acute lymphocytic leukemia. N Engl J Med 381:1240–1247. https://doi.org/10.1056/NEJMoa1817426

Yang Y et al (2016) A dual AAV system enables the Cas9-mediated correction of a metabolic liver disease in newborn mice. Nat Biotechnol 34:334–338. https://doi.org/10.1038/nbt.3469

Yang H et al (2020) Methods favoring homology-directed repair choice in response to CRISPR/Cas9 induced-double strand breaks. Int J Mol Sci 21:6461

Yip BH (2020) Recent advances in CRISPR/Cas9 delivery strategies. Biomol Ther 10:839. https://doi.org/10.3390/biom10060839

Zeng J et al (2020) Therapeutic base editing of human hematopoietic stem cells. Nat Med 26:535–541. https://doi.org/10.1038/s41591-020-0790-y

Zetsche B et al (2015) Cpf1 is a single RNA-guided endonuclease of a class 2 CRISPR-Cas system. Cell 163:759–771. https://doi.org/10.1016/j.cell.2015.09.038

Zhang F, Wen Y, Guo X (2014) CRISPR/Cas9 for genome editing: progress, implications and challenges. Hum Mol Genet 23:R40–R46

Zhang W et al (2020a) A high-throughput small molecule screen identifies farrerol as a potentiator of CRISPR/Cas9-mediated genome editing. elife 9:e56008

Zhang X et al (2020b) Dual base editor catalyzes both cytosine and adenine base conversions in human cells. Nat Biotechnol 38:856–860. https://doi.org/10.1038/s41587-020-0527-y

12 Aging, Causes, and Rejuvenation of Hematopoietic Stem Cells

Zhiyang Chen, Zhenyu Ju, and Yan Sun

Abstract

Hematopoietic stem cells (HSCs) undergo an age-related functional decline, which leads to a disruption of the blood system and contributes to the development of aging-associated hematopoietic diseases and malignancies. In this section, we provide a summary of the key hallmarks associated with HSC aging. We also examine the causal factors that contribute to HSC aging and emphasize potential approaches to mitigate HSC aging and age-related hematopoietic disorders.

Keywords

HSC · Aging · Niche · Therapeutic strategies

Z. Chen · Z. Ju
Key Laboratory of Regenerative Medicine of Ministry of Education, Guangzhou Regenerative Medicine and Health Guangdong Laboratory, Institute of Aging and Regenerative Medicine, Jinan University, Guangzhou, Guangdong, China
e-mail: zhiyangchen@jnu.edu.cn; zhenyuju@jnu.edu.cn

Y. Sun (✉)
Medical Research Institute, Guangdong Provincial People's Hospital (Guangdong Academy of Medical Sciences), Southern Medical University, Guangzhou, China
e-mail: sunyan@gdph.org.cn

12.1 Introduction

Aging is a spontaneous, inevitable, and complex natural process leading to a progressive functional decline of tissues and organs, accompanied by a decreased capacity to resist multiple stress and increased susceptibility to diseases, eventually resulting in death (Lopez-Otin et al. 2013). The aged blood has a diminished immune function and increased occurrences of anemia and myeloid and lymphoid malignancies, such as age-related clonal hematopoiesis, myelodysplastic syndromes, acute myeloid leukemia, chronic lymphocytic leukemia, multiple myeloma, and non-Hodgkin's lymphoma (de Haan and Lazare 2018). In particular, declining adaptive immune function and increased innate immune inflammation in the elderly populations lead to vaccine failure and increased risk and severity of infectious diseases (Weiskopf et al. 2009). It is well known that stem cell functionality declines with aging, contributing to age-associated pathologies and the whole organismal aging (Oh et al. 2014; Chandel et al. 2016). HSCs produce all blood cell types throughout an organism's life span and are defined by their inherent capacity for self-renewal and differentiation (Seita and Weissman 2010). HSC aging is associated with a gradual functional decline, leading to age-related hematopoietic disorders. Therefore, it is imperative to understand the biology of aging in HSC and develop new therapeutic approaches for treating and preventing

M. Zhao, P. Qian (eds.), *Hematopoietic Stem Cells*, Advances in Experimental Medicine and Biology 1442, https://doi.org/10.1007/978-981-99-7471-9_12

age-related hematopoietic and immune disorders and health-promoting strategies to support the quality of life for the elderly urgently needed. Most HSC aging studies have been carried out in mice and human, and accumulating evidences demonstrate that the unique aging properties of HSCs are genetically and epigenetically modulated, including both cell intrinsic and non-intrinsic factors. These factors have been involved in various physiological or pathological events such as oxidative stress, DNA damage, telomere shortening, and cell polarity and linked to decreased HSC self-renewal and regenerative capacity, increased HSC quantities, and lineage-biased differentiation. However, despite the aforementioned findings, the underlying mechanisms for HSC aging are yet to be fully understood. Identifying a more in-depth elucidation of the molecular and cellular mechanisms of the biological processes in HSC aging would facilitate understanding and intervention in the underlying mechanism adult HSC aging.

12.2 Cell Intrinsic Changes in HSCs During Aging

Aging of hematopoietic system is correlated with a high risk of anemia, impaired function of the adaptive immune system, and increased incidence of myeloid blood disorders (Verovskaya et al. 2019; Kurosawa and Iwama 2020). The aging-associated changes in blood system are most likely attributed to the alterations of HSCs upon aging. During the aging process, HSCs from older mice tend to exhibit an increased accumulation of DNA damage, as evidenced by the presence of γH2AX foci and comet assay (Moehrle and Geiger 2016). Aged HSCs exhibited a biased differentiation capacity toward myeloid lineages with a reduced lymphocyte output (Li et al. 2020). In line with these phenotypes, gene expression analyses have identified an upregulation of myeloid genes and a downregulation of lymphopoiesis-related genes in aged HSCs (Sun et al. 2014). In contrast to many other cell types that typically decrease in number during the aging process, it has been observed that the population of phenotypically defined HSCs actually increase with age in both mice and humans (Rossi et al. 2005; Chambers et al. 2007). Aged HSCs are deficient in their ability in regeneration and exhibit a diminished homing capacity (Liang et al. 2005). Furthermore, an increasing number of studies have demonstrated the presence of mutations, such as DNMT3A, TET2, ASXL1, JAK2, and others, in the aged hematopoietic system. These mutations are frequently observed at the stem and progenitor cell level and are strongly associated with the development of hematologic diseases during the aging process (Akunuru and Geiger 2016; Fujino et al. 2021; Chin et al. 2022). Furthermore, the epigenetic machineries that maintain the HSC transcription integrity decline during aging (Kramer and Challen 2017). The change of epigenetic programs controlling HSC function occurs at several levels, including DNA/histone methylation, histone acetylation, non-coding RNA, and high order of chromatin structure (Sun et al. 2014). In this section, we summarize the recent research identifying the changes in inflammation, metabolism, epigenetics, and also the clonal hematopoiesis in HSCs during aging.

12.2.1 Inflamm-Aging of HSCs

In addition to the function of HSCs in maintaining the homeostasis of hematopoietic process, they must also respond rapidly to hematopoietic challenges, including infection or blood loss. HSCs can be directly/indirectly activated and participate in hematopoiesis to meet the acute demands of inflammatory responses (Zhao et al. 2014a). Organism aging is characterized by increased inflammation and decreased stem cell function. Recent findings highlight the emerging role of inflammation signaling in shaping the hematopoietic system during HSC fate determination and aging. Aged HSCs exhibited increased responsiveness to inflammation-induced loss of self-renewal and enhanced differentiation (Chen et al. 2019). The alteration of microbiota and increased gut leakage are associated with the increase in the chronic low-grade inflammation during aging (O'Toole and Jeffery 2015; Nagpal et al. 2018). A recent study by Kovtonyuk et al. described for the first time the role of a microbiome/IL-1/IL1R1 axis as a self-sustaining driver of HSC aging (Kovtonyuk et al. 2022). Targeting IL-1 or antibiotic treatment attenuates the myeloid-biased differentiation phenotype of aged HSC (Kovtonyuk et al. 2022). Extensive attention has been paid to understand the relationship between the increased chronic inflammation and their role in aging-associated hematological malignancy. Mutations in Dnmt3a have been identified as drivers of clonal hematopoiesis, which is known to increase the risk of developing myeloid malignancies and adversely affect overall survival (Challen and Goodell 2020; Kandarakov and Belyavsky 2020). Recent studies have shown that the age-elevated inflammation promotes DNMT3a mutant HSC expansion leading to clonal hematopoiesis, raising the clinical value for treating myeloid malignancies with aging (Hormaechea-Agulla et al. 2021; Liao et al. 2022).

12.2.2 Clonal Hematopoiesis in Aging

Clonal hematopoiesis (CH) is a common finding among aging HSCs and is clearly linked to blood cancers, cardiovascular disease, and overall mortality. Thus, the potential health implications of CHIP (CH of indeterminate potential) are broad, and preventing CHIP progression undoubtedly benefits human health. However, these mechanisms by which the CHIP-associated mutations cause clonal expansion and enhanced pathogenesis of age-related diseases are elusive. Furthermore, charactering CH and understanding its clinical impact require the assembly of larger genetic cohorts and comprehensive genomic profiling, accompanied by extensive phenotypic information. These efforts can systematically uncover and potentially expand the links between CH and various clinical outcomes over time (Genovese et al. 2014; Jaiswal et al. 2014; Xie et al. 2014; Young et al. 2016; Zink et al. 2017). Chen et al. found that TP53 mutations drive clonal hematopoiesis and confer a competitive advantage to HSPCs by modulating epigenetic regulator EZH2, a promising target for preventing CHIP progression and treating hematological malignancies with TP53 mutations (Chen et al. 2018). PPM1D is a phosphatase that negatively regulates p53, a DNA damage response regulator frequently mutated in CH. Mutant Ppm1d hematopoietic cells outcompeted their wild-type counterparts following exposure to cytotoxic DNA-damaging agents (Hsu et al. 2018; Kahn et al. 2018); however, they do not recover from bone marrow transplantation. Both Tp53 and Ppm1d mutant HSC and other members of the DDR pathway can proliferate, leading to a stem cell competitive advantage in the setting of cytotoxic DNA-damaging agents (Bondar and Medzhitov 2010; Hsu et al. 2018; Kahn et al. 2018). While p53 and PPMID are involved in the DDR pathway, they may have distinct roles in accelerating HSPC expansion. It is unclear why mutations in TET2 or DNMT3A lead to clonal expansion. The recent findings reveal that the inflammatory cytokines after bacterial infection, which could be critical to the pre-leukemic expansion of somatic TET2 mutations (Meisel et al. 2018), indicate unidentified stressors. Therefore, it will be essential to understand the fitness landscapes that promote expansion of cells with distinct CH-associated mutations.

12.3 Aging of the Hematopoietic Stem Cell Niche

The HSC niche, a unique microenvironment, is organized in a complex three-dimensional architecture and composed of multiple hematopoietic and non-hematopoietic cells, extracellular matrix (ECM) components, and soluble factors to support HSC function (Yu and Scadden 2016). Three types of BM niches for HSCs have been described, these being designated the endosteal (Calvi et al. 2003; Zhang et al. 2003; Haylock et al. 2007; Guidi et al. 2017), arteriolar (Kunisaki et al. 2013), and sinusoid niches (Ding et al. 2012; Acar et al. 2015; Chen et al. 2016; Itkin et al. 2016). Although the enormous advances has been made in the understanding of the structure, function, and contribution of the BM niche in regulating HSCs, there are still numerous unexplored aspects that remain further clarification. However, the specific localization of HSCs and their niches remain challenging to determine and are still largely unknown (Calvi et al. 2003; Zhang et al. 2003; Kiel et al. 2005; Haylock et al. 2007; Mendez-Ferrer et al. 2010; Ding et al. 2012; Heazlewood et al. 2013; Kunisaki et al. 2013; Bruns et al. 2014; Nakamura-Ishizu et al. 2014; Zhao et al. 2014b; Acar et al. 2015; Chen et al. 2016; Itkin et al. 2016; Guidi et al. 2017). During aging, various specialized niche compartments undergo degeneration and remodeling, which in turn impacts the HSC behavior and function. Below, we will review the major niche cell types that have been identified to provide crucial support to HSCs.

12.3.1 Vascular Niche Upon Aging

During aging, the structure and function of BM vascular niche is profoundly alternated (Kusumbe et al. 2016; Sacma et al. 2019). Arteries and arterioles appear to be decreased in their length and diameter, which results in disorganization and loss of support for the maintenance of HSC quiescence (Sacma et al. 2019; Renders et al. 2021). Specifically, the vessels harboring type-H endothelial cells (ECs), which serve as a bridge between arteriolar and sinusoidal capillaries, undergo a significantly reduction over time. Meanwhile, smaller capillaries within the central bone marrow containing type-L ECs (with a diameter of less than 6 mm) exhibit expansion (Kusumbe et al. 2016), while small capillaries containing type L ECs (< 6 mm in diameter) in the central BM are expanded (Kusumbe et al. 2016; Ho et al. 2019). Sinusoids in the aged group maintain similar parameters of vessels compared to young samples. However, endothelial cell frequency (EC) decreases upon aging (Sacma et al. 2019). This vascular remodeling supports HSC functions and directly affects HSC behaviors (Kusumbe et al. 2016). During physiological aging, the impairment of Notch signaling in the arteriolar niche is because of the low expression of Dll1 in endosteal/arteriolar niches and Dll4 in type H EC. In addition, Jag2 is highly expressed at bone marrow ECs (Guo et al. 2017), and the expression level is retained within aging (Sacma et al. 2019). After chemotherapy, in the aged group, due to the lack of recovery of endothelial Jag2 at sinusoids, such treatment causes impaired reconstitution of sinusoid niches and hematopoiesis (Kusumbe et al. 2016, Sacma et al. 2019). These studies indicate that Notch signaling is critical in regulating aged endosteal/arteriolar and sinusoid niches (Sacma et al. 2019).

12.3.2 Sympathetic Adrenergic Signal Alterations

It has been reported that sympathetic adrenergic signals significantly influence the homing and egress of HSCs and granulocyte colony-stimulating factor-induced mobilization of HSC (Fielding and Mendez-Ferrer 2020), and the SNS innervation is strongly changed during aging (Maryanovich et al. 2018; Ho et al. 2019). However, the analyses on the changes in sympathetic adrenergic innervation showed conflicting findings. A recent study has suggested that noradrenergic nerve fibers by staining for tyrosine hydroxylase are reduced in old murine BM, coupled with a reduction of perivascular Nes-GFP bright cell innervation (Maryanovich et al. 2018). Taking advantage of surgical denervation of young BM to induce premature aging of the hematopoietic system, it has been shown that increased proliferation and lost polarity in HSC and $CD41^{+}$ HSCs are expanded (Maryanovich et al. 2018). Moreover, the removal of b3-ADR in young mice causes a premature aging phenotype of HSCs, and treatment with the b3-ADR could rejuvenate the aged phenotype after denervation. These findings suggest that b3-adrenergic activation is crucial in preserving the regenerative capacity of HSC (Maryanovich et al. 2018). Conversely, a more recent study demonstrated that sympathetic noradrenergic fibers increased in the mouse bone marrow (Ho et al. 2019). During aging, b2-adrenergic stimulation is increased and triggers MSC secretion of IL6, thus promoting myeloid cell expansion. Despite the absence of a consensus on the extent of adrenergic signaling alteration upon aging, it is clear that this adrenergic innervation strongly affects HSCs, contributing to the aging-associated phenotype through activation or inactivation of different AR. Hence, it is crucial to further investigate the potential contribution of other adrenergic and choligergic signaling that may influence hematopoietic aging (Maryanovich et al. 2018), which should be further investigated.

12.3.3 Endosteal Niche Degeneration

Aging greatly impacts the endosteal niche, making it one of the most affected compartments. Functionally, aged BM-derived MSCs demonstrate several qualitative changes. These changes include a diminished capacity for colony-forming unit fibroblasts (CFU-F) and mesensphere formation in vitro, decreased expression of HSC niche factors, and a reduction in the number of osteoblasts. Furthermore, there is a decrease in the release of osteopontin (OPN), which negatively regulates HSC proliferation (Chen et al. 2016; Maryanovich et al. 2018). Recent studies revealed that the numbers of nestin-GFP bright PDGFRβ^+NG2$^+$ MSCs in endosteal regions are reduced in the aged BM, while NES bright cells undergo a shift in their localization from arteries to sinusoids (Maryanovich et al. 2018; Ho et al. 2019; Sacma et al. 2019). Nes-GFP bright cell frequency decreases in the endosteum and increases near the central vein. Aged HSCs are more distant from the endosteum, arterioles, nestin-GFP high cells, and megakaryocytes. However, HSC distance from sinusoids and nestin-GFP low cells remains unaltered (Maryanovich et al. 2018, Ho et al. 2019, Sacma et al. 2019). These results strongly indicate that the BM microenvironment undergo changes with aging. HSCs lodge near perisinusoidal niches/non-endosteal (central) niches due to the contraction of endosteal niche. During aging, accumulation of BM adipocytes reduces hematopoietic repopulation capacity and disrupts bone regeneration (Ambrosi et al. 2017). As a consequence, an increase in adipocytes contributes to the high risk of osteoporosis and bone fracture in the elderly (Fazeli et al. 2013; Schwartz 2015). Recently, a study showed that maturing myeloid cell density increases surrounding adipocytes in the BM microenvironment, suggesting adipocytes may promote myeloid skewing of aging HSPCs (Aguilar-Navarro et al. 2020). The aging-related contraction of endosteal BM might cause lymphoid deficiency owing to the reduction in lymphocyte-specific niches. Recently, accumulating evidence has shed light on the dynamic interactions between B cell progenitors and MSCs in BM. These interactions play a vital role in supporting B lymphopoiesis and are mediated by key factors such as Cxcl12 and Il17. In addition, during aging, the frequency of LepR$^+$Osteolectin$^+$ osteogenic progenitors in BM niche decreases, which is important for osteogenesis and lymphopoiesis, contributing to the reduction of the amount of CLP within the BM in the elderly.

12.4 Perspective of Therapeutic Strategies for Anti-HSC Aging

Following a better understanding of the mechanisms underlying HSC aging, many approaches have been found to exert anti-HSC aging effect, including clearance of aged HSC by BCL-2 inhibitor ABT263 (Chang et al. 2016), reduction of ROS level by p38 MAPK inhibition (Ito et al. 2006; Jung et al. 2016), and decreasing mTOR activity by rapamycin (Chen et al. 2009). However, the effects of these methods on HSC rejuvenation still need to be fully revealed and optimized. Many other approaches that potentially reverse HSC aging are underway. In this section, we summarize the approaches that could potentially serve as therapeutic interventions to ameliorate HSC aging, including dietary restriction, prolonged fasting, and heterochronic parabiosis.

An increase in chronic inflammation is a driving force accelerating HSC aging (Caiado et al. 2021; Trowbridge and Starczynowski 2021; Kovtonyuk et al. 2022). Caloric restriction has been known to improve chronic inflammatory diseases via reducing circulating monocytes and pro-growth factors (Jordan et al. 2019). Many studies have shown caloric restriction improves animal health and elongates life span (Green et al. 2022; Spadaro et al. 2022). However, the underlying mechanism of caloric restriction on rejuvenation of HSC aging remains unclear. Cheng et al. show that prolonged fasting and refeeding cycles reduce circulating insulin-like growth factor-1 (IGF-1) levels and protein kinase A (PKA) activity, which enhance HSC self-renewal capacity and rejuvenate functional decline of aged HSC (Cheng et al. 2014). In line with this study, Tang et al. demonstrate that caloric restriction by 30% promotes maintenance of HSC quiescence and ameliorates aged HSC self-renewal by reducing IGF-1 (Tang et al. 2016). However, other study shows contradicting result that reduced IGF-1 signal in bone marrow microenvironment promotes HSC aging at middle age and that IFG-1 supplementation effectively rescues molecular and functional hallmarks of HSC aging (Young et al. 2021). In addition, rejuvenated old HSCs following prolonged fasting exhibit balanced lymphoid and myeloid differentiation (Cheng et al. 2014). Conversely, 30% reduction of calorie intake disturbs lymphopoiesis by reducing IL-7 and IL-6 levels (Tang et al. 2016). Thus, the exact role of IGF-1 and caloric restriction on HSC aging remains to be elucidated. In addition, the beneficial effects of caloric restriction on animal models have been investigated for a long time, yet the corresponding results in clinical trials have not been widely tested. A long-term caloric restriction could be a stressful procedure for vulnerable old individuals; it is therefore important to explore the molecule that targets the key signal underlying HSC aging to improve the therapeutic feasibility.

Parabiosis refers to the laboratory technique that combines two living organisms by surgical operation, allowing a shared blood circulation to be established. Heterochronic parabiosis, the pairing between a young and an aged mouse, has been widely used in aging research to assess the contribution of systemic factors that influence the processes of aging (Ashapkin et al. 2020). In the past decades, studies using heterochronic parabiosis convincingly claim the positive rejuvenation effects of a youthful systemic environment in several organs, including the skeletal muscle, heart, liver, and nervous system (Ashapkin et al. 2020; Ambrosi et al. 2021; Brioschi et al. 2021). A recent study shows that aged hematopoietic stem and progenitor cells (HSPCs) are also responsive to the changing systemic milieu after heterochronic parabiosis (Ma et al. 2022). Aged HSPCs express youthful transcriptomes as well as rejuvenating factors YY1 and CCL3, which are shown to reverse the functional decline and myeloid-biased differentiation of old HSPCs, respectively (Ma et al. 2022). In line with this finding, other study also indicates that HSCs are one of the cell types showing high responsiveness to heterochronic parabiosis (Palovics et al. 2022). However, it has also been claimed that neither parabiosis nor young serum transfusion rejuvenates the defect of aged HSC (Anon 2010; Kuribayashi et al. 2021). Wakako et al. have shown that transplantation of aged HSCs in young recipients without pre-conditioning allows successful engraftment of the aged HSCs in a young environment (Kuribayashi et al. 2021). Despite residing in a youthful niche, the old HSCs show myeloid-biased differentiation property and do not have any functional improvement (Kuribayashi et al. 2021). Therefore, the effect of blood-borne factors on rejuvenation of the HSC aging phenotype is inconclusive and requires further investigation.

12.5 Conclusive Remarks

In this chapter, we summarize the hallmarks of HSC aging with regard to self-renewal, myeloid-biased differentiation, and the connection of HSC aging to blood malignancies. HSC aging can be attributed to intrinsic changes, which encompass alterations in epigenetic patterns, metabolite levels, and an upregulation of inflammatory signaling. Extrinsic factors such as systemic factors and niche cells have also been known to affect HSC aging. As a result, functional decline of HSC during aging leads to defects in blood regeneration, immunosenescence, and an increase in inflammatory signals, which in turn accelerate aging process of blood system. In addition, clonal expansion of aged HSC caused by mutation accumulation has been shown to be associated with a higher risk of cardiovascular disease, suggesting a role of HSC aging in facilitating aging-related diseases in other organs (Pardali et al. 2020; Asada and Kitamura 2021). A better understanding of the mechanisms underlying HSC aging would promote discovery of novel factors that control HSC aging. This will not only pave a road for anti-HSC aging therapeutics but also lay the foundation for the treatment of HSC aging-related diseases.

References

Acar M, Kocherlakota KS, Murphy MM, Peyer JG, Oguro H, Inra CN, Jaiyeola C, Zhao Z, Luby-Phelps K, Morrison SJ (2015) Deep imaging of bone marrow shows non-dividing stem cells are mainly perisinusoidal. Nature 526(7571):126–130

Aguilar-Navarro AG, Meza-Leon B, Gratzinger D, Juarez-Aguilar FG, Chang Q, Ornatsky O, Tsui H, Esquivel-Gomez R, Hernandez-Ramirez A, Xie SZ, Dick JE, Flores-Figueroa E (2020) Human aging alters the spatial organization between CD34+ hematopoietic cells and adipocytes in bone marrow. Stem Cell Rep 15(2): 317–325

Akunuru S, Geiger H (2016) Aging, clonality, and rejuvenation of hematopoietic stem cells. Trends Mol Med 22(8):701–712

Ambrosi TH, Scialdone A, Graja A, Gohlke S, Jank AM, Bocian C, Woelk L, Fan H, Logan DW, Schurmann A, Saraiva LR, Schulz TJ (2017) Adipocyte accumulation in the bone marrow during obesity and aging impairs stem cell-based hematopoietic and bone regeneration. Cell Stem Cell 20(6):771–784.e776

Ambrosi TH, Marecic O, McArdle A, Sinha R, Gulati GS, Tong X, Wang Y, Steininger HM, Hoover MY, Koepke LS, Murphy MP, Sokol J, Seo EY, Tevlin R, Lopez M, Brewer RE, Mascharak S, Lu L, Ajanaku O, Conley SD, Seita J, Morri M, Neff NF, Sahoo D, Yang F, Weissman IL, Longaker MT, Chan CKF (2021) Aged skeletal stem cells generate an inflammatory degenerative niche. Nature 597(7875):256–262

Anon (2010) Retraction. Systemic signals regulate ageing and rejuvenation of blood stem cell niches. Nature 467(7317):872

Asada S, Kitamura T (2021) Clonal hematopoiesis and associated diseases: a review of recent findings. Cancer Sci 112(10):3962–3971

Ashapkin VV, Kutueva LI, Vanyushin BF (2020) The effects of parabiosis on aging and age-related diseases. Adv Exp Med Biol 1260:107–122

Bondar T, Medzhitov R (2010) p53-mediated hematopoietic stem and progenitor cell competition. Cell Stem Cell 6(4):309–322

Brioschi S, Wang WL, Peng V, Wang M, Shchukina I, Greenberg ZJ, Bando JK, Jaeger N, Czepielewski RS, Swain A, Mogilenko DA, Beatty WL, Bayguinov P, Fitzpatrick JAJ, Schuettpelz LG, Fronick CC, Smirnov I, Kipnis J, Shapiro VS, Wu GF, Gilfillan S, Cella M, Artyomov MN, Kleinstein SH, Colonna M (2021) Heterogeneity of meningeal B cells reveals a lymphopoietic niche at the CNS borders. Science 373(6553):eabf9277

Bruns I, Lucas D, Pinho S, Ahmed J, Lambert MP, Kunisaki Y, Scheiermann C, Schiff L, Poncz M, Bergman A, Frenette PS (2014) Megakaryocytes regulate hematopoietic stem cell quiescence through CXCL4 secretion. Nat Med 20(11):1315–1320

Caiado F, Pietras EM, Manz MG (2021) Inflammation as a regulator of hematopoietic stem cell function in disease, aging, and clonal selection. J Exp Med 218(7): e20201541

Calvi LM, Adams GB, Weibrecht KW, Weber JM, Olson DP, Knight MC, Martin RP, Schipani E, Divieti P, Bringhurst FR, Milner LA, Kronenberg HM, Scadden DT (2003) Osteoblastic cells regulate the haematopoietic stem cell niche. Nature 425(6960): 841–846

Challen GA, Goodell MA (2020) Clonal hematopoiesis: mechanisms driving dominance of stem cell clones. Blood 136(14):1590–1598

Chambers SM, Shaw CA, Gatza C, Fisk CJ, Donehower LA, Goodell MA (2007) Aging hematopoietic stem cells decline in function and exhibit epigenetic dysregulation. PLoS Biol 5(8):e201

Chandel NS, Jasper H, Ho TT, Passegue E (2016) Metabolic regulation of stem cell function in tissue homeostasis and organismal ageing. Nat Cell Biol 18(8): 823–832

Chang J, Wang Y, Shao L, Laberge RM, Demaria M, Campisi J, Janakiraman K, Sharpless NE, Ding S, Feng W, Luo Y, Wang X, Aykin-Burns N, Krager K, Ponnappan U, Hauer-Jensen M, Meng A, Zhou D (2016) Clearance of senescent cells by ABT263 rejuvenates aged hematopoietic stem cells in mice. Nat Med 22(1):78–83

Chen C, Liu Y, Liu Y, Zheng P (2009) mTOR regulation and therapeutic rejuvenation of aging hematopoietic stem cells. Sci Signal 2(98):ra75

Chen JY, Miyanishi M, Wang SK, Yamazaki S, Sinha R, Kao KS, Seita J, Sahoo D, Nakauchi H, Weissman IL (2016) Hoxb5 marks long-term haematopoietic stem cells and reveals a homogenous perivascular niche. Nature 530(7589):223–227

Chen S, Gao R, Yao C, Kobayashi M, Liu SZ, Yoder MC, Broxmeyer H, Kapur R, Boswell HS, Mayo LD, Liu Y (2018) Genotoxic stresses promote clonal expansion of hematopoietic stem cells expressing mutant p53. Leukemia 32(3):850–854

Chen Z, Amro EM, Becker F, Holzer M, Rasa SMM, Njeru SN, Han B, Di Sanzo S, Chen Y, Tang D, Tao S, Haenold R, Groth M, Romanov VS, Kirkpatrick JM, Kraus JM, Kestler HA, Marz M, Ori A, Neri F, Morita Y, Rudolph KL (2019) Cohesin-mediated NF-kappaB signaling limits hematopoietic stem cell self-renewal in aging and inflammation. J Exp Med 216(1):152–175

Cheng CW, Adams GB, Perin L, Wei M, Zhou X, Lam BS, Da Sacco S, Mirisola M, Quinn DI, Dorff TB, Kopchick JJ, Longo VD (2014) Prolonged fasting reduces IGF-1/PKA to promote hematopoietic-stem-cell-based regeneration and reverse immunosuppression. Cell Stem Cell 14(6):810–823

Chin DWL, Yoshizato T, Virding Culleton S, Grasso F, Barbachowska M, Ogawa S, Jacobsen SEW, Woll PS (2022) Aged healthy mice acquire clonal hematopoiesis mutations. Blood 139(4):629–634

de Haan G, Lazare SS (2018) Aging of hematopoietic stem cells. Blood 131(5):479–487

Ding L, Saunders TL, Enikolopov G, Morrison SJ (2012) Endothelial and perivascular cells maintain haematopoietic stem cells. Nature 481(7382):457–462

Fazeli PK, Horowitz MC, MacDougald OA, Scheller EL, Rodeheffer MS, Rosen CJ, Klibanski A (2013) Marrow fat and bone—new perspectives. J Clin Endocrinol Metab 98(3):935–945

Fielding C, Mendez-Ferrer S (2020) Neuronal regulation of bone marrow stem cell niches. F1000Res 9:F1000

Fujino T, Goyama S, Sugiura Y, Inoue D, Asada S, Yamasaki S, Matsumoto A, Yamaguchi K, Isobe Y, Tsuchiya A, Shikata S, Sato N, Morinaga H, Fukuyama T, Tanaka Y, Fukushima T, Takeda R, Yamamoto K, Honda H, Nishimura EK, Furukawa Y, Shibata T, Abdel-Wahab O, Suematsu M, Kitamura T (2021) Mutant ASXL1 induces age-related expansion of phenotypic hematopoietic stem cells through activation of Akt/mTOR pathway. Nat Commun 12(1):1826

Genovese G, Kähler AK, Handsaker RE, Lindberg J, Rose SA, Bakhoum SF, Chambert K, Mick E, Neale BM, Fromer M, Purcell SM, Svantesson O, Landén M, Höglund M, Lehmann S, Gabriel SB, Moran JL, Lander ES, Sullivan PF, Sklar P, Grönberg H, Hultman CM, McCarroll SA (2014) Clonal hematopoiesis and blood-cancer risk inferred from blood DNA sequence. N Engl J Med 371(26):2477–2487

Green CL, Lamming DW, Fontana L (2022) Molecular mechanisms of dietary restriction promoting health and longevity. Nat Rev Mol Cell Biol 23(1):56–73

Guidi N, Sacma M, Standker L, Soller K, Marka G, Eiwen K, Weiss JM, Kirchhoff F, Weil T, Cancelas JA, Florian MC, Geiger H (2017) Osteopontin attenuates aging-associated phenotypes of hematopoietic stem cells. EMBO J 36(10):1463

Guo P, Poulos MG, Palikuqi B, Badwe CR, Lis R, Kunar B, Ding BS, Rabbany SY, Shido K, Butler JM, Rafii S (2017) Endothelial jagged-2 sustains hematopoietic stem and progenitor reconstitution after myelosuppression. J Clin Invest 127(12):4242–4256

Haylock DN, Williams B, Johnston HM, Liu MC, Rutherford KE, Whitty GA, Simmons PJ, Bertoncello I, Nilsson SK (2007) Hemopoietic stem cells with higher hemopoietic potential reside at the bone marrow endosteum. Stem Cells 25(4):1062–1069

Heazlewood SY, Neaves RJ, Williams B, Haylock DN, Adams TE, Nilsson SK (2013) Megakaryocytes co-localise with hemopoietic stem cells and release cytokines that up-regulate stem cell proliferation. Stem Cell Res 11(2):782–792

Ho YH, Del Toro R, Rivera-Torres J, Rak J, Korn C, Garcia-Garcia A, Macias D, Gonzalez-Gomez C, Del Monte A, Wittner M, Waller AK, Foster HR, Lopez-Otin C, Johnson RS, Nerlov C, Ghevaert C, Vainchenker W, Louache F, Andres V, Mendez-Ferrer S (2019) Remodeling of bone marrow hematopoietic stem cell niches promotes myeloid cell expansion during premature or physiological aging. Cell Stem Cell 25(3):407–418.e406

Hormaechea-Agulla D, Matatall KA, Le DT, Kain B, Long X, Kus P, Jaksik R, Challen GA, Kimmel M, King KY (2021) Chronic infection drives Dnmt3a-loss-of-function clonal hematopoiesis via IFNgamma signaling. Cell Stem Cell 28(8):1428–1442.e1426

Hsu JI, Dayaram T, Tovy A, De Braekeleer E, Jeong M, Wang F, Zhang J, Heffernan TP, Gera S, Kovacs JJ, Marszalek JR, Bristow C, Yan Y, Garcia-Manero G, Kantarjian H, Vassiliou G, Futreal PA, Donehower LA, Takahashi K, Goodell MA (2018) PPM1D mutations drive clonal hematopoiesis in response to cytotoxic chemotherapy. Cell Stem Cell 23(5): 700–713.e706

Itkin T, Gur-Cohen S, Spencer JA, Schajnovitz A, Ramasamy SK, Kusumbe AP, Ledergor G, Jung Y, Milo I, Poulos MG, Kalinkovich A, Ludin A, Kollet O, Shakhar G, Butler JM, Rafii S, Adams RH, Scadden DT, Lin CP, Lapidot T (2016) Distinct bone marrow blood vessels differentially regulate haematopoiesis. Nature 532(7599):323–328

Ito K, Hirao A, Arai F, Takubo K, Matsuoka S, Miyamoto K, Ohmura M, Naka K, Hosokawa K, Ikeda Y, Suda T (2006) Reactive oxygen species act through p38 MAPK to limit the lifespan of hematopoietic stem cells. Nat Med 12(4):446–451

Jaiswal S, Fontanillas P, Flannick J, Manning A, Grauman PV, Mar BG, Lindsley RC, Mermel CH, Burtt N, Chavez A, Higgins JM, Moltchanov V, Kuo FC, Kluk MJ, Henderson B, Kinnunen L, Koistinen HA, Ladenvall C, Getz G, Correa A, Banahan BF, Gabriel S, Kathiresan S, Stringham HM, McCarthy MI, Boehnke M, Tuomilehto J, Haiman C, Groop L, Atzmon G, Wilson JG, Neuberg D, Altshuler D, Ebert BL (2014) Age-related clonal hematopoiesis associated with adverse outcomes. N Engl J Med 371(26):2488–2498

Jordan S, Tung N, Casanova-Acebes M, Chang C, Cantoni C, Zhang D, Wirtz TH, Naik S, Rose SA, Brocker CN, Gainullina A, Hornburg D, Horng S, Maier BB, Cravedi P, LeRoith D, Gonzalez FJ, Meissner F, Ochando J, Rahman A, Chipuk JE, Artyomov MN, Frenette PS, Piccio L, Berres ML, Gallagher EJ, Merad M (2019) Dietary intake regulates the circulating inflammatory monocyte pool. Cell 178(5):1102–1114.e1117

Jung H, Kim DO, Byun JE, Kim WS, Kim MJ, Song HY, Kim YK, Kang DK, Park YJ, Kim TD, Yoon SR, Lee HG, Choi EJ, Min SH, Choi I (2016) Thioredoxin-interacting protein regulates haematopoietic stem cell ageing and rejuvenation by inhibiting p38 kinase activity. Nat Commun 7:13674

Kahn JD, Miller PG, Silver AJ, Sellar RS, Bhatt S, Gibson C, McConkey M, Adams D, Mar B, Mertins P, Fereshetian S, Krug K, Zhu H, Letai A, Carr SA, Doench J, Jaiswal S, Ebert BL (2018) PPM1D-truncating mutations confer resistance to chemotherapy and sensitivity to PPM1D inhibition in hematopoietic cells. Blood 132(11):1095–1105

Kandarakov O, Belyavsky A (2020) Clonal hematopoiesis, cardiovascular diseases and hematopoietic stem cells. Int J Mol Sci 21(21):7902

Kiel MJ, Yilmaz OH, Iwashita T, Yilmaz OH, Terhorst C, Morrison SJ (2005) SLAM family receptors distinguish hematopoietic stem and progenitor cells and reveal endothelial niches for stem cells. Cell 121(7): 1109–1121

Kovtonyuk LV, Caiado F, Garcia-Martin S, Manz EM, Helbling P, Takizawa H, Boettcher S, Al-Shahrour F, Nombela-Arrieta C, Slack E, Manz MG (2022) IL-1 mediates microbiome-induced inflammaging of hematopoietic stem cells in mice. Blood 139(1):44–58

Kramer A, Challen GA (2017) The epigenetic basis of hematopoietic stem cell aging. Semin Hematol 54(1): 19–24

Kunisaki Y, Bruns I, Scheiermann C, Ahmed J, Pinho S, Zhang D, Mizoguchi T, Wei Q, Lucas D, Ito K, Mar JC, Bergman A, Frenette PS (2013) Arteriolar niches maintain haematopoietic stem cell quiescence. Nature 502(7473):637–643

Kuribayashi W, Oshima M, Itokawa N, Koide S, Nakajima-Takagi Y, Yamashita M, Yamazaki S, Rahmutulla B, Miura F, Ito T, Kaneda A, Iwama A (2021) Limited rejuvenation of aged hematopoietic stem cells in young bone marrow niche. J Exp Med 218(3):e20192283

Kurosawa S, Iwama A (2020) Aging and leukemic evolution of hematopoietic stem cells under various stress conditions. Inflamm Regen 40(1):29

Kusumbe AP, Ramasamy SK, Itkin T, Mae MA, Langen UH, Betsholtz C, Lapidot T, Adams RH (2016) Age-dependent modulation of vascular niches for haematopoietic stem cells. Nature 532(7599):380–384

Li X, Zeng X, Xu Y, Wang B, Zhao Y, Lai X, Qian P, Huang H (2020) Mechanisms and rejuvenation strategies for aged hematopoietic stem cells. J Hematol Oncol 13(1):31

Liang Y, Van Zant G, Szilvassy SJ (2005) Effects of aging on the homing and engraftment of murine hematopoietic stem and progenitor cells. Blood 106(4):1479–1487

Liao M, Chen R, Yang Y, He H, Xu L, Jiang Y, Guo Z, He W, Jiang H, Wang J (2022) Aging-elevated inflammation promotes DNMT3A R878H-driven clonal hematopoiesis. Acta Pharm Sin B 12(2):678–691

Lopez-Otin C, Blasco MA, Partridge L, Serrano M, Kroemer G (2013) The hallmarks of aging. Cell 153(6):1194–1217

Ma S, Wang S, Ye Y, Ren J, Chen R, Li W, Li J, Zhao L, Zhao Q, Sun G, Jing Y, Zuo Y, Xiong M, Yang Y, Wang Q, Lei J, Sun S, Long X, Song M, Yu S, Chan P, Wang J, Zhou Q, Belmonte JCI, Qu J, Zhang W, Liu GH (2022) Heterochronic parabiosis induces stem cell revitalization and systemic rejuvenation across aged tissues. Cell Stem Cell 29(6):990–1005.e1010

Maryanovich M, Zahalka AH, Pierce H, Pinho S, Nakahara F, Asada N, Wei Q, Wang X, Ciero P, Xu J, Leftin A, Frenette PS (2018) Adrenergic nerve degeneration in bone marrow drives aging of the hematopoietic stem cell niche. Nat Med 24(6):782–791

Meisel M, Hinterleitner R, Pacis A, Chen L, Earley ZM, Mayassi T, Pierre JF, Ernest JD, Galipeau HJ, Thuille N, Bouziat R, Buscarlet M, Ringus DL, Wang Y, Li Y, Dinh V, Kim SM, McDonald BD, Zurenski MA, Musch MW, Furtado GC, Lira SA, Baier G, Chang EB, Eren AM, Weber CR, Busque L, Godley LA, Verdú EF, Barreiro LB, Jabri B (2018) Microbial signals drive pre-leukaemic myeloproliferation in a Tet2-deficient host. Nature 557(7706):580–584. https://doi.org/10.1038/s41586-018-0125-z

Mendez-Ferrer S, Michurina TV, Ferraro F, Mazloom AR, Macarthur BD, Lira SA, Scadden DT, Ma'ayan A, Enikolopov GN, Frenette PS (2010) Mesenchymal and haematopoietic stem cells form a unique bone marrow niche. Nature 466(7308):829–834

Moehrle BM, Geiger H (2016) Aging of hematopoietic stem cells: DNA damage and mutations? Exp Hematol 44(10):895–901

Nagpal R, Mainali R, Ahmadi S, Wang S, Singh R, Kavanagh K, Kitzman DW, Kushugulova A, Marotta F, Yadav H (2018) Gut microbiome and aging: physiological and mechanistic insights. Nutr Healthy Aging 4(4):267–285

Nakamura-Ishizu A, Takubo K, Fujioka M, Suda T (2014) Megakaryocytes are essential for HSC quiescence through the production of thrombopoietin. Biochem Biophys Res Commun 454(2):353–357

O'Toole PW, Jeffery IB (2015) Gut microbiota and aging. Science 350(6265):1214–1215

Oh J, Lee YD, Wagers AJ (2014) Stem cell aging: mechanisms, regulators and therapeutic opportunities. Nat Med 20(8):870–880

Palovics R, Keller A, Schaum N, Tan W, Fehlmann T, Borja M, Kern F, Bonanno L, Calcuttawala K, Webber J, McGeever A, Tabula Muris C, Luo J, Pisco AO, Karkanias J, Neff NF, Darmanis S, Quake SR, Wyss-Coray T (2022) Molecular hallmarks of heterochronic parabiosis at single-cell resolution. Nature 603(7900):309–314

Pardali E, Dimmeler S, Zeiher AM, Rieger MA (2020) Clonal hematopoiesis, aging, and cardiovascular diseases. Exp Hematol 83:95–104

Renders S, Svendsen AF, Panten J, Rama N, Maryanovich M, Sommerkamp P, Ladel L, Redavid AR, Gibert B, Lazare S, Ducarouge B, Schonberger K, Narr A, Tourbez M, Dethmers-Ausema B, Zwart E, Hotz-Wagenblatt A, Zhang D, Korn C, Zeisberger P, Przybylla A, Sohn M, Mendez-Ferrer S, Heikenwalder M, Brune M, Klimmeck D, Bystrykh L, Frenette PS, Mehlen P, de Haan G, Cabezas-Wallscheid N, Trumpp A (2021) Niche derived netrin-1 regulates hematopoietic stem cell dormancy via its receptor neogenin-1. Nat Commun 12(1): 608

Rossi DJ, Bryder D, Zahn JM, Ahlenius H, Sonu R, Wagers AJ, Weissman IL (2005) Cell intrinsic alterations underlie hematopoietic stem cell aging. Proc Natl Acad Sci U S A 102(26):9194–9199

Sacma M, Pospiech J, Bogeska R, de Back W, Mallm JP, Sakk V, Soller K, Marka G, Vollmer A, Karns R, Cabezas-Wallscheid N, Trumpp A, Mendez-Ferrer S, Milsom MD, Mulaw MA, Geiger H, Florian MC (2019) Haematopoietic stem cells in perisinusoidal niches are protected from ageing. Nat Cell Biol 21(11):1309–1320

Schwartz AV (2015) Marrow fat and bone: review of clinical findings. Front Endocrinol (Lausanne) 6:40

Seita J, Weissman IL (2010) Hematopoietic stem cell: self-renewal versus differentiation. Wiley Interdiscip Rev Syst Biol Med 2(6):640–653

Spadaro O, Youm Y, Shchukina I, Ryu S, Sidorov S, Ravussin A, Nguyen K, Aladyeva E, Predeus AN, Smith SR, Ravussin E, Galban C, Artyomov MN, Dixit VD (2022) Caloric restriction in humans reveals immunometabolic regulators of health span. Science 375(6581):671–677

Sun D, Luo M, Jeong M, Rodriguez B, Xia Z, Hannah R, Wang H, Le T, Faull KF, Chen R, Gu H, Bock C, Meissner A, Gottgens B, Darlington GJ, Li W, Goodell MA (2014) Epigenomic profiling of young and aged HSCs reveals concerted changes during aging that reinforce self-renewal. Cell Stem Cell 14(5):673–688

Tang D, Tao S, Chen Z, Koliesnik IO, Calmes PG, Hoerr V, Han B, Gebert N, Zornig M, Loffler B, Morita Y, Rudolph KL (2016) Dietary restriction improves repopulation but impairs lymphoid differentiation capacity of hematopoietic stem cells in early aging. J Exp Med 213(4):535–553

Trowbridge JJ, Starczynowski DT (2021) Innate immune pathways and inflammation in hematopoietic aging, clonal hematopoiesis, and MDS. J Exp Med 218(7): e20201544

Verovskaya EV, Dellorusso PV, Passegue E (2019) Losing sense of self and surroundings: hematopoietic stem cell aging and leukemic transformation. Trends Mol Med 25(6):494–515

Weiskopf D, Weinberger B, Grubeck-Loebenstein B (2009) The aging of the immune system. Transpl Int 22(11):1041–1050

Xie M, Lu C, Wang J, McLellan MD (2014) Age-related mutations associated with clonal hematopoietic expansion and malignancies. Nat Med 20(12):1472–1478

Young AL, Challen GA, Birmann BM, Druley TE (2016) Clonal haematopoiesis harbouring AML-associated mutations is ubiquitous in healthy adults. Nat Commun 7:12484

Young K, Eudy E, Bell R, Loberg MA, Stearns T, Sharma D, Velten L, Haas S, Filippi MD, Trowbridge JJ (2021) Decline in IGF1 in the bone marrow microenvironment initiates hematopoietic stem cell aging. Cell Stem Cell 28(8):1473–1482.e1477

Yu VW, Scadden DT (2016) Heterogeneity of the bone marrow niche. Curr Opin Hematol 23(4):331–338

Zhang J, Niu C, Ye L, Huang H, He X, Tong WG, Ross J, Haug J, Johnson T, Feng JQ, Harris S, Wiedemann LM, Mishina Y, Li L (2003) Identification of the haematopoietic stem cell niche and control of the niche size. Nature 425(6960):836–841

Zhao JL, Ma C, O'Connell RM, Mehta A, DiLoreto R, Heath JR, Baltimore D (2014a) Conversion of danger signals into cytokine signals by hematopoietic stem and progenitor cells for regulation of stress-induced hematopoiesis. Cell Stem Cell 14(4):445–459

Zhao M, Perry JM, Marshall H, Venkatraman A, Qian P, He XC, Ahamed J, Li L (2014b) Megakaryocytes maintain homeostatic quiescence and promote post-injury regeneration of hematopoietic stem cells. Nat Med 20(11):1321–1326

Zink F, Stacey SN, Norddahl GL, Frigge ML, Magnusson OT, Jonsdottir I, Thorgeirsson TE, Sigurdsson A, Gudjonsson SA, Gudmundsson J, Jonasson JG, Tryggvadottir L, Jonsson T, Helgason A, Gylfason A, Sulem P, Rafnar T, Thorsteinsdottir U, Gudbjartsson DF, Masson G, Kong A, Stefansson K (2017) Clonal hematopoiesis, with and without candidate driver mutations, is common in the elderly. Blood 130(6): 742–752

The manufacturer's authorised representative in the EU is Springer
Nature Customer Service Centre GmbH, Europaplatz 3, 69115 Heidelberg,
Germany. If you have any concerns regarding our products, please
contact ProductSafety@springernature.com

Printed and bound by CPI Group (UK) Ltd, Croydon, CR0 4YY
01/10/2026
02225899-0020